STRUCTURAL STEELWORK DESIGN to BS 5

2nd Ed.

STRUCTURAL STEELWORK DESIGN to BS 5950

2nd EDITION

L J Morris

•

D R Plum

LONGMAN

© Addison Wesley Longman 1996
Addison Wesley Longman Limited
Edinburgh Gate
Harlow
Essex
CM20 2JE
England

and Associated Companies throughout the world.

The rights of Linden J Morris and David R Plum to be identified as authors of this Work has been asserted by them in accordance with the Copyright, Designs and Patents Act 1988.

All rights reserved. No part of this publication may be reproduced, stored in a retrieval system, or transmitted in any form or by any means, electronic, mechanical, photocopying, recording or otherwise, without prior written permission of the publisher or a licence permitting restricted copying in the United Kingdom issued by the Copyright Licensing Agency Ltd, 90 Tottenham Court Road, London W1P 9HE.

Cover designed by Chris Eley Design, Reading
and printed by The Riverside Printing Co. (Reading) Ltd
Cover photograph of Hay's Galleria, London, courtesy of British Steel
Typeset by Tradespools Ltd, Frome, Somerset
Printed and bound at The Bath Press, Bath

First printed 1996

ISBN 0-582-23088-8

British Library Cataloguing-in-Publication Data
A catalogue record for this book is available from the British Library.

The publishers wish to thank Ward Building Components for permission to use their material.

Preface

Structural steelwork design is usually taught in degree and diploma courses after an initial grounding in the theory of structures and strength of materials. The design teaching usually covers both simple structural elements and complete buildings. More complex elements and buildings are often covered in postgraduate courses, but the ideas and concepts outlined in this text still provide the basis for more complicated structures. This book has been prepared primarily for the student, but also for those engineers in practice who are not familiar with BS 5950: *Structural use of steelwork in buildings*.

This book falls naturally into two parts. Part I sets out in detail the design of elements (beams, columns, etc.) frequently found in a structural steel framework. Part II shows how these elements are combined to form a building frame, and should prove especially useful to the engineer in the context of practical design. Part II also develops other considerations such as the overall stability of building structures. Those with some experience of element design may prefer to start with Part II, using the cross-references to re-examine element design as necessary. A final chapter considers detailing practice, and the effects of a number of practical considerations such as fabrication and fire protection.

It is assumed that the reader has some knowledge of structural analysis and that a basic understanding of metallurgy has been gained elsewhere. The design examples concentrate on manual methods to ensure a proper understanding of steelwork behaviour, with suggestions where computing could be used. Detailed programs for specific microcomputers are increasingly being written, and a number of complete design packages are available commercially.

The principal documents required by the reader are:

BS 5950: Part 1 (1985): *Design in simple construction; hot rolled sections.*
   British Standards Institution.
Steelwork design. Vol 1: *Section properties; member capacities.* Steel
   Construction Institute.

The first of these documents is also available in abridged extract form from British Standards Institution as:

Extracts from British Standards for students of structural design.

The second document is available in extract form (dimensions and properties only) in the following two publications:

A check list for designers. Steel Construction Institute.
Structural Sections to BS 4: Part 1 & BS 4848: Part 4. British Steel Corporation.

The latter two extracts are updated regularly and the latest edition should be used. It should be noted that since 1995, the symbols used in the BSC publication for the main dimensions of rolled sections have been changed to reflect the Eurocode 3 nomenclature. The relevant changes are noted at the foot of this page.

Throughout the book, clause references and notation follow those given in BS 5950: Part 1 (*design in simple and continuous construction*), except for those chapters which deal specifically with composite construction when the clause references and notation follow those in BS 5950: Part 3.1 (*design of composite beams*) and Part 4 (*design of floors with profiled steel sheeting*).

The main change in the second edition has been the introduction of a chapter on the design of plate girders. The authors have also taken the opportunity to update the text in the light of current practice and latest design information.

While every effort has been made to check both calculations and interpretation of BS 5950 the authors cannot accept any responsibility for inadvertent errors.

LJM
DRP

Acknowledgements

The authors acknowledge the assistance of many structural engineers in industry and teaching, to whom details of both interpretation and current practice have been submitted, and whose helpful comments have been incorporated into the text. In particular, the collaboration with P.A. Rutter in the initial drafting of the first edition proved invaluable.

Extracts from BS 5950 are reproduced by permission of the British Standards Institution. Complete copies can be obtained from BSI at Linford Wood, Milton Keynes, MK14 6LE.

The equivalent Eurocode 3 notation for the relevant notation given in BS 5950: Part 1.

$D \equiv h$; $B \equiv b$; $t \equiv s$; $T \equiv t$
$b \equiv b/2t$; $d/t \equiv d/s$; $d \equiv A$ (**T-section only**)

CONTENTS

Preface   *v*

Foreword to first edition   *xiii*

PART I   THE DESIGN OF STRUCTURAL STEEL ELEMENTS   1

1 INTRODUCTION TO DESIGN IN STEELWORK   3

1.1	Design requirements	5
1.2	Scope of BS 5950 *Structural use of steelwork in buildings*	5
1.3	Limit state design	6
1.4	Partial safety factors	6
1.5	Loading	7
1.6	Internal forces and moments	9
1.7	Stresses and deformations	9
1.8	Layout of calculations	11
1.9	Structural theory	12
1.10	Format of chapters	13
	Study references	13

2 LOADING AND LOAD COMBINATIONS   15

2.1	Dead loads	15
2.2	Imposed loads	16
2.3	Wind loads	16
2.4	Load combinations	16
2.5	Example 1. Loading of a simply supported gantry girder	17
2.6	Example 2. Loading of continuous spans	18
2.7	Example 3. Loading of a portal frame	21
	Study references	26

3 BEAMS IN BUILDINGS   28

3.1	Beams with full lateral restraint	29
3.2	Beams without full lateral restraint	29
3.3	Simplified design procedures	30

3.4	Moment capacity of members (local capacity check)	30
3.5	Buckling resistance (member buckling check)	31
3.6	Other considerations	31
3.7	Example 4. Beam supporting concrete floor slab (restrained beam)	32
3.8	Example 5. Beam supporting plant loads (unrestrained beam)	36
	Study references	40

4 PURLINS AND SIDE RAILS — 42

4.1	Design requirements for purlins and side rails	42
4.2	Example 6. Purlin on sloping roof	43
4.3	Example 7. Design of side rail	47
4.4	Example 8. Design of multi-span purlin	49

5 CRANE GIRDERS — 52

5.1	Crane wheel loads	52
5.2	Maximum load effects	54
5.3	Example 9. Crane girder without lateral restraint along span	55
5.4	Example 10. Crane girder with lateral restraint	61
	Study references	64

6 TRUSSES — 65

6.1	Types of truss and their use	65
6.2	Loading and analysis	66
6.3	Slenderness of members	67
6.4	Compression resistance	69
6.5	Tension capacity	69
6.6	Connections	70
6.7	Example 11. Roof truss with sloping rafter	71
6.8	Example 12. Lattice girder	76
	Study references	78

7 SIMPLE AND COMPOUND COLUMNS — 80

7.1	Types of column	80
7.2	Axial compression	81
7.3	Slenderness	82
7.4	Bending and eccentricity	83
7.5	Local capacity	84
7.6	Overall buckling	85

7.7	Example 13. Column for industrial building	85
7.8	Example 14. Laced column for industrial building	91
	Study references	93

8 COLUMN BASES & BRACKETS — 94

8.1	Column bases	94
8.2	Design of column bases	95
8.3	Brackets	96
8.4	Design of brackets	97
8.5	Example 15. Design of slab base	99
8.6	Example 16. Design of crane girder bracket (face)	101
8.7	Example 17. Design of crane girder bracket (lapped)	102
	Study references	103

9 COMPOSITE BEAMS & SLABS — 105

9.1	Composite beams	106
9.2	Shear and moment capacity of composite beams	106
9.3	Shear connectors	107
9.4	Local shear in concrete	108
9.5	Deflections	109
9.6	Composite slabs	110
9.7	Example 18. Composite beam in building	110
	Study references	112

10 BRACING — 113

10.1	Loading resisted by bracing	113
10.2	Sway stability	113
10.3	Multi-storey bracing	114
10.4	Single-storey bracing	115
10.5	Beam, truss and column bracing	116
10.6	Example 19. Gable wind girder and side bracing	116
10.7	Example 20. Multi-storey wind bracing	119
	Study references	120

11 PLATE GIRDERS — 121

| 11.1 | Introduction | 121 |
| 11.2 | Design of unstiffened plate girder | 124 |

	11.3	Example 21. Design of unstiffened plate girder – thick webs	125
	11.4	Example 22. Design of unstiffened plate girder – thin webs	129
	11.5	Design of stiffened plate girder	135
	11.6	Example 23. Design of stiffened plate girder – excluding tension field action	136
	11.7	Design of girder including tension field action	143
	11.8	Example 24. Design of stiffened plate girder – utilizing tension field action	145
	11.9	Other considerations	155
		Study references	159

PART II THE DESIGN OF STRUCTURAL STEEL FRAMEWORKS — 161

12 DESIGN OF SINGLE-STOREY BUILDING – LATTICE GIRDER AND COLUMN CONSTRUCTION — 163

	12.1	Introduction	163
	12.2	Design brief	165
	12.3	Preliminary design decisions	165
	12.4	Loading	166
	12.5	Design of purlins and sheeting rails	170
	12.6	Design of lattice girder	174
	12.7	Design of column members	191
	12.8	Overall stability of building	196
	12.9	Design of gable posts	205
	12.10	Design of connections	206
	12.11	Design of foundation block	219
	12.12	Other considerations	222
		Study references	224

13 DESIGN OF SINGLE-STOREY BUILDING – PORTAL FRAME CONSTRUCTION — 226

	13.1	Introduction	226
	13.2	Design brief	227
	13.3	Design information	227
	13.4	Design of purlins and sheeting rails	228
	13.5	Spacing of secondary members	228
	13.6	Design of portal frame	229
	13.7	Frame stability	242
	13.8	Member stability – lateral torsional buckling	243

13.9	Design of connections	256
13.10	Gable framing	265
13.11	Overall stability of building	273
13.12	Design of main column base	275
13.13	Design of foundation block	277
13.14	Other considerations	279
	Study references	279

14 DESIGN OF AN OFFICE BLOCK – COMPOSITE CONSTRUCTION — 281

14.1	Layout and basic choices	281
14.2	Loading	284
14.3	Roof beam design	286
14.4	Typical floor beam design	291
14.5	Column design	296
14.6	Connections	300
14.7	Wind bracing	301
14.8	Wind resistance by frame action	303
	Study references	304

15 DETAILING PRACTICE AND OTHER REQUIREMENTS — 305

15.1	Fabrication processes	305
15.2	Steelwork drawings	307
15.3	Cost considerations	312
15.4	Fire protection	312
15.5	Corrosion protection	313
	Study references	313

Appendix A	315
Appendix B	317
Index	319

Foreword to First Edition

In 1969, the British Standards Institution Committee B/20 responsible for BS 449, a permissible stress structural steelwork design code, instigated the preparation of a new draft code based on limit state principles and incorporating the latest research into the behaviour of structural components and complete structures. The draft was issued for public comment in 1977 and attracted considerable adverse comment from an industry long acquainted with the simpler design methods in BS 449.

It was realized by the newly constituted BSI Committee, CSB/27, that a redraft of the B/20 document would be necessary before it would be acceptable to the construction industry. The work of redrafting was undertaken by Constrado, partly funded by the European Coal and Steel Community, and the task was guided by a small steering group representing the interests of consulting engineers, steelwork fabricators and the Department of the Environment.

Prior to the completion of the redraft, calibration was carried out by the Building Research Establishment to derive suitable values for load and material factors, and design exercises to compare the design of whole structures to the draft code with designs to BS 449 were directed by Constrado. The object of these studies was to assess whether the recommendations to be contained within the new code would produce structural designs which would be no less safe than designs to BS 449 but would give an improvement in overall economy.

The resulting code of practice, BS 5950: Part 1, published in 1985, covering the design in simple and continuous construction of hot rolled steel sections, and Part 2, dealing with the specification of materials, fabrication and erection, achieved the greater simplicity sought by industry while allowing the design of building structures to be based on the more rational approach of limit state theory than the permissible stress method of BS 449. Part 3, which is in the course of preparation, will give recommendations for design in composite construction.

While BS 5950: Part 1 is explicit in its design recommendations, the code is intended to be used by appropriately qualified persons who have experience in structural steelwork design and construction. There is a need, therefore, for a text for university and college students engaged on courses in civil and structural engineering which gives clear guidance on the application of the code to typical building structures by worked examples which set out the calculations undertaken in the design office. This is achieved in this book through explanatory text, full calculations and reference to the BS 5950: Part 1 clauses.

The first part of the book deals with the design of various types of structural members and the second part deals with complete designs of the most commonly encountered structures, namely, single-storey industrial buildings and multi-storey office blocks. Apart from the relevant codes and standards, references are given to well established publications commonly found in the designer's office. The book should also be useful to the design engineer requiring an understanding of the application of the limit state code.

P.A. Rutter
Partner, Scott Wilson Kirkpatrick and Partners
Member of BS 5950 Committee

PART I

THE DESIGN OF STRUCTURAL STEEL ELEMENTS

A simple basis for design is to consider a structural framework composed of a number of elements connected together. Loads are sustained by the element, and its reactions transferred to other elements via the connections. In this simple concept for design it is essential that the overall action of the framework is considered. Therefore an introduction to the concept of overall stability of the structure is given in terms of bracing systems.

1

INTRODUCTION TO DESIGN IN STEELWORK

Structural steelwork can be either a single member or an assembly of a number of steel sections connected together in such a way that they perform a specified function. The function required by a client or owner will vary enormously but may include:

- **building frames** by which loads must be supported safely and without undue movement, and to which a weatherproof envelope must be attached;
- **chemical plant supports** by which loads must be supported but which commonly require no external envelope;
- **containers** which will retain liquids, granular materials or gases, and which may also be elevated as a further structural function;
- **masts** which must safely support mechanical or electrical equipment at specified heights, and in which the deflections, vibrations and fatigue must be controlled;
- **chimneys** which will support flues carrying waste gases to safe heights;
- **bridges** which must support traffic and other loads over greatly varying spans, and for which degrees of movement may be permitted;
- **temporary supports** used during the construction of some part of a structure, which may be of steelwork, concrete, brickwork, etc., in which safety for short periods and speed of assembly are important.

It will be noticed that both safety and movement of the structures described are important for proper function and, together with economy, these will be the main considerations when discussing the design method later. It should be noted that the design of only some of the above structures is covered by BS 5950 and hence discussed in the later chapters.

Steel sections are rolled or formed into a variety of cross-sections, a selection of which is shown in Fig. 1.1, together with their common descriptions. The majority of these cross-sections are obtained by the hot rolling of steel billets in a rolling mill, while a minority, sometimes involving complex shapes, are cold formed from steel sheet. Hollow sections are obtained by extrusion or by bending plates to the required cross-section, and seaming (welding) them to form tubes. The sections are usually produced in a variety of grades of steel having different strengths and other properties. The commonest grade is known as 43A, referred to sometimes as mild steel, having a yield strength in the range 245–275 N/mm^2. In some types of structure other grades (43B, 43C, etc.), having the same yield strength, are more suitable owing to their higher resistance to brittle fracture.

Fig 1.1 Steel sections

Description	Universal beam	Universal column	Channel
Typical size	305 × 127 × 48 UB	254 × 254 × 107 UC	203 × 89 RSC

Description	Equal angle	Rectangular hollow section
Typical size	150 × 150 × 18 Angle	200 × 100 × 8.0 RHS

In addition to cross-section, the shape of a steel section will include reference to its length, curvature if required, cutting and drilling for connections, etc., all of which are needed to ensure that each part fits accurately into the finished structure. These further shaping processes are known as fabrication and are carried out in a fabricating works or 'shop'. It is for this stage that drawings giving precise dimensions of the steelwork will be required, showing what the designer intends. In many cases these drawings are produced by the fabricators rather than by the designer of the structure.

The final stage of producing a structure in steelwork is the erection or putting together of the various elements on site to form the required framework. At this final stage the safety of the partly finished structure must be checked, and prior thought given as to how the framework is to be erected in order to define the location of each part with precision. The steelwork usually forms only a skeleton to which other building elements (floors, walls, etc.) are fixed, but in frameworks supporting chemical or mechanical plant the steelwork may be the sole structural medium.

1.1 DESIGN REQUIREMENTS

The design of any structure must be judged by whether it fulfils the required function safely, can be built with economy and can maintain an acceptable appearance for its specified lifetime. It follows that the design of structural steelwork also will be assessed by these criteria of safety, economy and appearance.

Safety is assessed by considering the strength of the structure relative to the loads which it is expected to carry. In practice, this assessment is applied to each structural element in turn, but these individual element checks are not sufficient without considering the overall safety of the framework. The strength of the structural element must always exceed the effects of the loads by a margin which is known as a factor of safety. The method of providing the factor of safety is discussed in Section 1.4. In the general sense assessment of the structure includes all the criteria by which its performance will be judged, e.g. strength, deflection, vibration, etc.

While in practice economy of the design is of great importance to the owner of the finished structure, students are rarely required to make a full economic assessment. However, two basic matters should be taken into account. Firstly, the finished design should match, without excessively exceeding, as many of the design criteria as possible. Clearly, the provision of excess strength in a structural element without reason will not be judged economic. Secondly, in structural steelwork construction only part of the cost is contained in the rolled steel sections, and a large part of the cost results from the fabrication and erection process. Consequently, economic design does not result from finding the smallest structural size and weight without considering the difficulties of fabrication. In many cases repetition of a member size and standardization of components can lead to substantial overall savings.

The appearance of the finished structure is generally of great importance owing to the very size and impact of frames in structural steel. The achievement of an elegant design is desirable not only in complete structures but also in small design details. It is here that the student should try to achieve stylish, neat and balanced solutions to problems set. In many cases these will prove to be the strongest and most economic solutions also.

1.2 SCOPE OF BS 5950 *STRUCTURAL USE OF STEELWORK IN BUILDINGS*

BS 5950 is subdivided into nine parts, each being published separately. Parts 3 and 5 to 9 inclusive are awaiting publication.

- Part 1: *Design in simple and continuous construction: hot rolled sections* (1985)
- Part 2: *Specification for materials, fabrication and erection: hot rolled sections* (1985)
- Part 3: *Design in composite construction*
- Part 4: *Design of floors with profiled steel sheeting* (1982)
- Part 5: *Design of cold formed sections*
- Part 6: *Design in light gauge sheeting, decking and cladding*

Part 7: *Specification for materials and workmanship: cold formed sections*
Part 8: *Design of fire protection for structural steelwork*
Part 9: *Stressed skin design*

The purpose of BS 5950 is to define common criteria for the design of structural steelwork in buildings and allied structures, and to give guidance to designers on methods of assessing compliance with those criteria. Part 1 of this British Standard deals with design in simple and continuous construction for hot rolled sections. Part 2 covers the specification for materials, fabrication and erection. The following chapters give examples of the design of buildings principally covered by Part 1 and Part 2 of BS 5950.

Use is also made of other parts for particular design requirements such as composite construction, and these are referred to where appropriate in the following chapters. BS 5400 is the appropriate code for the design of bridges, and may also be a more appropriate basis for the design of other types of plated structure, e.g. bunkers.

1.3 LIMIT STATE DESIGN

In common with most current UK codes of practice, BS 5950 adopts a limit state approach to design. In this approach, the designer selects a number of criteria by which to assess the proper functioning of the structure and then checks whether they have been satisfied. The criteria are divided into two main groups based on whether assessment is made of the collapse (ultimate) condition, or normal working (serviceability) condition.

Ultimate limit state includes:

- strength (safety)
- stability (overturning)
- fatigue fracture (not normally considered in buildings)
- brittle fracture
- structural integrity (including accidental damage)

Serviceability limit state includes:

- deflection
- durability
- vibration

Limiting values for each criterion are given in BS 5950: Part 1 and their use is demonstrated in the following chapters. The designer should, however, always be aware of the need for additional or varied criteria.

1.4 PARTIAL SAFETY FACTORS

Safety factors are used in all designs to allow for variabilities of load, material, workmanship and so on, which cannot be assessed with absolute certainty. They must be sufficient to cover:

INTRODUCTION TO DESIGN IN STEELWORK

1. load variations;
2. load combinations;
3. design and detailing procedures;
4. fabrication and erection procedures;
5. material variations.

The safety factor can be applied at one point in the design (global or overall safety factor), or at several points (partial safety factors). In steelwork design a partial safety factor γ_f (the load factor) is applied to the loads (variations 1 to 3 above) and another factor γ_m to the material strengths (variations 4 and 5). BS 5950: Part 1 includes a value of γ_l for structural performance within the value of γ_f, and assumes a value of 1.0 for γ_m. The use of $\gamma_m = 1.0$ does not imply that no margin of safety for material has been included, but rather that a suitable allowance has been made in the design strengths given in e.g. Table 1.2 of this chapter. Typical values of γ_f are given in Table 1.1 with further values given in BS 5950: Part 1, table 2. Application of the factors to different loads in combination is given in Chapter 2 and throughout the design examples. The value of each load factor reflects the accuracy with which a load can be estimated, and the likelihood of the simultaneous occurrence of a given combination of loads.

Table 1.1 Partial safety factor for loads

Loading	Load factor γ_f
Dead load W_d	1.4
Dead load restraining uplift	1.0
Imposed load W_i	1.6
Wind load W_w	1.4
Combined loads $(W_d + W_i + W_w)$	1.2

1.5 LOADING

In most cases design begins with as accurate as possible an assessment of the loads to be carried. These may be given, or obtained from a British Standard[1] or other appropriate source. They will be used in idealized forms as either distributed loads or point (concentrated) loads. Chapter 2 sets out typical loadings and gives examples of how they are combined in design.

These external loads, sometimes called actions, form only part of the total forces on a structure, or on a structural element. The reactions to the loads on each element must be obtained as design proceeds. These reactions must be carried through to supporting elements, so that all external loads, including self weight of members, are transferred through the structure by the shortest load path, until the foundation is reached. This process is essential to safe design. A simple example of this process is shown in Fig. 1.2, in which the load path for the external load (snow) on a section of roof cladding is traced to the foundation. The cladding (sheeting) carries the snow load as well as

Fig 1.2 Load transfer

INTRODUCTION TO DESIGN IN STEELWORK 9

self weight, and this combined load produces a reaction from a typical purlin. This constitutes a load W on the purlin, producing reactions P on the rafter. Similar purlin reactions together with the rafter self weight constitute the rafter load, producing reactions R from the roof truss (at each node). Loading the roof truss produces vertical reaction T from the column, and also a horizontal reaction S if the loading is non-vertical. These in turn, acting on the column, produce foundation reactions H, V and M.

1.6 INTERNAL FORCES AND MOMENTS

Loads with their reactions may be used to find internal forces and moments within any structural element. The usual method is to draw shear force and bending moment diagrams[2,3]. These diagrams are graphs showing how the internal forces vary along a structural element as a result of a set of stationary (static) loads. Influence lines and moment/force envelopes may be needed in cases of moving (dynamic) loads (see Chapter 2). It is also necessary to find the axial force present in a structural element, particularly vertical members, and in some cases the torsion as well. Diagrams may be used to advantage for these forces also.

In many cases design concentrates on specific values of maximum bending moment or shear force at a known position, e.g. mid-span. In these cases formulae or coefficients may be useful and can be obtained from standard tables or charts[4].

1.7 STRESSES AND DEFORMATIONS

In the design of steelwork to BS 5950: Part 1 stresses are used to obtain the capacities of structural sections in bending, shear, axial force, etc., and any combinations of these forces. Stresses used are generally based on the yield stresses appropriate to the steel quality and maximum thickness required by the designer, and detailed in table 6, BS 5950: Part 1, which is based on values stipulated in BS 4360[5]. The design strength p_y is used, for example, to calculate the moment capacity of a steel section.

$$\text{Moment capacity } M_c = p_y S_x$$

where S_x is the major axis plastic modulus of the section.

Strength is used to define an ultimate stress for a particular situation (bending, shear, etc.), and will include an adjustment for partial safety factor (material) and buckling (local or overall).

Capacity refers to a local moment of resistance (or shear or axial force) at a section based on the given strength but disregarding overall (member) buckling. **Resistance** refers to maximum moment with due regard to overall (member) buckling.

In some parts of the design it may be necessary to assess stresses when the steel is in the linear elastic condition. In this case the linear elastic bending theory may be used[6,7] in which

$$M/I = \sigma/y = E/R$$

Deformations are usually required in the design and are derived from elastic bending theory. The commonest requirement is the calculation of deflections. These are found using formulae derived from bending theory[8,9] but in more complex cases may require the use of moment-area methods[10,11] or computer programs. In general, the deflection due to unfactored imposed loads only is required.

Strains are not normally calculated in steelwork design and excessive strains are avoided by limiting stresses and other design parameters.

Stresses which should be used in steelwork design are given in detail in BS 5950: Part 1, clause 3.1.1. Some common values of design strengths are given in Table 1.2.

Table 1.2 Steel design strengths

Steel grade BS 4360	Maximum thickness (mm)	Design strength p_y (N/mm²)
43A, 43B, 43C	16	275
	40	265
	63	255
	100	245
50B, 50C	16	355
	40	345
	63	340
	100	325

Note that the steel grades 43A, etc., are specified in BS 4360[5], which defines the mechanical and other properties of the steel. The most important properties for structural use are yield strength, tensile strength and impact test values. The designations A, B, C indicate increasing resistance to impact and brittle fracture, with no significant change in the other mechanical properties.

The cross-section of a structural member needs to be classified according to BS 5950: Part 1, clause 3.5 in order to assess the resistance to local buckling of the section. Cross-sections are classified as plastic, compact, semi-compact and slender by reference to the breadth/thickness ratios of flange outstands and webs (Fig. 1.1), and also to the design strength. Details are given in clause 3.5 and table 7 of BS 5950: Part 1. The classifications of most hot rolled sections in grades 43 and 50 are given with their section properties in reference (12).

Recommended values of maximum deflections are given in BS 5950: Part 1, table 5. Some common values are reproduced in Table 1.3. In cases where the steelwork structure is to support machinery, cranes and other moving loads, more stringent limitations on deflection may be necessary. Values of maximum deflections should be checked with the manufacturers of any machinery to be used.

Table 1.3 Deflection limits

Structural element	Deflection limit due to unfactored imposed load
Cantilever	Length/180
Beam (brittle finishes)	Span/360
Other beam	Span/200
Purlin or sheeting rail	To suit cladding (but span/200 may be used)
Crane girder (vertical)	Span/600
Crane girder (horizontal)	Span/500

1.8 LAYOUT OF CALCULATIONS

Before design calculations are started, the designer must first interpret the client's drawings so that a structural arrangement can be decided to carry the loads down to the foundations. This structural arrangement must avoid intrusion into space required by the client's processes or operations. It is broken down into simple structural elements which are each given an individual code number by the steel designer (see Chapter 15). Calculations for an individual element can thus be identified, as in the design examples.

Clarity is essential in setting out calculations, and the designer should make sure that they can be checked without constant reference to that designer. Designers develop their own individual styles for setting out their calculations. Design offices of consulting engineers, local authorities and contractors often use one particular format as a house style. The student should start using a basic format such as that given in the text, but adapting it to suit the particular structure.

1.8.1 Subdivision

Subdivide the calculations into appropriate sections using subheadings such as 'dimensions', 'loading', 'moment capacity'. This makes checking of the basic assumptions and the results much easier, and helps the designer achieve a neat presentation.

1.8.2 Sketches

Engineers think pictorially, and should develop a spatial awareness. A sketch will clearly indicate what the designer intended, while in a string of numbers a serious omission can be overlooked. Sketches in the following chapters are placed in the left-hand margin where convenient.

1.8.3 References

Sources of information must be quoted as:

- loading — British Standard; manufacturer's catalogue; client's brief
- dimensions — drawing number
- stresses — BS 5950 clause number; research paper

This ensures that future queries about the calculations can be answered quickly and that subsequent alterations can be easily detected. It will also assist the designer when carrying out a similar design at a later date and this can be of great value to a student who will one day design in earnest.

1.8.4 Results

In design, results (or output), such as member size, load, moment, from one stage of the calculations may be used as input at a further stage. It is important therefore that such information is easily obtained from the calculations. Such results may be highlighted by placing in an output margin (on the right-hand side), or by placing in a 'box', or merely by underlining or using coloured marker pens; in this book, bold type has been used.

The student should attempt to maintain realism in calculation, and avoid quoting the eight or more digits produced by calculators and micro-computers. Loading is commonly no more than two-figure accurate, and section properties are given to only three figures. No amount of calculation will give results of higher accuracy.

1.8.5 Relationship with drawings

In most cases, the final results of design calculations are member sizes, bolt numbers, connection layouts and so on, all of which information must be conveyed to the fabricator/contractor on drawings. It is, however, common in steelwork design for some of the drawings to be carried out by persons other than the design engineer, such as detailing draughtsmen, technician engineers, or even the fabricator. It is therefore essential that the final output should be clearly marked in the calculations. Specific requirements such as connection details must be clearly sketched.

1.9 STRUCTURAL THEORY

It is assumed in the following chapters that the reader will have available a copy of BS 5950: Part 1 or extracts from it. Tables and charts for design will not be reproduced in full in the text but extracts will be given where appropriate. In addition, properties of steelwork sections will be required. These are available from the Steel Construction Institute[12]. The meaning of the properties given in these publications must be understood and may be studied in, for example, references (6, 7).

It may be useful at some point for the student to examine the background to the steelwork design method and BS 5950. Reference is therefore made to the Steel Construction Institute publication[13] which is intended as explanatory to BS 5950: Part 1.

The designer of steelwork elements and structures must have a clear understanding of the theory of structures and strength of materials. The student is referred to the relevant sections of textbooks such as those given in references (6) to (11) and further explanation is avoided.

1.10 FORMAT OF CHAPTERS

The following chapters provide design examples of structural steelwork elements (Part I) and structural steelwork frameworks (Part II).

The chapter order is intended to guide a student with a basic knowledge of the theory of structures and strength of materials into steelwork design. It therefore starts (in Part I) with loading, including combination effects, and proceeds to simple elements with which the student is probably already familiar. Later, more complex elements and those requiring special treatment are introduced. In Part II the simple elements are combined to form complete structures. While this chapter arrangement is preferred for teaching, in the actual design of structural steel elements, the calculations are usually arranged in load order, i.e. as indicated in Section 1.5 and Fig. 1.2. This is sometimes known as reverse construction order.

Each chapter (in Part I) starts with basic definitions of structural members and how they act. General notes on the design of the element/frame follow and then the design calculations are set out for one or more examples demonstrating the main variations.

The calculations follow the layout suggested in Section 1.8. References to BS 5950: Part 1 are given merely by quoting the appropriate clause, e.g. 'clause 2.4.1', or table, e.g. 'BS table 13'. References for the structural theory required by the student, or for background to BS 5950, are given as study topics at the end of each chapter with a numerical reference in the text, e.g. (3).

STUDY REFERENCES

Topic	References
1. Loading	BS 6399 *Design Loading for Buildings* Part 1: *Dead and imposed loads* (1984) Part 2: *Wind loads* (1995) Part 3: *Imposed roof loads* (1988)
2. BM and SF diagrams	**Marshall W.T. & Nelson H.M.** (1990) Examples of bending moment and shear force diagrams, *Structures*, pp. 23–4. Longman

3.	BM and SF diagrams	**Coates R.C., Coutie M.G. & Kong F.K.** (1988) Shear forces and bending moments, *Structural Analysis*, pp. 58–71. Van Nostrand Reinhold
4.	BM and SF coefficients	(1992) *Steel Designers' Manual*, pp. 1026–53. Blackwell
5.	Steel quality	BS 4360 (1990) *Specification for Weldable Structural Steels*
6.	Theory of bending	**Marshall W.T. & Nelson H.M.** (1990) Bending stress analysis, *Structures*, pp. 134–43. Longman
7.	Theory of bending	**Hearn E.J.** (1985) Bending, *Mechanics of Materials*, vol. 1, pp. 62–8. Pergamon
8.	Deflections	**Marshall W.T. & Nelson H.M.** (1990) Bending deformation, *Structures*, pp. 203–18. Longman
9.	Deflections	**Hearn E.J.** (1985) Slope and deflection of beams, *Mechanics of Materials* vol. 1, pp. 92–107. Pergamon
10.	Moment-area	**Croxton P.L.C. & Martin L.H.** (1990) Area-moment method of analysis, *Solving Problems in Structures*, vol. 2, pp. 25–47. Longman
11	Moment-area method	**Coates R.C., Coutie M.G. & Kong F.K.** (1988) Moment-area methods, *Structural Analysis*, pp. 176–81. Van Nostrand Reinhold
12.	Section properties	(1987) *Steelwork Design* vol. 1, Section properties member capacities. Steel Construction Institute
13.	Background to BS 5950	**Dowling P.J., Knowles P. & Owens G.W.** (1988) *Structural Steel Design*. Steel Construction Institute

2

LOADING AND LOAD COMBINATIONS

The loading for most structures is obtained from the appropriate British Standard[1,2], the manufacturer's data and similar sources. The loads obtained must, however, be combined to simulate what is perceived by the designer to occur in practice, and be multiplied by appropriate load factors. The process of combining loads and including the load factors is carried out for simple structural elements when deriving the bending moments, shear forces, etc., which will occur. For more complex structures it is advantageous to include the load factors after deriving the bending moments etc., so that specific combinations can be examined more readily.

2.1 DEAD LOADS

These will include the following:
- own weight of steel member (kg/m of steel section)
- other permanent parts of building, etc., not normally moveable (e.g. concrete floor slabs, brick/block walls, finishes, cladding)

They are calculated either from density of material (kg/m^3) or specific weight (kN/m^3), or from manufacturers' data contained in catalogues or manuals. Table 2.1 shows some typical values; these are all permanent loads and are combined with the appropriate dead load partial safety factor (see Section 1.4).

Table 2.1 Typical values of common structural materials

Material	Density (kg/m^3)	Specific weight (kN/m^3)
Steel	7850	77
Reinforced concrete	2420	23.7
Brickwork	2000–2300	20–23
Timber	500–900	5–9

2.2 IMPOSED LOADS

These will include the following temporary loads:

- snow on roofs[1]
- people
- furniture
- equipment such as cranes and other machinery
- semi-permanent partitions which are moveable

Imposed loads vary with the function of the room or building[1], and some typical values are shown in Table 2.2. All imposed loads are based on experience within the construction industry and the statistical analyses of observed cases. These are all temporary loads and are combined with the imposed load partial safety factor (see Section 1.4).

Table 2.2 Typical values of imposed loads

Building usage	Imposed load (kN/m^2)
Residential (self-contained dwellings)	1.5
Offices (depending on room usage)	2.5–5.0
Educational (classrooms)	3.0
Theatres (areas with fixed seating)	4.0
Warehousing (general storage)	2.4 per m height
Industrial workshops	5.0

2.3 WIND LOADS

The wind loads used in the text are based on CP3, ChV, Part 2[2] (as BS 6399): Part 2[1] was issued too late to be incorporated). Basic wind speed, appropriate to the location of the building, is selected and reduced to a design wind speed using factors which take into consideration topography, surrounding buildings, height above ground level, component size and period of exposure. The design wind speed is equated to a dynamic pressure q (kN/m^2). Owing to building and roof shape, openings in walls, etc., pressures and suctions, both external and internal, will arise. Pressure coefficients external (C_{pe}) and internal (C_{pi}) may be used as shown in Section 2.7.

$$\text{Force on any element} = (C_{pe} - C_{pi})q \times \text{area of element}$$

Wind data with suitable factors and coefficients are given in reference (1). The method of obtaining the quasi-static wind load used in design is given in greater detail in Sections 2.7 and 12.4.3.

2.4 LOAD COMBINATIONS

Loads on any structure must be arranged in design so that the maximum force or moment is achieved at the point in the structure being considered. Hence all realistic load combinations must be considered to ensure that all peak

2.5 EXAMPLE I. LOADING OF A SIMPLY SUPPORTED GANTRY GIRDER

(a) Dimensions

Simply supported span   6.0 m
Crane wheels' centres   3.6 m
(See further description in Section 5.1)

(b) Loads specified

Self weight of girder (uniformly distributed)   1.5 kN/m
Maximum crane wheel load (static)   220 kN
Weight of crab   60 kN
Hook load to be lifted   200 kN

Dynamic effects to be included in accordance with BS 6399: Part 1[1] at 25%. (For further details of dynamic effects and the derivation of maximum wheel loads see Section 5.1.)

(c) Moments and forces (due to u.d.l.)

(See Fig. 2.1.)

Fig. 2.1

Self weight W_d (factored by γ_f) $= 1.4 \times 1.5 \times 6.0 = 12.6$ kN
Ultimate midspan BM $= W_d L/8$ $= 12.6 \times 6/8$ $= 9.0$ kNm
Ultimate reactions $R_A = R_B = WL/2 = 12.6/2$ $= 6.0$ kN

                                                                 6.3kN

(d) Moments and forces (due to vertical wheel load)

Wheel load W_c (including γ_f and 25% impact)
$= 1.6 \times 1.25 \times 220 = 440$ kN

The positions of moving loads to give maximum values of moment and shear force are given in Section 5.2 and references (3, 4). The maximum values of each case are now given (and see Fig. 2.2).

Ultimate BM under wheel (case 1)
$= 2W_c(L/2 - c/4)^2/L$
$= 2 \times 440(6.0/2 - 3.6/4)^2/6.0 = 647$ kNm

Ultimate BM under wheel (case 2)
$= W_c L/4 = 440 \times 6.0/4$ $= 660$ kNm

Case 2 gives maximum ultimate BM of 660 kNm.

Fig. 2.2

Ultimate reaction R_A (case 3)
$$= W_c(2 - c/L)$$
$$= 440(2 - 3.6/6.0) = 616\,\text{kN}$$
Total ultimate BM $\quad= 9 + 660 =$ **669 kNm**
Total ultimate reaction $= 6 + 616 =$ **622 kN**

(e) Moments and forces (due to horizontal wheel load)

In addition to the vertical loads and forces calculated above, horizontal surge loading due to movement of the crab and hook load gives rise to horizontal moments and forces (see Section 5.1) equal to 10% of these loads.

Horizontal surge load W_{hc} (incl. γ_f) $= 1.6 \times 0.10(200 + 60)$
$$= 41.6\,\text{kN}$$

This is divided between 4 wheels (assuming double-flanged wheels):
Horizontal wheel load $\quad W_{hc} = 41.6/4 = 10.4\,\text{kN}$

Using calculations similar to those for vertical moments and forces:
Ultimate horizontal BM (case 2)
$$= W_{hc}L/4$$
$$= 10.4 \times 6.0/4 \quad = \textbf{15.6 kNm}$$
Ultimate horizontal reaction (case 3)
$$= W_{hc}(2 - c/L)$$
$$= 10.4(2 - 3.6/6.0) = \textbf{14.6 kN}$$

Note that the γ_f value of 1.6 may be reduced to 1.4 where vertical and horizontal crane loads act together, and this combination must be checked in practice (see Section 5.3).

2.6 EXAMPLE 2. LOADING OF CONTINUOUS SPANS

Obtain the maximum values of bending moment, shear force and reaction for a continuous beam at the positions noted in (c), (d), (e) and (f).

(a) Dimensions

Main beams, spaced at 4.5 m centres, supporting a concrete slab (spanning one way only) for office accommodation. Steel beams are to have four continuous spans of 8.0 m. Assume uniform section properties.

(b) Loading

Self weight of beam	1.0 kN/m
Concrete slab and finish	5.4 kN/m^2
Imposed loading (offices)	5.0 kN/m^2

The dead loading (self weight, slab and finishes) is fixed, but the imposed loading is moveable. The dead loading must be present ($\gamma_f = 1.4$), while the imposed loading may be present ($\gamma_f = 1.6$) or absent.

LOADING AND LOAD COMBINATIONS 19

For one span:

Self weight	$= 1.0 \times 8.0$	$= 8$ kN
Slab + finishes	$= 5.4 \times 4.5 \times 8.0$	$= 194$ kN
Dead load W_d (total)		$= 202$ kN
Imposed load W_i	$= 5.0 \times 4.5 \times 8.0$	$= 180$ kN
Maximum span load	$= 1.4W_d + 1.6W_i$	$= 571$ kN
Minimum span load	$= 1.4W_d$	$= 283$ kN

Arrangements of loading to produce maximum moments and forces may be found by inspection of the appropriate influence lines. The use of influence lines to give the required arrangements (patterns) of loading is described in references (5, 6). An influence line shows the effect, say of bending moment, due to a moving unit load. Hence the maximum BM is obtained by placing the imposed loads where the influence is of one sign only, e.g. in Fig. 2.3 the maximum BM due to imposed load at the middle of span 1 is obtained by placing the imposed load on spans 1 and 3. Load on spans 2 and 4 would produce a BM of opposite sign at the point considered.

In the design office, standard loading arrangements are used to speed up this process of selection, with influence lines used for more complex cases.

(c) Load pattern for mid-span moment M_1

Maximum value of M_1 is produced when spans 1 and 3 have the maximum loading and spans 2 and 4 have the minimum loading. Using the influence lines shown, the loading patterns producing maximum effect may be obtained, and are summarized below.

Fig. 2.3 Influence line for M_1

(d) Load pattern for support moment M_b

Maximum value of M_b is produced when spans 1, 2 and 4 have the maximum loading and span 3 has the minimum loading (Fig. 2.4).

Fig. 2.4 Influence line for M_b

(e) Load pattern for reaction R_b

Maximum as in (d) (Fig. 2.5).

Fig. 2.5 Influence line for R_b

(f) Load pattern for shear force S_{ab}

Maximum as in (c) (Fig. 2.6).

Fig. 2.6 Influence line for S_{ab}

Arrangements of loading for maximum moments and forces:

	Span loading (kN)			
	W_1	W_2	W_3	W_4
For moment M_1	571	283	571	283
For moment M_b	571	571	283	571
For reaction R_b	571	571	283	571
For shear force S_{ab}	571	283	571	571

(g) Moments and forces

Moments and forces may be found by any analytical method, but for equal span cases coefficients are available[7], where, for example

$$M_1 = (\alpha_{11}W_1 + \alpha_{12}W_2 + \alpha_{13}W_3 + \alpha_{14}W_4)L$$

and M_b, R_b and S_{ab} are given by similar expressions. The coefficients α may be summarized:

Span loaded	Moment or force coefficients α			
	M_1	M_b	R_b	S_{ab}
1	0.094	0.068	0.652	0.433
2	− 0.024	− 0.048	0.545	− 0.049
3	0.006	0.014	− 0.080	0.013
4	− 0.003	− 0.005	0.027	− 0.005

Hence the maximum value of M_1 occurs with the load arrangement shown and

$$M_1 = (0.094 \times 571 - 0.024 \times 283 + 0.006 \times 571 - 0.003 \times 283)8.0$$
$$= 396 \text{ kNm}$$

Similarly,

$$M_b = -521 \text{ kNm}$$
$$R_b = 676 \text{ kN}$$
$$S_{ab} = 238 \text{ kN}$$

2.7 EXAMPLE 3. LOADING OF A PORTAL FRAME

(a) Frame

See Fig. 2.7: pitched portal with pinned feet; span 38 m, spaced at 6 m centres.

Fig. 2.7 Portal frame

(b) Loading

Self weight of frame	0.90 kN/m
Cladding (roof and walls)	0.09 kN/m²
Snow and services	0.75 kN/m²
Wind pressure q (walls)	1.20 kN/m²
Wind pressure q (roof)	1.20 kN/m²

Wind pressures are based on a basic wind speed of 50 m/s for a location in Scotland.

Factors[2] S_1 and S_2 are both taken as 1.0 and factor S_3 as 0.88 for a height of 10 m.

The design wind speed is hence 44 m/s, giving a dynamic pressure of 1.20 kN/m².

22 STRUCTURAL STEELWORK DESIGN TO BS 5950

It is possible to use a lower wind pressure below a height of 5 m but this makes the analysis more complex for very little reduction in frame moments.

(c) Pressure coefficients

External pressure coefficients (C_{pe}) may be found and are summarized in the following table. These values are obtained from reference (2).

	C_{pe} for frame member			
	AB	BC	CD	DE
Wind on side	0.7	−1.2	−0.4	−0.25
Wind on end	−0.5	−0.6	−0.6	−0.5

Internal pressure coefficients (C_{pi}) should be obtained[1], and are taken in this example as +0.2 (maximum) and −0.3 (minimum) which are combined algebraically with the values of C_{pe} above.

Figure 2.8 shows the individual pressure coefficients, and Fig. 2.9 shows the various combination cases.

(a) Wind on side   (b) Wind on end   (c) Internal pressure   (d) Internal suction

Fig. 2.8

Case 1 (a + b)   Case 2 (a + d)   Case 3 (b + c)   Case 4 (b + d)

Fig. 2.9

	($C_{pe} - C_{pi}$) for frame member			
Case	AB	BC	CD	DE
1. Wind on side + internal pressure	0.5	−1.4	−0.6	−0.45
2. Wind on side + internal suction	1.0	−0.9	−0.1	0.05
3. Wind on end + internal pressure	−0.7	−0.8	−0.8	−0.7
4. Wind on end + internal suction	−0.2	−0.3	−0.3	−0.2

It may be noted that case 4 is similar to case 3, but has lower values and may be discarded.

(d) Member loads

Dead load on roof is calculated on the projected area (Fig. 2.10):
Self weight $= 0.9 \times 19.3$ $= 17.5\,\text{kN}$
Cladding $= 0.09 \times 19.3 \times 6.0 = 10.5\,\text{kN}$
Total W_{dr} $= 28.0\,\text{kN}$

Dead load on walls (Fig. 2.11):
Self weight $= 0.9 \times 5.5$ $= 5.0\,\text{kN}$
Cladding $= 0.09 \times 5.5 \times 6.0$ $= 3.0\,\text{kN}$
Total W_{dw} $= 8.0\,\text{kN}$

Fig. 2.10

Fig. 2.11

Imposed load on roof given on plan area (and services) (Fig. 2.12):
Snow load $W_i = 0.75 \times 19.0 \times 6.0 = 86\,\text{kN}$

Wind loads for case (1) – wind on side + internal pressure:

Wind load on wall W_{ww} (Fig. 2.13) $= 0.5 \times 1.20 \times 6.0 \times 5.5 = 15\,\text{kN}$

Wind pressure on roof is divided into vertical and horizontal components (Fig. 2.14),

Vertical component $W_{wv} = -1.4 \times 1.20 \times 6.0 \times 19.0 = -192\,\text{kN}$
Horizontal component $W_{wh} = -1.4 \times 1.20 \times 6.0 \times 3.4 = -34\,\text{kN}$

Fig. 2.12

Fig. 2.13

Fig. 2.14

In the same manner wind load for each member and each case may be calculated. The loading due to dead, imposed and wind loads may be summarized using the positive notation in Fig. 2.15.

Fig. 2.15 Loading summary

	Unfactored loads (kN)					
	W_1	W_2	W_3	W_4	W_5	W_6
Dead load W_d	0	0	28	28	0	0
Imposed load W_i	0	0	86	86	0	0
Wind case (1) W_{w1}	20	−34	−192	−82	−15	−18
Wind case (2) W_{w2}	40	−22	−123	−14	−2	−2
Wind case (3) W_{w3}	−28	−20	−109	−109	−20	−28

Note that the wall dead load is not included at this stage, as it does not produce a bending moment. Values of γ_f are not yet included.

(e) Forces and moments

The loads given in Fig. 2.15 may be used to obtain moments and forces by any analytical method or by use of charts or coefficients. Each load W has an appropriate set of coefficients α giving the required moment or force[6]. For any load case, the effects of all six loads must be summed.

For dead load:

$$H_a = \gamma_{11} W_1 + \gamma_{21} W_2 + \gamma_{31} W_3 + \gamma_{41} W_4 + \gamma_{51} W_5 + \gamma_{61} W_6$$

and similarly for each force or moment.

The coefficients for elastic analyses may be calculated and tabulated:

Load	Coefficient α for moment or force						
	H_a (kN)	H_e (kN)	V_a (kN)	V_e (kN)	M_b (kNm)	M_c (kNm)	M_d (kNm)
W_1	−0.806	0.194	−0.072	0.072	1.683	−0.351	−1.067
W_2	−0.541	0.459	−0.189	0.189	2.975	−0.485	−2.525
W_3	0.434	0.434	0.750	0.250	−2.385	0.890	−2.385
W_4	0.434	0.434	0.250	0.750	−2.385	0.890	−2.385
W_5	0.459	−0.541	0.189	−0.189	−2.525	0.485	−2.975
W_6	0.194	−0.806	0.072	−0.072	−1.067	−0.351	1.683

Each force or moment may be calculated using equations similar to that given for H_a above.

The expression of coefficients (α) and loads (W) as arrays allows for combining by use of a computer program. This would be of particular advantage if matrix multiplication was available.

$$\{W\}[\alpha] = \{F\}$$

By computer or by hand the moments and forces are calculated and tabulated:

	H_a (kN)	H_e (kN)	V_a (kN)	V_e (kN)	M_b (kNm)	M_c (kNm)	M_d (kNm)
Value of moment or force due to unfactored loads							
W_d	24.3	24.3	28.0	28.0	−133.6	49.8	−133.6
W_i	74.6	74.6	86.0	86.0	−410.2	153.1	−410.2
W_{w1}	−127.0	−108.0	−163.6	−110.4	643.1	−220.8	643.1
W_{w2}	−81.1	−59.1	−95.0	−42.0	335.8	−123.6	330.3
W_{w3}	−75.8	−75.8	−109.0	−109.0	493.7	−155.0	493.7

(f) Combinations and load factors

Combinations of load must now be considered and at this stage the values of γ_f may be included. Possible combinations are:

Group 1 Dead + imposed $1.4W_d + 1.6W_i$
Group 2 Dead + wind $1.4W_d + 1.4W_{w1}$
                                                                $1.4W_d + 1.4W_{w2}$
                                                                $1.4W_d + 1.4W_{w3}$
Group 3 Dead + imposed + wind $1.2W_d + 1.2W_i + 1.2W_{w1}$
                                                               $1.2W_d + 1.2W_i + 1.2W_{w2}$
                                                                $1.2W_d + 1.2W_i + 1.2W_{w3}$
Group 4 Dead (restraining uplift) + wind $1.0W_d + 1.4W_{w1}$
                                                               $1.0W_d + 1.4W_{w3}$

Group 4 is intended for use when considering restraint against uplift or overturning, i.e. maximum wind plus minimum dead load. Some combinations may be eliminated by inspection, but care must be taken to retain combinations giving maximum values of opposite sign. Some of the combinations in Groups 2 and 3 have been discarded in the following table.

For group 1 combination ($1.4W_d + 1.6W_i$):

$$H_a = 1.4 \times 24.3 + 1.6 \times 74.6 = 153\,\text{kN}$$

Frame forces (kN) and moments (kNm) for factored loads:

26 STRUCTURAL STEELWORK DESIGN TO BS 5950

Group		H_a	H_e	V_a	V_e	M_b	M_c	M_d
1.	$1.4W_d$ $+1.6W_i$	153	153	177	177	−843	315	−843
2.	$1.4W_d$ $+1.4W_{w1}$	−134	−117	−190	−116	713	−239	713
3.	$1.2W_d$ $+1.2W_i$ $+1.2W_{w2}$	21	48	23	86	−250	95	−256
4.	$1.0W_d$ $+1.4W_{w1}$	−154	−127	−201	−127	764*	−259*	764*
	$1.0W_d$ $+1.4W_{w3}$	−82	−82	−125	−125	557	−167	557

* While these values are maxima (opposite sign) the group 4 combinations (BS 5950) are to cover uplift and overturning only. It is necessary, however, for all the effects of a load combination to be considered (see also Chapter 12).

Maximum and minimum values may be selected from the table. Bending moment diagrams may be drawn as shown in Figs. 2.16 and 2.17.

Fig. 2.16 Bending moments for combination 1

Fig. 2.17 Bending moments for combination 2

Finally the effect of wall dead load can be added, if axial force in the columns is required (combination 1):

Maximum axial force at A = $177 + 1.4 \times 8.2 =$ **190 kN**
Minimum axial force at A = $-209 + 1.0 \times 8.2 =$ **−201 kN**

The alternative method of analysis is by application of plastic theory (see Chapter 13). In this elastic analysis of a portal frame, the loading combinations can be added, but in Chapter 13 each load combination produces its own unique collapse mechanism, i.e. each load combination must be analysed independently when applying plastic theory.

STUDY REFERENCES

Topic	References
1. Loading	BS 6399 *Design Loading for Buildings* Part 1: *Dead and imposed loads* (1984) Part 2: *Wind loads* (1995) Part 3: *Imposed roof loads* (1988)

2. Wind Loading   British Standard Institute CP3, Chapter V, Part 2

3. Moving loads effects   **Marshall W.T. & Nelson H.M.** (1990) Moving loads and influence lines, *Structures*, pp. 79–106. Longman

4. Moving load effects   **Wang C.K.** (1983) Influence lines for statically determinate beams, *Intermediate Structural Analysis*, pp. 459–67. McGraw-Hill

5. Influence lines   **Coates R.C., Coutie M.G. & Kong F.K.** (1988) Mueller-Breslau's principle. Model analysis, *Structural Analysis*, pp. 127–31. Van Nostrand Reinhold

6. Influence lines   **Wang C.K.** (1983) Influence lines for statically determinate beams, *Intermediate Structural Analysis*, pp. 496–503. McGraw-Hill

7. BM and SF coefficients   (1992) Design theory, *Steel Designer's Manual*, pp. 1051–4. Blackwell

8. BM and SF coefficients   (1992) Design theory, *Steel Designer's Manual*, pp. 1080–97. Blackwell

3
BEAMS IN BUILDINGS

Most buildings are intended to provide load carrying floors, and to contain these within a weathertight envelope. In some structural frameworks the weathertight function is not needed, while the load carrying function only is required, e.g. supporting chemical plant. The loads to be supported are often placed on floor slabs of concrete, or on steel or timber grid floors, and these are in turn supported on the steelwork beams. In some cases, especially in industrial buildings, loads from equipment may be placed directly on to the beams without the use of a floor slab. Wind loads also must be carried to the beams by provision of cladding of adequate strength, and by secondary members such as purlins and side rails.

Beams which carry loads from floors or other beams to the columns are generally called main beams. Secondary beams will be provided to transfer load to the main beams, or in some cases just to give lateral stability to columns, while themselves carrying only their self weight. The manner in which loads are distributed from the floors on to the beams needs careful consideration so that each beam is designed for a realistic proportion of the total load. Examples of load distribution for one-way and two-way spanning slabs are shown in Fig. 3.1.

Fig. 3.1 Load distribution (excluding self weight)

3.1 BEAMS WITH FULL LATERAL RESTRAINT

Many beams in a steel framework will be restrained laterally by the floors which transmit the loads to them. Concrete floor slabs, and wall or roof cladding, are generally able to give this lateral support or restraint. Timber floors and open steel floors are less certain in providing restraint. The degree of attachment of the flange to the floor may need to be assessed[1].

Alternatively, lateral restraint may be provided by bracing members at specific points along the beams[1]. If adequate bracing or floor slab restraint is present then lateral torsional instability will be prevented. In addition, this need not be considered for beams in which:

- the section is bent about a minor axis, or
- the section has a high torsional stiffness, e.g. a rectangular hollow section.

Beams, in which lateral torsional instability will not occur, are classed as restrained and are designed as illustrated in Section 3.7.

3.2 BEAMS WITHOUT FULL LATERAL RESTRAINT

An understanding of the behaviour of struts[2,3] will be useful in appreciating the behaviour of beams where full lateral restraint is not provided. The compression flange of such members will show a tendency to fail by buckling sideways (laterally) in the most flexible plane. Design factors which will influence the lateral stability can be summarized as:

- the length of the member between adequate lateral restraints;
- the shape of the cross-section;
- the variation of moment along the beam;
- the form of end restraint provided;
- the manner in which the load is applied, i.e. to tension or compression flange.

These factors and their effects are discussed in detail in reference (1), and are set out in clause 4.3.7 of BS 5950. The buckling resistance (M_b) of a beam may be found by use of a number of parameters and factors:

- **Effective length** (L_E), which allows for the effects of end restraint as well as type of beam, and the existence of destabilizing forces.
- **Minor axis slenderness** (λ), which includes lateral stiffness in the form of r_y, and is defined by $\lambda = L_E/r_y$.
- **Torsional index** (x), which is a measure of the torsional stiffness of a cross-section.
- **Slenderness factor** (v), which allows for torsional stiffness and includes the ratio of λ/x.
- **Slenderness correction factor** (n), which is dependent on the moment variation along the beam.
- **Buckling parameter** (u), which allows for section type and includes a factor for warping.

- **Equivalent slenderness** (λ_{LT}), which combines the above parameters and from which the bending strength (p_b) may be derived:

$$\lambda_{LT} = nuv\lambda$$

In addition, an equivalent moment factor (m) is used which allows for the effect of moment variation along the beam.

3.3 SIMPLIFIED DESIGN PROCEDURES

The design of many simple beams will not require the calculation of all the above parameters. In particular, simply supported beams carrying distributed loads and not subjected to destabilizing loads will use:

$m = 1.0$
$n = 0.94$
$u = 0.9$ (for I, H and channel sections) or from published tables
$ = 1.0$ (for other sections)

A conservative approach is allowed in BS 5950 and may be used in the design of I and H sections only, basing the bending strength (p_b) on the design strength, the minor axis slenderness (λ) and the torsional index (x), which for this method may be approximated to D/T. This approach is useful in the preliminary sizing of members and is given in clause 4.3.7.7.

3.4 MOMENT CAPACITY OF MEMBERS (LOCAL CAPACITY CHECK)

The local moment capacity (M_c) at any critical point along a member must not be less than the applied bending moment at that point. The moment capacity will depend on:

- the design strength and the elastic or plastic modulus;
- the co-existent shear;
- the possibility of local buckling of the cross-section.

Providing the applied shear force is not more than 0.6 of the shear capacity, no reduction in moment capacity is needed, and

$$M_c = p_y S$$
but $M_c \not> 1.2 p_y Z$

where S is the plastic modulus
 Z is the elastic modulus

Note that the limitation $1.2 p_y Z$ is to prevent the onset of plasticity below working load. For UB, UC and joist sections the ratio S/Z is less than 1.2 and the plastic moment capacity governs design. For sections where $S/Z > 1.2$, the constant 1.2 is replaced by the ratio (γ_o) of factored load/unfactored load. The limitation $1.2 p_y Z$ is therefore purely notional and becomes in practice $\gamma_o p_y Z$.

If 0.6 of shear capacity is exceeded, some reduction in M_c will occur as set out in clause 4.2.6.

Local buckling can be avoided by applying a limitation to the width/thickness ratios of elements of the cross-section. This leads to the classification of cross-sections discussed in Section 1.7.

Where members are subjected to bending about both axes a combination relationship must be satisfied:

(a)   For plastic and compact sections:
For UB, UC and joist sections $\quad (M_x/M_{cx})^2 + M_y/(M_{cy}) \not> 1$
For RHS, CHS and solid sections $\quad (M_x/M_{cx})^{5/3} + M_y(M_{cy})^{5/3} \not> 1$
For channel, angle and all other sections $\quad M_x/M_{cx} + M_y/M_{cy} \not> 1$
where M_x, M_y are applied moments about x and y axes
M_{cx}, M_{cy} are moment capacities about x and y axes

(b)   For semi-compact and slender sections:
(and as a simplified method for compact sections in *(a)* above)
$$M_x/M_{cx} + M_y/M_{cy} \not> 1$$

3.5   BUCKLING RESISTANCE (MEMBER BUCKLING CHECK)

Members not provided with full lateral restraint (Section 3.2) must be checked for lateral torsional buckling resistance (M_b) as well as moment capacity. The buckling resistance depends on the bending strength (Section 3.2) and the **plastic** modulus:

$$M_b = p_b S_x$$

Where members are subjected to bending about both axes (without axial load) a combination relationship must be satisfied:

$$m_x M_x/M_b + m_y M_y/M_{cy} \not> 1$$

where m_x, m_y are equivalent uniform moment factors
M_{cy} is the moment capacity about the y axis but without the restriction of $1.2p_y Z$ (as in Section 3.4).

This is described as a 'more exact' approach (clause 4.8.3.3.2) which is less conservative than the 'simplified' approach (clause 4.8.3.3.1), in which M_{cy} is defined as $p_y Z_y$. Also, a simplified approach for bending about two axes (without axial load) does not reduce the calculations.

3.6   OTHER CONSIDERATIONS

In addition to the above requirements for moment capacity and buckling resistance, a member is usually required to meet some deflection criteria. These are outlined in Section 1.5 and reference (1).

The application of heavy loads or reactions to a member may produce high local stresses and it is necessary to check that the web bearing and web buckling requirements are satisfied. These requirements are generally significant only in beams carrying heavy point loads such as crane girders (Chapter 5) or beams supporting column members within the span.

32 STRUCTURAL STEELWORK DESIGN TO BS 5950

Connections must be provided at junctions between members and must safely transmit the calculated loads from one member to the next. A variety of connection details exist for most common situations and are fully described in the SCI handbook[4]. Design information for connections is given in section 6, BS 5950, which includes some guidance on bolt spacing and edge distances. Bolt and weld sizes and capacities are given in reference (5).

3.7 EXAMPLE 4. BEAM SUPPORTING CONCRETE FLOOR SLAB (RESTRAINED BEAM)

(a) Dimensions

(See Fig. 3.2.)

Beams centres	6.0 m
Span (simply supported)	7.4 m
Concrete slab (spanning in two directions)	250 mm thick
Finishing screed	40 mm thick

Fig. 3.2 Slab and beams

(b) Loading

Concrete slab	23.7 kN/m^3
Screed (40 mm)	0.9 kN/m^2
Imposed load	5.0 kN/m^2

For preliminary calculation, an estimated self weight is included. Assume beam to be **533 × 210 × 92 UB** (grade 43A). It is sufficiently accurate to take beam weight of 92 kg/m = 0.92 kN/m. Member size must be finally confirmed after all the design checks have been carried out.

BEAMS IN BUILDINGS **33**

Fig. 3.3

Fig. 3.4

Fig. 3.5

Fig. 3.6

Self weight $\quad 0.92 \times 7.4 \quad = 7 \text{ kN}$
Dead load slab $\quad 23.7 \times 0.250 = 5.93 \text{ kN/m}^2$
Dead load screed $\quad\quad\quad\quad\quad = 0.90 \text{ kN/m}^2$
$\quad\quad\quad\quad\quad\quad\quad\quad\quad\quad\quad\quad\quad\quad\; 6.83 \text{ kN/m}^2$

Area of slab supported by beam (Fig. 3.3):

$\quad$ rectangles $\quad 2 \times 1.4 \times 3.0 \quad = 8.4 \text{ m}^2$
$\quad$ triangles $\quad\; 4 \times 3.0 \times 3.0/2 = 18.0 \text{ m}^2$

Dead load W_d:

$\quad$ on rectangles $\quad 6.83 \times 8.4 \;\; = 57 \text{ kN}$
$\quad$ on triangles $\quad\;\; 6.83 \times 18.0 = 123 \text{ kN}$

Imposed load W_i:

$\quad$ on rectangles $\quad 5.0 \times 8.4 \;\; = 42 \text{ kN}$
$\quad$ on triangles $\quad\;\; 5.0 \times 18.0 = 90 \text{ kN}$

Ultimate load (factored) (Fig. 3.4):

$\quad$ uniformly distributed $\quad\quad\quad\quad 1.4 \times 7 = 10 \text{ kN}$
$\quad$ on rectangles $\quad\quad\quad\quad 1.4 \times 57 + 1.6 \times 42 = 147 \text{ kN}$
$\quad$ on triangles $\quad\quad\quad\quad\; 1.4 \times 123 + 1.6 \times 90 = 316 \text{ kN}$

[handwritten annotation: UDL, Live load]
$\dfrac{10 \times 7.4}{8} + 232 \times 3.7 \; - \; 158 \times 1.7 - 74 \times 0.35$
$= 9.25 \; + \; 858.4 \; - \; 268.6 \; - \; 25.9 = 573 \text{ kN}$

(c) BM and SF

(See Fig. 3.5.)

Maximum ultimate moment M_x (mid-span)
$= 10 \times 7.4/8 + 232 \times 3.0 - 158 \times 1.7 - 74 \times 10.35 = \mathbf{573 \text{ kNm}}$
Maximum ultimate shear force $F_x = 10/2 + 232 = \mathbf{237 \text{ kN}}$

(d) Shear capacity

Using the design strength from Table 1.2 for grade 43A steel, noting that maximum thickness of section is 15.6 mm:

$$p_y = 275 \text{ N/mm}^2$$

clause 4.2.3 $\quad$ Shear capacity $P_v = 0.6 \, p_y \, A_v$
$\quad\quad\quad\quad\quad\quad\quad = 0.6 \times 0.275 \times 531.1 \times 10.2 = \mathbf{897 \text{ kN}}$
Shear force $F_x/P_v = 0.26$

Therefore, as $F_x/P_v < 0.6$, there will be no reduction in moment capacity (see clause 4.2.5).

(e) Moment capacity

The concrete slab provides full restraint to the compression flange (Fig. 3.6), and lateral torsional buckling is not considered. The chosen UB is

34 STRUCTURAL STEELWORK DESIGN TO BS 5950

clause 3.5.2
clause 4.2.5

a plastic section ($b/T = 6.7$) and at mid-span the shear force is zero.

$$M_{cx} = p_y S_x = 275 \times 2370 \times 10^{-3} = \mathbf{651 \text{ kNm}}$$

Note that because the units are p_y (N/mm²) and S_x (cm³) then the 10^{-3} must be included for M_{cx} in order to obtain the correct unit of kNm. Alternatively, there is no need for 10^{-3} if 0.275 kN/mm² is used for p_y. But

$$M_{cx} \not> 1.2\, p_y Z_y = 1.2 \times 275 \times 2080 \times 10^{-3} = 686 \text{ kNm}$$

Note that for I and H sections bent about the x axis the expression $p_y S_x$ governs the design. For bending about the y axis, however, the expression $1.2\, p_y Z_y$ governs the design. The factor 1.2 in this expression may be increased to the ratio factored load/unfactored load (clause 4.2.5):

$$M_x / M_{cx} = 573/652 = \mathbf{0.88}$$

Section is satisfactory.

(f) Deflection

Deflection (which is a serviceability limit state) must be calculated on the basis of the unfactored imposed loads:

$$W_x = 90 + 42 = 132 \text{ kN}$$

Assume the load is approximately triangular and hence formulae are available for deflection calculations[6]

$$\begin{aligned}\delta_x &= W_x L^3 / 60 E I_x \\ &= 132 \times 7400^3 / (60 \times 205 \times 55\,400 \times 10^4) \\ &= \mathbf{8.4 \text{ mm}} \qquad 7.85mm\end{aligned}$$

BS table 5 Deflection limit $= 7400/360 = 20.6$ mm

(g) Connection

The design of connections which are both robust and practicable, yet economic, is developed by experience. Typical examples may be found in references (4, 7).

The connection at each end of the beam must be able to transmit the ultimate shear force of 237 kN to the column or other support. The connection forms part of the beam, i.e. the point of support is the column to cleat interface. Design practice assumes that the column bolts support shear force only, while the beam bolts carry shear force, together with a small bending moment.

$$BM = 237 \times 0.05 = 12.1 \text{ kNm}$$

Assume **9 bolts, 22 mm diameter** (grade 4.6) as shown in Fig. 3.7.

2 no. 90 × 90 × 10 Ls
400 long

$e = 50$
Beam
M22 bolts

Fig. 3.7

BEAMS IN BUILDINGS

(i) **COLUMN BOLTS**

Vertical shear/bolt = 237/6 = 39.5 kN (single shear)

clause 6.3.2 Shear capacity $P_s = p_s A_s$, where A_s is the cross-sectional area at the root of the bolt thread:

$$P_s = 0.160 \times 303 = 48.5 \text{ kN/bolt}$$

clause 6.3.3.2 Bearing capacity of bolts $P_{bb} = dtp_{bb}$:

$$P_{bb} = 22 \times 10 \times 0.460 = 101 \text{ kN/bolt}$$

clause 6.3.3.3 Bearing capacity of the angles (dtp_{bs}) is the same as that for bolts, because p_{bs} for grade 43A steel has the same value as p_{bb} for grade 4.6 bolts, i.e. 460 N/mm². In addition, the bearing capacity of the angles must comply with the criterion, $P_{be} \not> etp_{bs}/2$ (e defined in Fig. 3.7).

$$P_{bs} = 50 \times 10 \times 0.460/2 = 115 \text{ kN/bolt}$$

Note that the column flange will also require checking if it is less than 10 mm thick. Column bolt connection is satisfactory. Capacities of bolts and bearing values may alternatively be obtained from reference (5).

(ii) **BEAM BOLTS**

(See Fig. 3.8.)

Double shear capacity/bolt $P_s = 2 \times 0.160 \times 303 = 97.0$ kN

Vertical shear/bolt $= 242/3 = 80.7$ kN (double shear)

Maximum horizontal bolt forces $= (Md_{max}/\Sigma d^2)$ due to bending moment are discussed in Section 8.4.

Horizontal shear/bolt $= 11.9 \times 0.15/(2 \times 0.15^2) = 39.7$ kN

Resultant shear/bolt $= \sqrt{(80.7^2 + 39.7^2)}$ $= \mathbf{88.4}$ **kN**

Bearing capacity of bolt, $P_{bb} = dtp_{bb}$
$$= 22 \times 10.2 \times 460 = 97.6 \text{ kN}$$

clause 6.3.3.3 As $p_{bs} = p_{bb}$, then bearing capacity of the web plate is the same as for the bolt.

Also, $P_{bs} \not> etp_{bs}/2 = 89 \times 10.2 \times 460/2 = 209$ kN

(iii) **ANGLE CLEATS**

clause 4.2.3 Shear area of cleats (allowing for 24 mm holes)
$$= 0.9(400 \times 10 \times 2 - 3 \times 2 \times 10 \times 24) = 5904 \text{ mm}^2$$

Shear capacity $P_v = 0.6 \times 0.275 \times 5904$ $= 974$ kN

A check for bending may also be carried out but will generally give a high bending capacity relative to the applied moment[3].

Fig. 3.8

3.8 EXAMPLE 5. BEAM SUPPORTING PLANT LOADS (UNRESTRAINED BEAM)

(a) Dimensions

Main beam is simply supported and spans 9.0 m (Fig. 3.9). Boilers are supported symmetrically on secondary beams (A and B) of span 6.0 m, which are at 5.0 m centres.

Fig. 3.9 Plant loads

Fig. 3.10

(b) Loading

Boiler loading (each)	400 kN
Open steel flooring (carried on beams A and B)	0.3 kN/m²
Imposed load (outside boiler area)	4.5 kN/m²

The boilers produce reactions of 100 kN at the end of each secondary beam. Allow 4.0 kN for the self weight of each secondary beam (not designed here). Assume **610 × 305 × 149 UB** (grade 43) for main beam.

Self weight (ultimate) = $1.4 \times 9.0 \times 1.49 = 18.8$ kN

With reference to Fig. 3.10:

Flooring on beam A = $0.3 \times 4.0 \times 6.0 = 7.2$ kN
Imposed load on beam A = $4.5 \times 3.0 \times 4.0 = 54.0$ kN

Imposed load area excludes boiler area.

With reference to Fig. 3.11:

$$\text{Flooring on beam B} = 5.4 \text{ kN}$$
$$\text{Imposed load on beam B} = 40.5 \text{ kN}$$

Ultimate point load $W_A = 1.4(4.0 + 7.2) + 1.6(200 + 54) = 422$ kN
Ultimate point load $W_B = 1.4(4.0 + 5.4) + 1.6(200 + 40.5) = 398$ kN

Fig. 3.11

(c) BM and SF

Ultimate shear force
$$F_1 = 19/2 \times 422 \times 6.0/9.0 + 398 \times 1.0/9.0 = \mathbf{335 \text{ kN}}$$
Ultimate shear force
$$F_2 = 19/2 \times 398 \times 8.0/9.0 + 422 \times 3.0/9.0 = \mathbf{504 \text{ kN}}$$

With reference to Fig. 3.12:

Ultimate moment $M_A = 335 \times 3.0 = \mathbf{1005 \text{ kNm}}$
Ultimate moment $M_B = 504 \times 1.0 = \mathbf{504 \text{ kNm}}$

Note that in calculating the moments, the small reduction due to the self weight is ignored.

Fig. 3.12

(d) Shear capacity

Design strength p_y for steel 19.7 mm thick (grade 43A) = 265 N/mm² (see Table 1.2).

clause 4.2.3 Shear capacity $P_v = 0.6\, p_y A_v$
$$= 0.6 \times 265 \times 609.6 \times 11.9 \times 10^{-3} = 1153 \text{ kN}$$

Shear force $F_1 = 335$ kN
$$F_1/P_v = 335/1153 = 0.29$$

Shear force $F_2 = 504$ kN
$$F_2/P_v = 504/1153 = 0.44 \leq 0.6$$

This maximum coexistent shear force is present at point B, while the maximum moment occurs at point A.

(e) Moment capacity

clause 3.5.2 The chosen section is a plastic section ($b/T = 7.7$).

Moment capacity $M_c = p_y S_x$
$$= 265 \times 4570 \times 10^{-3} = \mathbf{1210 \text{ kNm}}$$
Moment ratio $M_A/M_c = 1005/1210 = 0.83 < 1$

Section is satisfactory.

(f) Buckling resistance

The buckling resistance moment of the beam in the part-span AB must be found, and in this part the moment varies from 504 kNm to 1005 kNm. It is assumed that the steel flooring does not provide lateral restraint, but that the secondary beams give positional and rotational restraint at 5.0 m spacing. Loading between the restraints is of a minor nature (self weight only) and is ignored for use of BS table 13 as it would affect the moments by less than 10% (see also BS table 16).

clause 4.3.5

$L_E = 5.0$ m
Slenderness $\lambda = L_E/r_y$
$= 5000/69.9 = 72$ (both r_y and L_E given in mm)
Torsional index $x = 32.5$
$\lambda/x = 2.2$

BS table 14
BS table 13
clause 4.3.7.5

$v = 0.94$ (for $N = 0.5$, i.e. equal flanges)
$n = 1.0$ (for member not loaded between restraints)
$u = 0.886$
$\lambda_{LT} = nuv\lambda$
$= 1.0 \times 0.886 \times 0.94 \times 72 = 60$

BS table 11

Bending strength $p_b = 207$ N/mm^2
Buckling resistance $M_b = p_b S_x$
$= 207 \times 4570 \times 10^{-3} = \mathbf{946 \text{ kNm}}$

clause 4.3.7.2,
BS tables 13, 18

Equivalent uniform moment factor m needs to be obtained for a member not loaded between restraints:

$\beta = 504/1005 = 0.50$
hence $m = 0.76$

Equivalent uniform moment $\overline{M} = mM_A$
$= 0.76 \times 1005 = \mathbf{764 \text{ kNm}}$

hence $\overline{M}/M_b = 764/946 = \mathbf{0.81}$ and section is satisfactory.

Try a smaller section (610 × 229 × 140 UB):

$S_x = 4150$ cm^4
$\lambda = 99.4$
$P_b = 161$ N/mm^2
$M_b = 668$ kNm
$\overline{M}/M_b = \mathbf{1.14}$

which is not satisfactory.

This comparison indicates the sensitivity of the buckling resistance moment to small changes in section properties, particularly to a reduction in flange width.

(g) Deflection

Calculation of the deflection for the serviceability imposed loading cannot be carried easily by the use of formulae, which become complex for non-standard cases. Serviceability point load

$$W_A = 200 + 54 = 254 \text{ kN}$$
$$W_B = 200 + 40.5 = 240 \text{ kN}$$

With reference to Fig. 3.13, mid-span deflection may be found by the moment-area method[8,9]:

$$\delta = \int (M(x)/EI)dx$$
$$= (149 \times 0.667 + 1045 \times 2.75 + 355 \times 3.333)/$$
$$(205 \times 124660 \times 10^{-5}) = \mathbf{16.3 \text{ mm}}$$

Calculation by Macaulay's method[10,11] gives the point of maximum deflection 4.65 m from support 2 with a value of 17.6 mm.

Deflection limit = 9000/200 = 45 mm

An approximate estimate of deflection is often obtained by treating the load as an equivalent u.d.l.

Fig. 3.13 BM diagram (areas in **bold**)

(h) Connection

The connection at support 2 of the main beam must transmit an ultimate shear force of 504 kN and follows the method given in Section 3.7(g):

$$M = 504 \times 0.05 = 25.3 \text{ kNm}$$

Assume 22 mm bolts (grade 8.8) as shown in Fig. 3.14.

Fig. 3.14

(i) COLUMN BOLTS

clause 6.2.3
clause 6.3.3.3

Vertical shear/bolt $F_s = 504/10 = 50.4$ kN
Shear capacity/bolt $P_s = 0.375 \times 303 = 114$ kN
Bearing capacity of angles/bolt $P_{bs} = 22 \times 12 \times 0.460 = 121$ kN
but P_{bs}/bolt $\not> 50 \times 12 \times 0.460/2 = 138$ kN

Note that the column flange will require checking if less than 12 mm thick. Column bolt connection is satisfactory.

(ii) BEAM BOLTS

(See Fig. 3.15.)

Vertical shear/bolt	$= 504/5$	$= 100.8$ kN
Horizontal shear/bolt	$= M d_{max}/\Sigma d^2$	
	$= 25.3 \times 0.20/2(0.10^2 + 0.20^2)$	
		$= 50.6$ kN
Resultant shear/bolt	$= \sqrt{(100.8^2 + 50.6^2)}$	$= \mathbf{113}$ **kN**
Shear capacity (double shear)	$= 0.375 \times 2 \times 303$	$= 227$ kN
Bearing capacity/bolt	$P_{bb} = 22 \times 11.9 \times 1.035$	$= 271$ kN
Bearing capacity of web plate	$P_{bs} = 22 \times 11.9 \times 0.460$	$= 120$ kN
but	$P_{bs} \ngtr 89 \times 11.9 \times 0.460/2 = 243$ kN	

clause 6.3.3.3

Beam bolt connection is satisfactory.

(iii) ANGLE CLEATS

clause 4.2.3

Shear area of cleats (24 mm holes)
$$= 0.9(500 \times 12 \times 2 - 5 \times 2 \times 12 \times 24) = 8208 \text{ mm}^2$$

Shear capacity $P_v = 0.6 \times 0.275 \times 8208 = 1350$ kN

$F_v/P_v = 504/1350 = 0.37$

Connection cleat is satisfactory.

Fig. 3.15

STUDY REFERENCES

Topic	References
1. Lateral restraint BS 5950	**Dowling P.J., Knowles P. & Owens G.W.** (1988) *Structural Steel Design.* Steel Construction Institute
2. Strut behaviour	**Marshall W.T. & Nelson H.M.** (1990) Elastic stability analysis, *Structure*, pp. 420–52. Longman
3. Strut behaviour	**Coates R.C., Coutie M.G. & Kong F.K.** (1988) Instability of struts and frameworks, *Structural Analysis*, pp. 58–71. Van Nostrand Reinhold
4. Connections	(1993) *Joints in Simple Connections* vol. 1. Steel Construction Institute
5. Bolt details	(1985) *Steelwork Design* vol. 1, Section properties, member capacities. Steel Construction Institute
6. Deflection formulae	(1992) Design theory, *Steel Designers' Manual*, pp. 1026–50. Blackwell
7. Connections	(1992) *Joints in Simple Connections* vol. 2. Steel Construction Institute
8. Moment-area method	**Croxton P.L.C. & Martin L.H.** (1990) Area-moment methods of analysis, *Solving Problems in Structures* vol. 2. pp. 25–47. Longman

9. Moment-area method	**Coates R.C., Coutie M.G. & Kong F.K.** (1988) Moment-area methods, *Structural Analysis*, pp. 176–81. Van Nostrand Reinhold
10. Deflection	**Marshall W.T. & Nelson H.M.** (1990) Singularity functions, *Structures*, pp. 233–8. Longman
11. Deflection	**Hearn E.J.** (1985) Slope and deflection of beams, *Mechanics of Materials* vol. 1, pp. 102–7. Pergamon

4

PURLINS AND SIDE RAILS

In the UK purlins and side rails used in the construction of industrial buildings are often fabricated from cold formed sections. These sections can be designed in accordance with Part 5 of BS 5950, but the load tables for these sections are frequently based on test data. The sections are marketed by companies specializing in this field who will normally give the appropriate spans and allowed loadings in their catalogues. Sections of this kind are commonly of channel or zed form as illustrated in Fig. 4.1. Although their design is not covered in this chapter, the selection of cold formed sections is discussed in Chapter 12.

Fig. 4.1 Cold formed purlins

Sigma   Zed   Zeta

Hot rolled sections may be used as an alternative, and in some situations may be preferred to cold formed sections. The design of angles and hollow sections may be carried out by empirical methods which are covered by clause 4.12.4. The full design procedure (i.e. non-empirical) is set out in this chapter.

4.1 DESIGN REQUIREMENTS FOR PURLINS AND SIDE RAILS

The design of steelwork in bending is dependent on the degree of lateral restraint given to the compression flange and the torsional restraint of the beam, and also on the degree of lateral/torsional restraint given at the beam supports. These restraints are given in detail in clause 4.3 and have been discussed and demonstrated in Chapter 3. Side rails and purlins may be considered to have lateral restraint of the compression flange owing to the presence of the cladding, based on adequate fixings (clause 4.12.1). Loads will be transferred to the steel member via the cladding (see Fig. 4.2), and the

Fig. 4.2

dead, imposed and wind pressure loads will cause the flange restrained by the cladding to be in compression. Wind suction load can, however, reverse this arrangement, i.e. the unrestrained flange will be in compression. Torsional restraint to a beam involves both flanges being held in position and for purlins and side rails this will be true only at the supports.

Side rails are subjected to both vertical loading (cladding) and horizontal loading (wind pressure/suction), but in general the vertical loading is considered to be taken by the cladding acting as a deep girder. Consequently, only moments in the horizontal plane (due to wind) are considered in design. In the design of new construction where the cladding is penetrated by holes for access, ductwork or conveyors, the design engineer should be satisfied that the cladding and fixings are capable of acting in this manner.

Sag rods are sometimes used to reduce the effective length of purlins and side rails, and result in continuous beam design (see Section 4.4). Where sag rods are used, provision must be made for the end reaction on eaves or apex beams. As is shown in the following examples, there is no reason why purlins and side rails should not be designed as beams subject to biaxial bending in accordance with the normal design rules.

4.2 EXAMPLE 6. PURLIN ON SLOPING ROOF

(a) Dimensions

See Fig. 4.2: purlins at 2.0 m centres; span 6.0 m simply supported; rafter slope 20°.

(b) Loading

Dead load (cladding + insulation panels) 0.15 kN/m²
Imposed load   0.75 kN/m² (on plan)
Wind load   0.40 kN/m² (suction)

Reference should be made to Chapter 2 for the derivation of loads, the direction in which each will act, and the area appropriate to each load. Maximum values of bending moment and shear force must be found at the ultimate limit state making due allowance for the slope angle and including the γ_f factors.

Assume purlin to be **152 × 76** channel section, grade 43A steel (see Fig. 4.3).

Fig. 4.3 W_d (or W_i or W_w) 6.0 m at 2.0 m centres

Cladding $2.0 \times 6.0 \times 0.15 = 1.80$ kN
Self weight $6.0 \times 0.18 = \underline{1.08\text{ kN}}$
Total dead load $W_d = 2.88$ kN
Imposed load $= 2.0 \cos 20° \times 6.0 \times 0.75$
$W_i = 8.46$ kN
Wind load $= 2.0 \times 6.0 \times (-0.40)$
$W_w = -4.80$ kN

The rafter slope of 20° results in purlins at the same angle. Components of load are used to calculate moments about the x and y axes, i.e. normal and tangential to the rafter (Fig. 4.4). As with side rails, it would be possible to ignore bending in the plane of the cladding, but in practice, biaxial bending is usually considered in purlin design.

$W_{dx} = 2.88 \cos 20° = 2.71$ kN
$W_{dy} = 2.88 \sin 20° = 0.99$ kN
$W_{ix} = 8.46 \cos 20° = 7.95$ kN
$W_{iy} = 8.46 \sin 20° = 2.89$ kN
$W_{wx} = -4.80$ kN

Note that W_{wy} is zero as wind pressure is perpendicular to the surface on which it acts, i.e. normal to the rafter.

Ultimate load $W_x = 1.4 \times 2.71 + 1.6 \times 7.95 = 16.5$ kN

where 1.4 and 1.6 are the appropriate λ_f factors (Section 1.7).

Fig. 4.4

(c) BM and SF

Max. ultimate moment $M_x = 16.5 \times 6.0/8 = 12.4$ kNm
Max. ultimate shear force $F_x = 16.5/2 = 8.3$ kN
Similarly, $W_y = 6.0$ kN
$M_y = 4.5$ kNm
$F_y = 3.0$ kN

Fig. 4.5

(d) Shear capacity

Design strength p_y is given in Table 1.2 and for the selected purlin section is 275 N/mm².

clause 4.2.3

Shear area $A_{vx} = 152.4 \times 6.4 = 975$ mm²
Shear capacity $P_{vx} = 0.6\, p_y A_{vx}$
$= 0.6 \times 275 \times 975 \times 10^{-3} = 161$ kN
Shear area $A_{vy} = 0.9 A_o$
$= 0.9 \times 2 \times 76.2 \times 9.0 = 1234$ mm²
Shear capacity $P_{vy} = 0.6 \times 275 \times 1234 \times 10^{-3} = 204$ kN

==It may be noted that in purlin design, shear capacity is usually high relative to shear force.==

PURLINS AND SIDE RAILS

(e) Moment capacity

The section classification of a channel subject to biaxial bending depends on b/T and d/t which in this case are 8.47 and 16.5, respectively. The channel is therefore a plastic section. Hence, the moment capacity

BS table 7

clause 4.2.5
$$M_{cx} = p_y S_x = 275 \times 130 \times 10^{-3} = 35.8 \text{ kNm}$$

but M_{cx} must not exceed $1.2 \, p_y Z_x$
$$= 1.2 \times 275 \times 112 \times 10^{-3} = 37.0 \text{ kNm}$$

Note that 10^{-3} must be included to give M_{cx} in kNm, when N/mm² for p_y and cm³ for S_x. Alternatively, p_y may be expressed as 0.275 kN/mm², but this requires care later when axial forces and stresses are used.

$$M_{cy} = p_y S_y = 0.275 \times 41.3 = 11.4 \text{ kN m}$$

The ratio S_y/Z_y is greater than 1.2 and hence the constant 1.2 is replaced by the ratio factored load/unfactored load $(6.0/[0.99 + 2.89] = 1.55)$.

M_{cy} must not exceed $1.55 \, p_y Z_y$
$$= 1.55 \times 0.275 \times 21.0 = 9.0 \text{ kNm}$$

The local capacity check may now be carried out (Section 3.4):

$M_x/M_{cx} + M_y/M_{cy} \not> 1$ (for a channel section)
$12.4/35.8 + 4.5/9.0 = \textbf{0.85}$

The local capacity of the section is therefore adequate.

(f) Buckling resistance

The buckling resistance moment M_b of the section does not need to be found because the beam is restrained by the cladding in the x plane (Fig. 4.6) and instability is not considered for a moment about the minor axis (Fig. 4.7) (Section 3.1).

Fig. 4.6   **Fig. 4.7**

(g) Wind suction

The effect of the wind suction load has so far not been considered, and in some situations it could be critical. In combination with loads W_d and W_i, a lower total load W is clearly produced.

$$\begin{aligned}
\text{Ultimate load } W_x &= 1.0 \times 2.71 + 1.4 \times 4.8 = -4.01\,\text{kN}\\
M_x &= -4.01 \times 6.0/8 &= -3.01\,\text{kNm}\\
W_y &= 1.0 \times 0.99 &= 0.99\,\text{kN}\\
M_y &= 0.99 \times 6.0/8 &= 0.75\,\text{kNm}
\end{aligned}$$

The value of M_x is much lower than the value 12.8 kNm used earlier, but the negative sign indicates that the lower flange of the channel is in compression and this flange is not restrained. The buckling resistance M_b must therefore be found.

The effective length L_E of the purlin may be found from BS table 9.

clause 4.3.5

$$L_E = 1.0 \times 6.0 = 6.0\,\text{m}$$

Slenderness $\lambda = L_E/r_y = 6000/22.4 = 268$ (which is less than 350 as required by clause 4.7.3.2) where L_E and r_y are in mm. Equivalent slenderness λ_{LT} allowing for lateral torsional buckling is given by:

$$\lambda_{LT} = nuv\lambda$$

Torsional index $x = 14.5$
$\lambda/x = 268/14.5 = 18$

BS table 14

$v = 0.49$
$u = 0.902$

BS table 16

$n = 0.94$ (for $\beta = 0$ and $\gamma = 0$)
$\lambda_{LT} = 0.94 \times 0.902 \times 0.49 \times 268 = 111$

Clause 4.3.7.4
BS table 11

Bending strength p_b may be obtained: $p_b = 108\,\text{N/mm}^2$

Buckling resistance $M_b = p_b S_x$
$= 108 \times 130 \times 10^{-3} = 14.0\,\text{kNm}$

BS table 13

The overall buckling check may now be carried out using an equivalent uniform moment factor (m) equal to 1.0 (member loaded between restraints):

$$mM_x/M_b + mM_y/M_{cy} \not> 1$$

$$3.01/14.0 + 0.75/(275 \times 41.3 \times 10^{-3}) = \mathbf{0.28}$$

The overall buckling of the section is therefore satisfactory.

The diagrams for bending moment and shear force shown in Fig. 4.5 indicate that maximum values are not coincident and it is not therefore necessary to check moment capacity in the presence of shear load. Purlin design does not normally need a check on web bearing and buckling as the applied concentrated loads are low – note the low values of shear force. The check for bearing and buckling of the web is particularly needed where heavy concentrated loads occur, and reference may be made to Chapter 5 for the relevant calculations.

PURLINS AND SIDE RAILS

(h) Deflection

Deflection limits for purlins are not specified in BS table 5 but a limit of span/200 is commonly adopted.

Deflection $\delta_x = 5W_x L^3/384EI_x$

where W_x is the serviceability imposed load, i.e. 7.95 kN and E is 205 kN/mm²

$\delta_x = 5 \times 7.95 \times 6000^3/(384 \times 205 \times 852 \times 10^4) = \mathbf{12.8\,mm}$
$\delta_y = \mathbf{28.5\,mm}$
Deflection limit $= 6000/200 = 30\,mm$

(i) Connections

The connection of the purlin to the rafter may be made by bolting it to a cleat as shown in Fig. 4.8. The design of these connections is usually nominal due to the low reactions at the end of the purlins. However, the transfer of forces between the purlin and rafter should be considered. For the channel section chosen, W_x and W_y transfer to the rafter through a cleat. Bolts must be provided but will be nominal due to the low reactions involved (8.0 kN and 3.0 kN). Chapter 3 gives calculations for a bolted connection in more detail.

Multi-span (continuous) purlins may be used and minor changes in design are considered in Section 4.4.

Fig. 4.8 Purlin connection

4.3 EXAMPLE 7. DESIGN OF SIDE RAIL

(a) Dimensions

(See Fig. 4.9): side rails at 2.0 m centres; span 5.0 m simply supported.

(b) Loading

Dead load (cladding/insulation panels)   0.18 kN/m²
Wind load (pressure)   0.80 kN/m²

48 STRUCTURAL STEELWORK DESIGN TO BS 5950

Fig. 4.9

Maximum values of bending moment and shear force must be found allowing for the wind loading (horizontal) only (Fig. 4.9) and including the safety factor γ_f.

Assume the side rail to be **125 × 75 × 10** unequal angle, grade 43 steel. An angle, such as that chosen, provides greater resistance to bending (higher section properties) about the x axis than the y axis, compared to that for an equal angle of the same area (weight).

$$
\begin{aligned}
&\text{Cladding } 2.0 \times 5.0 \times 0.18 &&= 1.80\,\text{kN} \\
&\text{Self weight } 5.0 \times 0.15 &&= \underline{0.75\,\text{kN}} \\
&\text{Total dead load } W_d &&= 2.55\,\text{kN} \\
&\text{Wind load } W_w = 2.0 \times 5.0 \times 0.80 &&= 8.0\ \text{kN}
\end{aligned}
$$

The loads W_w and W_d act in planes at right angles producing moments about x and y axes of the steel section, but only moments about x are used in design (as discussed in Section 4.1).

$$
\begin{aligned}
\text{Ultimate load } W_x &= 1.4 W_w \\
&= 1.4 \times 8.0 = 11.2\,\text{kN}
\end{aligned}
$$

(c) BM and SF

With reference to Fig. 4.10:

Maximum moment $\quad M_x = 11.2 \times 5.0/8 = \mathbf{7.0\,kNm}$
Maximum shear force $F_x = 11.2/2 \quad\quad\ = \mathbf{5.6\,kN}$

Fig. 4.10

(d) Shear capacity

Design strength p_y is 275 N/mm² (Section 1.7).

clause 4.2.3

Shear area $\quad A_v = 0.9 \times 125 \times 10 = 1125\,\text{mm}^2$
Shear capacity $P_v = 0.6 \times 275 \times 1125 \times 10^{-3} = 186\,\text{kN}$

The shear capacity is clearly very large relative to the shear force.

Fig. 4.11

BS table 7

(e) Moment capacity

For single angles, lateral restraint is provided by the cladding, which also ensures bending about the x axis, rather than about a weaker axis (Fig. 4.11). The moment capacity only of the section is therefore checked. The section chosen is defined as semi-compact having $b/T = 7.5$ and $d/T = 12.5$ (both < 15), and $(b+d)/T = 20$ (< 23), hence:

$$M_{cx} = p_y Z_x$$
$$= 275 \times 36.5 \times 10^{-3} = 10.0 \, \text{kNm}$$
$$M_x/M_{cx} = 7.0/10/.0 = \mathbf{0.70}$$

Section is satisfactory.

The design of side rails does not normally include a check on web bearing and buckling, as discussed in Section 4.2(g).

(f) Deflection

Calculation of deflection is based on the serviceability condition, i.e. with unfactored loads.

$$W_w = 8.0 \, \text{kN}$$
$$\delta_y = 5 W_w L^3 / 384 E I_x$$
$$= 5 \times 8.0 \times 5000^3 / (384 \times 205 \times 302 \times 104)$$
$$= \mathbf{21.0 \, mm}$$

Although clause 4.12.2 avoids specifying any value, use a deflection limit of, say, $L/200 = 25$ mm.

4.4 EXAMPLE 8. DESIGN OF MULTI-SPAN PURLIN

Continuity of a structural element over two or more spans may be useful in order to reduce the maximum moments to be resisted, and hence the section size, and to improve the buckling resistance of the member.

1. In general, the bending moments in a continuous beam are less than those in simply supported beams of the same span. It should be noted, however, that a two-span beam has the same moment ($WL/8$) at the middle support as the mid-span moment of a simply supported beam.
2. The resistance of a member of lateral torsional buckling is improved by continuity and this is reflected in BS table 16.

Continuity may be achieved by fabricating members of length equal to two or more spans. Length will, however, be limited by requirements for delivery and flexibility during site erection. For the purlin designed in Section 4.2 a length of not more than two spans (12 m) would be acceptable (they can be delivered bundled together to reduce flexibility). Continuity can also be arranged by use of site connections capable of transmitting bending moments. Such connections are costly to fabricate and to assemble and are rarely used in small structural elements such as purlins and side rails. Using the same example as in Section 4.2, the design is repeated for a purlin continuous over two spans of 6.0 m.

(a) Dimensions

As Section 4.2(a).

(b) Loading

As Section 4.2; assume purlin to be **152 × 76** channel, grade 43 steel:

$W_d = 2.88$ kN
$W_{dx} = 2.71$ kN
$W_{dy} = 0.99$ kN
Ultimate load $W_x = 16.5$ kN
Ultimate load $W_y = 6.0$ kN

(c) BM and SF

With reference to Fig. 4.12:

Maximum ultimate moment (at central support)

$M_x = 16.5 \times 6.0/8 = 12.4$ kNm
$F_x = 0.65 \times 16.5 = 10.4$ kN
$M_y = 4.5$ kNm
$F_y = 3.8$ kN

Fig. 4.12

(d) Shear capacity

Shear force is less than 0.6 shear capacity, as Section 4.1(d).

(e) Moment capacity

BS table 7   The 152 × 76 channel is a compact section ($b/T = 8.45$), hence

$M_{cx} = 275 \times 130 \times 10^{-3} = 35.8$ kNm
$M_{cy} = 9.0$ kNm

$M_x/M_{cx} + M_y/M_{cy} \not> 1$

$12.4/35.8 + 4.5/9.0 = \mathbf{0.85}$

PURLINS AND SIDE RAILS

(f) Buckling resistance

$$\lambda = L_E/r_y = 6000/22.4 = 268$$

The factor n is obtained from BS table 16 for $\beta = 0$ and $\gamma = M/M_o = -1.0$, as the end moment M and simply supported moment M_o are equal (but opposite sign), hence

$n \ \ = 0.66$
$m \ \ = 1.0$
$u \ \ = 0.902$
$\lambda/x = 268/14.5 = 18.5$

where x is the torsional index.

BS table 14
$v \ \ = 0.49$
$\lambda_{LT} = 0.66 \times 0.902 \times 0.49 \times 268 = 78$

BS table 11
$p_b \ \ = 170 \text{ N/mm}^2$
$M_b \ \ = 170 \times 130 \times 10^{-3} = 22.1 \text{ kNm}$

$M_x/M_b + M_y/M_{cy} \not> 1$

$12.4/22.1 + 4.5/(275 \times 41.3 \times 10^{-3}) = \mathbf{0.96}$

(g) Deflection

From Example 6, the imposed load at serviceability limit state is

$W_{ix} = 7.95 \text{ kN}$
$\delta_x \ \ = 7.95 \times 6000^3/(185 \times 205 \times 852 \times 10^4) = \mathbf{5.3 \text{ mm}}$
$W^{iy} = 2.89 \text{ kN}$
$\delta^y \ \ = \mathbf{14.5 \text{ mm}}$
Deflection limit $6000/200 = 30 \text{ mm}$

5

CRANE GIRDERS

Fig. 5.1 Crane gantry girder

(diagram labels: Crane rail; Crane gantry girder (UB and plate welded together))

Industrial buildings commonly house manufacturing processes which involve heavy items being moved from one point to another during assembly, fabrication or plant maintenance. In some cases overhead cranes are the best way of providing a heavy lifting facility covering virtually the whole area of the building. These cranes are usually electrically operated, and are provided by specialist suppliers. The crane is usually supported on four wheels running on special crane rails. These rails are not considered to have significant bending strength, and each is supported on a crane beam or girder (Fig. 5.1). The design of this girder, but not the rail, is part of the steelwork designer's brief. However, the position and attachment of the rail on the crane girder must be considered, as a bad detail can led to fatigue problems, particularly for heavy duty cranes. The attachment of the rail should allow future adjustment to be carried out, as continuous movement of the crane can cause lateral movement of the rail.

5.1 CRANE WHEEL LOADS

Parts of a typical overhead crane are shown in Fig. 5.2. The weight or load associated with each part should be obtained from the crane supplier's data, and then be combined to give the crane wheel loads. Alternative wheel loads may be given directly by the crane manufacturer. Reference may be made to BS 6399: Part 1[1] for full details of loading effects. The following notes apply to single crane operation only.

The crab with the hook load may occupy any position on the crane frame up to the minimum approach shown in Fig. 5.2. Hence the vertical load on the nearer pair of wheels can be calculated, adding an amount for the crane frame, which is usually divided equally between the wheels. Maximum wheel loads are often provided by the crane manufacturer.

An allowance for impact of 25% is made for most light/medium duty cranes (classes Q1 and Q2), and this is added to each vertical wheel load. For heavy duty cranes (classes Q3 and Q4) reference should be made to BS 2573[2] and to supplier's data for appropriate impact values.

In addition to the vertical loads transferred from the wheels to the crane rail, horizontal loads can also develop. The first of these is called surge and acts at right angles (laterally) to the girder and at the level of the rail. This surge load covers the acceleration and braking of the crab when moving

Fig. 5.2 Overhead crane

along the crane frame, together with the effects of non-vertical lifting. The value of this load is assessed in BS 6399[1] at 10% of the sum of the crab weight and hook load. It is divided equally between the four crane wheels when the wheels are double flanged and can act in either direction.

The second horizontal load (longitudinal) is the braking load of the whole crane, and in this case acts along the crane girder at the level of the top flange. The value of this load is assessed at 5% of each wheel load, and is therefore a maximum when the wheel load is a maximum. As before, the braking load covers acceleration as well as non-vertical lifting.

The loads are summarized in Fig. 5.3. In addition, gantry girders intended to carry class Q3 and Q4 cranes (as defined in BS 2573: Part 1) should be designed for the crabbing forces given in clause 4.11.2.

Fig. 5.3 Crane loads

The safety factor γ_f for crane loads (ultimate limit state) is taken as 1.6, i.e. as for imposed loads generally (Section 1.7). Whenever the vertical load and the surge load are combined in the design of a member, the safety factor should however be taken as 1.4 for both loads (BS table 2). Further detailed provisions for gantry girders are given in clauses 4.11 and 2.4.1.2.

5.2 MAXIMUM LOAD EFFECTS

Moving loads, such as crane wheels, will result in bending moments and shear forces which vary as the loads travel along the supporting girder. In simply supported beams the maximum shear force will occur immediately adjacent to a support, while the maximum bending moment will occur near, but not necessarily at, mid-span. In general, influence lines[3,4,5] should be used to find the load positions producing maximum values of shear force and bending moment.

The maximum effects of two moving loads may be found from formulae[3] as demonstrated in Section 2.5. For a simply supported beam the load positions shown in Fig. 5.4 give maximum values:

Shear force (max) $= W(2 - c/L)$
Bending moment (max) $= WL/4$
or $= 2W(L/2 - c/4)^2/L$

The greater of the bending moment values should be adopted.

The design of the bracket supporting a crane girder uses the value of maximum reaction from adjacent simply supported beams, as in Fig. 5.4. Where adjacent spans are equal, the reaction is equal to the shear force, i.e.

Reaction (max) $= W(2 - c/L)$

Fig. 5.4 Maximum BM, SF and R

CRANE GIRDERS 55

In all cases the effect of self weight (uniformly distributed) of the girder must be added.

5.3 EXAMPLE 9. CRANE GIRDER WITHOUT LATERAL RESTRAINT ALONG SPAN

(a) Dimensions

Span of crane	15.0 m
Wheel centres	3.5 m
Minimum hook approach	0.7 m
Span of crane girder	6.5 m (simply supported)

(b) Loading

Class Q2 (no crabbing forces need be calculated)
Hook load	200 kN
Weight of crab	60 kN
Weight of crane (excluding crab)	270 kN

(c) Wheel loads

Vertical wheel load from:
 hook load $200(15.0 - 0.7)/(15.0 \times 2)$ = 95.3 kN
 crab load $60(15.0 - 0.7)/(15.0 \times 2)$ = 28.6 kN
 crane load $270/4$ = 67.5 kN
 Total vertical load = 191.4 kN per wheel
Vertical load W_c (including allowance for impact and γ_f)
 $= 1.25 \times 1.4 \times 191.4 = 335$ kN
Where vertical load is considered acting alone then γ_f is 1.6 and W_c becomes 383 kN.

Lateral (horizontal) surge load is 10% of hook+crab load:
 $= 0.10(200 + 60) = 26.0$ kN
Total lateral load $= 26/4$ = 6.5 kN per wheel
Surge load W_{hc} (including γ_f) $= 1.4 \times 6.5$ = 9.1 kN
Longitudinal (horizontal) braking load is 5% of wheel load and including γ_f is:

$0.05 \times 1.6 \times 191.4 = 15.3$ kN per wheel

In the following design it will become clear that the critical considerations are lateral buckling, and web bearing at the support. Hence first sizing of the girder would be based on these criteria. Assume a $610 \times 305 \times 179$ UB (grade 43) with extra plate (grade 43) welded to top flange. This plate is used to give additional strength to the top flange which is assumed to act alone to resist the lateral (surge) loading. A channel section may be preferred instead of the flat plate; this type of section was commonly used in the past.

Dead load due to self weight of girder (1.79+0.36 kN/m) and rail (0.25 kN/m) including γ_f is

$$W_d = 1.4 \times 2.40 \times 6.5 = 21.8 \text{ kN}$$

Fig. 5.5

(d) BM and SF

Moment due to vertical wheel loads is
either $W_c L/4$ $= 335 \times 6.5/4 = 544$ kN
or $2W_c(L/2 - c/4)^2/L = 2 \times 335(6.5/2 - 3.5/4)^2/6.5$
$= 581$ kNm (664 kNm when acting alone)
Moment due to dead load $= 21.5 \times 6.5/8 = 18$ kNm
Max. ultimate moment $M_x = 581 + 18$
$= \mathbf{599}$ **kNm (682 kNm when acting alone)**

Although the dead load maximum BM occurs at mid-span, and the wheel maximum occurs a distance $c/4$ away, it is usual to assume the value of M_x to be the sum of the maxima as shown.

Moment due to surge load $= 2 \times 9.1(6.5/2 - 3.5/4)^2/6.5$
$= 15.8$ kNm
Max. ultimate moment $M_y = \mathbf{15.8}$ **kNm**

Shear force due to vertical wheel loads is:
$W_c(2 - c/L) = 335(2 - 3.5/6.5)$
$= 490$ kN (560 kN when acting alone)
Vertical shear force due to dead load $= 21.8/2 = 11$ kN
Max. ultimate shear force $F_x = 490 + 11$
$= \mathbf{501}$ **kN (571 kN when acting alone)**
Lateral shear force due to surge load $= 9.1(2 - 3.5/6.5) = 13.3$ kN
Max. ultimate shear force $F_y = \mathbf{13.3}$ **kN**
Max. ultimate reaction $R_x = 490 + 21.8 = 512$ kN
$R_y = 13.3$ kN

CRANE GIRDERS 57

(e) Shear capacity

clause 4.2.3 Design strength for chosen section with flange 22 mm thick:
Table 1.2
$$p_y = 265 \text{ N/mm}^2$$

Shear capacity $P_{vx} = 0.6 p_y A_v$
$$= 0.6 \times 0.265 \times 617.5 \times 14.1 = 1380 \text{ kN}$$
$$F_x/P_{vx} = \mathbf{0.41} < 0.60$$
Shear capacity $P_{vy} = 0.6 \times 0.265(307 \times 23.6 + 280 \times 120) = 2030 \text{ kN}$
$$F_y/P_{vy} = \mathbf{0.01} < 0.60$$

(f) Moment capacity

clause 3.5.5 The chosen section (Fig. 5.6) is a plastic section with

b/T (internal) $= 280/20 = 14$
b/T (external) $= 75.5/10.4 = 6.5$

Fig. 5.6

For the built-up section chosen, the designer may need to calculate the plastic modulus (S_x)[6,7], if this is not available in published tables. The properties may be obtained from formulae given in Appendix A.

Area of plate $A_p = 280 \times 20 = 5600 \text{ mm}^2$
Total area $A = 228 + 5600 \times 10^{-2} = 284 \text{ cm}^2$

Plastic section properties:

$d_p = 5600/(2 \times 14.1) = 198.6 \text{ mm}$ (i.e. 130 mm below top face)
$S_x = 5520 + [14.1 \times 198.6^2 + 5600(617.5/2 + 20/2 - 198.6)]10^{-3}$
$\quad = 6750 \text{ cm}^3$
S_y (for top flange only) $= (23.6 \times 307^2/4 + 20 \times 280^2/4)10^{-3}$
$\quad = 948 \text{ cm}^3$

Elastic section properties:

$d_e = 5600(617.5 + 20)/(284 \times 10^2)$
$I_x = 152\,000 + 228 \times 62.8^2 \times 10^{-2}$
$\quad + 5600(617.5/2 + 20/2 - 62.8)^2 \times 10^{-4}$
$\quad = 198\,000 \text{ cm}^4$
$Z_x = 198\,000/(617.5/2 + 62.8)\,10^{-1} = 5320 \text{ cm}^3$

$I_y = 11\,400 + 20 \times 280^3/(12 \times 10^4) = 15100 \text{ cm}^4$
$r_y = \sqrt{(I_y/A)} = \sqrt{(15\,100/284)} = 7.28 \text{ cm}$

for tension flange about y-y axis
$$I_{tf} = 23.6 \times 307^3/(12 \times 10^4) = \quad 5690 \text{ cm}^4$$

for compression flange about y-y axis
$$I_{cf} = 5690 + 20 \times 280^3/(12 \times 10^4) = 9350 \text{ cm}^4$$
$$Z_y \text{ (for top flange only)} = 2\, I_{cf}/B$$
$$= 2 \times 9350/(307 \times 10^{-1}) = 609 \text{ cm}^3$$

Torsional index may be calculated using the appropriate method in BS appendix B.2.5.1(c).

$$h_s = 617.5 + 20 - 23.6/2 - 43.6/2 = 604 \text{ mm}$$
$$\Sigma b_t + h_w t_w = \text{total area} = 28\,400 \text{ mm}^2$$
$$\Sigma b_t^3 + h_w t_w^3 = 2 \times 307 \times 23.6^3 + 280 \times 20.0^3 + 570.3 \times 14.1^3$$
$$= 11.91 \times 10^6 \text{ mm}^4$$
$$x = 604[28\,400/(11.91 \times 10^6)]^{1/2} = 29.5 \text{ mm}$$

Local moment capacity
$$M_{cx} = p_y\, S_x$$
$$= 265 \times 6750 \times 10^{-3} = \mathbf{1790} \text{ kNm}$$

To prevent plasticity at working load (see Section 3.4), $M_{cx} \not> 1.4\, p_y\, Z_x$ where factored load/unfactored load = 1.4

$$M_{cx} \not> 1.4 \times 265 \times 5320 \times 10^{-3} = 1970 \text{ kNm}$$
$$M_{cy} = p_y\, S_y \text{ where } S_y \text{ is for top flange only}$$
$$= 265 \times 948 \times 10^{-3} = \quad 251 \text{ kNm}$$

But $M_{cy} \not> 1.4 p_y\, Z_y$ using constant 1.4 as noted above and Z_y for top flange only,

$$M_{cy} \not> 1.4 \times 265 \times 609 \times 10^{-3} = \mathbf{226} \text{ kNm}$$

Combined local capacity check
$$M_x/M_{cx} + M_y/M_{cy} \not> 1$$
$$599/1790 + 15.8/226 = \mathbf{0.40}$$

Acting alone without surge
$$M_x/M_{cx} = 683/1790 = \mathbf{0.38}$$

Hence section chosen is satisfactory.

Fig. 5.7

(g) Buckling resistance

The buckling resistance may be found in the same way as in Section 3.8(f), but allowing for the destabilizing effect of the surge load. Hence

clause 4.11.3
BS table 13

$$m = n = 1.0$$

No restraint is provided between the ends of the girder. At the supports the diaphragm gives partial restraint against torsion, but the compression flange is not restrained (Fig. 5.7).

BS table 9
$$L_E = 1.2(L + 2D)$$
$$= 1.2(6.5 + 2 \times 0.637) = 9.33 \text{ m}$$

|||BS table 14||Slenderness $\lambda = L_E/r_y$
$= 9330/72.8 = 128$
$\lambda/x = 128/29.5 = 4.3$
$N = I_{cf}/(I_{cf} + I_{tf})$
$= 9350/(9350 + 5690) = 0.62$
$v = 0.80$|
|BS table 13
clause 4.3.7.5||$u = 1.0$ (conservatively)
$\gamma_{LT} = nuv\lambda$
$= 1.0 \times 0.9 \times 0.80 \times 128 = 102$|
|BS table 12||Bending strength $\quad p_b = 150 \text{ N/mm}^2$
Buckling resistance $M_{bx} = p_b S_x$
$= 128 \times 6750 \times 10^{-3} = 689 \text{ kNm}$|

clause 4.11.3   Equivalent uniform moment factor $m = 1.0$
Overall buckling check

$$mM_x/M_{bx} + mM_y/p_y Z_y \ngtr 1$$

$$599/864 + 15.8/(265 \times 609 \times 10^{-3}) = \mathbf{0.97}$$

Acting alone without surge

$$M_x/M_{bx} = 683/689 = \mathbf{0.99}$$

Hence the section is satisfactory.

(h) Web buckling

At points of concentrated load (wheel loads or reactions) the web of the girder must be checked for local buckling[8] (see Fig. 5.8). If necessary, load carrying stiffeners must be introduced to prevent local buckling of the web.

Fig. 5.8

		Dispersion length **under wheel**
		$b_1 = 2 \times 75 \quad = 150 \text{ mm}$
$n_1 = 617 + 2 \times 20 \quad = 657 \text{ mm}$
clause 4.11.5		Web slenderness $\lambda \quad = 2.5d/t$
clause 4.5.2.1		$= 2.5 \times 537/14.1 = 195$
Compressive strength $p_c = 131 \text{ N/mm}^2$
BS table 27c		Buckling resistance $P_w = (b_1 + n_1) t p_c$
$= (150 + 657)14.1 \times 0.131 = \mathbf{1490 \text{ kN}}$
Max. wheel load $\quad = 383 \text{ kN}$

Hence buckling resistance is satisfactory.

Minimum stiff bearing length required **at support**

$b_1 = F_x/(tp_c) - n_1 \text{ mm}$
$F_x = 571 \text{ kN}$ (support reaction)
$n_1 = 309 \text{ mm}$
$b_1 = 571/(14.1 \times 0.131) - 309 = -10 \text{ mm}$

i.e. no stiff bearing is required at support for web bearing.

Fig. 5.9

(i) Web bearing

At the same points the web of the girder must be checked for local crushing[8] (see Fig. 5.10). If necessary, bearing stiffeners must be introduced to prevent local crushing of the web.

clause 4.11.5 Load dispersion under wheel $= 2(75+43.6+16.5) = 270$ mm

Bearing capacity $P_{crip} = 270 \times 14.1 \times 0.265 = \mathbf{1009\ kN}$
Maximum wheel load $= 383$ kN

Fig. 5.10

Load dispersion at support:

Minimum stiff bearing $= F_x/(tp_{yw}) - n_2$
$n_2 = (23.6 + 16.5)2.5 = 100$ mm
$F_x = 571$ kN (support reaction)
$b_1 = 571/(14.1 \times 0.265) - 100 = 53$ mm

Web bearing at the support requires a minimum stiff bearing of 53 mm; check that the supports provide this minimum stiff bearing to prevent web bearing capacity of the girder from being exceeded (see Fig 5.10).

(j) Deflection

Serviceability vertical wheel load excluding impact

$W_c = 191.4$ kN

Max. deflection for position given[9]
$\delta_c = W_c L^3 (3a/4L - a^3/L^3)/6EI$
$a = (L - c)/2 = 1.5$ m

$\delta_c = 3.5$ mm
Deflection limit $= 6500/600 = 10.8$ mm

(k) Connection

The vertical forces are transmitted to the supporting bracket by direct bearing (Fig. 5.11). Horizontal reactions are present from surge load (13.3 kN) and horizontal braking (15.3 kN). The surge load is transmitted to the column by the diaphragm (Fig. 5.7). The braking force will be transmitted by nominal bolts; provide, say two M20 bolts (grade 4.6).

Fig. 5.11

Fig. 5.12

5.4 EXAMPLE 10. CRANE GIRDER WITH LATERAL RESTRAINT

The design in Section 5.3 may be repeated but including a lattice restraint to the compression flange. In practice such a lattice girder may have been provided specifically for this purpose, or to support access platforms or walkways (Fig. 5.12). In modern UK practice it is rarely economic to include such a restraint in order to reduce the beam size, except in heavy industrial buildings.

(a) Dimensions

As Section 5.3.

(b) Loading

As Section 5.3 (making no correction for the changed self weight).

(c) Wheel loads

As Section 5.3.

(d) BM and SF

As Section 5.3, i.e.

Max. $M_x = 599$ kNm (683 kNm when acting alone)
Max. $M_y = 15.8$ kNm

(e) Shear capacity

A smaller UB may be chosen as lateral restraint is provided. Assume a **533 × 210 × 122** UB with no plate added. Note that in the design in Section 5.3 web bearing and buckling were critical, hence a revised section having a similar web thickness is chosen.

clause 4.2.3   Shear capacity $p_x = 0.6 p_y A_v$
$$= 0.6 \times 0.265 \times 544.6 \times 12.8 = 1108 \text{ kN}$$
$$F_x/P_x = 571/1108 = \mathbf{0.52}$$

(f) Moment capacity

The chosen UB is a plastic section ($b/T = 5.0$)
$$\text{Moment capacity } M_{cx} = p_y S_x$$
$$= 0.265 \times 3200 = 849 \text{ kNm}$$
But
$$M_{cx} = 1.2 p_y Z_y$$
$$= 1.2 \times 0.265 \times 2799 = 890 \text{ kNm}$$
$$M_x/M_{cx} = 683/849 = \mathbf{0.80}$$

Section is satisfactory.

clause 4.3.2   The effect of the lateral moment M_y is considered later in part (j) in combination with a restraint force equal to 2.5% of the flange force. Note that the value of 1% in the initial version of clause 4.3.2 was considered to be too small.

Flange force may be estimated as: $599/0.523 = \mathbf{1145 \text{ kN}}$ (1306 kN when acting alone); see Fig. 5.13.
Restraint force $= \mathbf{28.6 \text{ kN}}$ (32.6 kN when acting alone)

Fig. 5.13

(g) Web buckling/bearing

Minimum stiff bearing length at support to resist reaction of 571 kN.
$$n_1 = 544.6/2 = 272 \text{ mm}$$
$$\lambda = 2.5 \times 477/12.8 = 93$$

BS table 27c $p_c = 134 \text{ N/mm}^2$
Min. bearing length $b_l = 571/(12.8 \times 0.134) - 272 = 61 \text{ mm}$

$$n_2 = (21.3 + 12.7)2.5 = 85 \text{ mm}$$
Min. bearing length $b_l = 571/(12.8 \times 0.265) - 85 = 84 \text{ mm}$

Therefore, a stiff bearing length of at least 84 mm is needed at the supports to give adequate bearing strength for the girder.

(h) Deflection

Wheel load $= 191.4 \text{ kN}$
Calculation as in Section 5.3(j) gives $\delta = 9.0 \text{ mm}$
Limit $= 6500/600 = 10.5 \text{ mm}$

(i) Connection

As Section 5.3 but the lattice transmits horizontal forces to the support: hence the diaphragm is not needed.

(j) Lattice

The lattice girder which provides restraint to the crane girder top flange is loaded by

(i) the restraint force of 11.45 kN (13.06 kN when acting alone) which is considered to be distributed between the nodes of the lattice;
(ii) the surge load of 9.1 kN per wheel (10.4 kN when acting alone).

Fig. 5.14

The truss is analysed by graphics, calculation, or computer, giving member forces as tabulated (kN):

Panel	Top chord	Bottom chord	Diagonal	Post
1	0	21.2	30.0	0
2	21.2	38.3	24.2	17.1
3	38.3	51.3	18.4	13.0
4	51.3	51.0	0.3	0.3
5	46.6	51.0	6.1	0
6	38.2	46.6	11.9	4.3
7	25.7	38.2	11.7	8.4
8	0	25.7	36.3	12.5
9				0

The applied forces may act in either direction and hence the member forces may be either tension or compression. Designing the chord members for a compression of 51.3 kN; use a **45 × 45 × 6 equal angle**

$$\lambda = L_E/r_{rr} = 1.0 \times 813/8.67 = 94$$
$$p_c = 135 \text{ N/mm}^2$$

BS table 27c   Compression resistance $P_c = A_g p_c$
$$= 5.09 \times 135 \times 10^{-1} = \mathbf{69 \text{ kN}}$$

Designing the diagonals for a compression of 36.2 kN using the same single angle:

Effective length $L_e = 1.0 \times 813/\cos 45° = 1150$ mm
$$\lambda = 1150/8.67 = 133$$

BS table 27c
$$p_c = 83 \text{ N/mm}^2$$
$$P_c = 5.09 \times 83 \times 10^{-1} = \mathbf{42 \text{ kN}}$$

The basic lattice member is therefore a 45 × 45 × 6 equal angle and might be fabricated as a welded truss. For a more detailed consideration of truss design reference should be made to Chapters 6 and 12.

STUDY REFERENCES

Topic	References
1. Crane loading	BS 6399 *Design Loading for Buildings* Part 1: *Dead and imposed loads* (1984)
2. Crane types	BS 2573: Part 1 *Rules for the Design of Cranes: Specification for classification, stress calculation and design criteria of structures* (1983)
3. Influence lines	**Marshall W.T. & Nelson H.M.** (1990) Moving loads and influence lines, *Structures*, pp. 79–106. Longman
4. Influence lines	**Coates R.C., Coutie M.G. & Kong F.K.** (1988) Mueller-Breslau's principle. Model analysis, *Structural Analysis*, pp. 127–31. Van Nostrand Reinhold
5. Influence lines	**Wang C.K.** (1983) Influence lines for statically determinate beams, *Intermediate Structural Analysis*, pp. 459–67. McGraw-Hill
6. Plastic modulus	**Marshall W.T. & Nelson H.M.** (1990) Plastic bending, *Structures*, pp. 532–6. Longman
7. Plastic modulus	**Horne M.R., & Morris L.J.** (1981) *Plastic Design of Low-Rise Frames*. Collins
8. Web buckling and bearing	**Dowling P.J., Knowles P. & Owens G.W.** (1988) *Structural Steel Design*. Steel Construction Institute
9. Deflection formulae	(1992) Design theory, *Steel Designers' Manual* pp. 1026–50. Blackwell

6

TRUSSES

Trusses and lattice girders are fabricated from the various steel sections available, joined together by welding or by bolting usually via gusset (connecting) plates. Generally the trusses act in one plane and are usually designed as pin-jointed frames, although some main members may be designed as continuous. Where members lie in three dimensions the truss is known as a space frame. Trusses and lattice girders are particularly suited to long spans, as they can be made to any overall depth, and are commonly used in bridge construction. In buildings they have particular application for roof structures, and for members supporting heavy loads (columns from floors above) and for members having longer spans.

The use of a greater overall depth leads to a large saving in weight of steel compared with a universal beam. This saving of material cost can offset the extra fabrication costs in certain cases.

6.1 TYPES OF TRUSS AND THEIR USE

A selection of roof trusses is shown in Fig. 6.1, where the roof slopes and spans dictate the shape of the truss and the layout of the members. Hipped trusses are used for small spans, economically up to 6 m, the lattice girders for medium spans, and the mansard for large spans. Such trusses are lightly loaded by snow and wind load, together with a small allowance for services. It is unusual for lifting facilities to be supported from the roof trusses. The resulting member and connection sizes are therefore relatively small.

Heavy trusses may be used in multi-storey buildings where column loads from the floors above need to be carried. Examples of these are shown in Fig. 6.2. Trusses of this type carry very heavy loads, and are similar in layout and member size to bridge structures.

A common method of providing stability to a building, whether single or multi-storey, is to use an arrangement of bracing members. These are essentially formed into a truss and carry the horizontal loads, such as wind, to the foundations by acting as horizontal and vertical frameworks. Examples are shown in Fig. 6.3. Bracing is considered in more detail in Chapter 10, and graphically illustrated in Chapters 12 and 13 in terms of the overall stability of single-storey buildings.

Fig. 6.1 Roof trusses

Hipped

Lattice

Mansard

Fig. 6.2 Support trusses

Column loads

N-truss

Column loads

Vierendeel

Fig. 6.3 Bracing

Multi-storey building

Single-storey building

6.2 LOADING AND ANALYSIS

Loading will consist of dead, imposed and wind loads as described in Chapter 2. Combinations of loads giving maximum effects in individual members must be considered (see Section 2.4) and safety factors γ_f must be included (Section 1.7).

The loads will usually be transferred to the truss by other members such as purlins (Fig. 1.2) or by beams in the case of a floor truss. A wind bracing will be loaded by the gable posts, or by side members such as the eaves beam. It is ideal if the loads can be transferred to the truss at the node points, but commonly (as shown in Fig. 1.2) this is not possible. In roof truss design the purlin positions may not be known initially, and allowing for the possibility of purlin changes during future re-roofing, a random position for loads is often allowed.

The analysis therefore involves several stages:

(a) Analysis of the truss assuming pin-joints (except Vierendeel trusses) and loading at the nodes.
This may proceed using manual methods – joint resolution, method of

sections and graphical means are all suitable[1,2] – or computer techniques. Several analyses may be needed where different arrangements of dead, imposed and wind loading must be considered.

(b) Analysis of the load bearing member such as the rafter as a continuous beam supported at the nodes and loaded by the purlins. In cases where the load positions are uncertain the rafter moment may be taken as $WL/6$ (clause 4.10c), where W is the purlin load and L is the node to node length perpendicular to W.

(c) Assessment of stresses due to eccentricity of the connections. Ideally the centroidal axes of members should meet at the nodes. Where this is not possible the members and connections should be designed for the moments due to the eccentricity, if significant.

(d) Assessments of the effects of joint rigidity and deflections. Secondary stresses become important in some trusses having short thick members, but may be neglected where more slender members are used (clause 4.10).

The overall analysis of the truss will therefore involve the summation of two or more effects. Analyses (a) and (b) must always be considered, while (c) and (d) may be avoided by meeting certain conditions.

6.3 SLENDERNESS OF MEMBERS

The slenderness λ of a compression member (a strut) is given by

$$\lambda = L_E/r$$

where L_E is the effective length of the strut about the appropriate axis
r is the radius of gyration about the appropriate axis

The requirements of clauses 4.7.2 and 4.7.10 define the effective lengths of the chord and internal members of trusses and are illustrated in Figs. 6.4 and 6.5. These requirements take into account the effect of the nodes (joints) which divide the top chord into a number of in-plane effective lengths. In the lateral (out-of-plane) direction the purlins restrain the top chord. The radius of gyration appropriate to a given strut depends on the possible axis of buckling, and these are shown in Figs. 6.4 and 6.5.

The effective lengths of discontinuous struts are increased where single bolted connections are used, resulting in reduced compressive strengths. Hence single bolted connections usually result in less economic trusses.

Where double angles are used as shown in Figs. 6.4 and 6.5 it is necessary to reduce the slenderness of the individual component (single angle) by interconnecting the angles at points between the joints. These connections are usually a single bolt (minimum 16 mm diameter), with a packing between the angles equal to the gusset thickness, and are commonly placed at third or quarter points along the member.

slenderness $\lambda_b = \sqrt{(\lambda_m^2 + \lambda_c^2)}$
where $\lambda_m = s_2/r_{yy}$
$\lambda_c = s_2/3r_{vv}$ for members having connections which divide s_2 into three equal parts

68 STRUCTURAL STEELWORK DESIGN TO BS 5950

Fig. 6.4 λ for continuous chords

Double angle
RHS

λ = maximum of
$0.85\, s_1/r_{xx}$ or s_2/r_{yy}

Single angle

λ = maximum of
$0.7\, s_1/r_{vv}$ or s_2/r_{vv}

Fig. 6.5 λ for discontinuous struts

Double angle

λ = maximum of
$0.85\, L/r_{min}$ or $0.7\, L/r_{xx} + 30$

Single angle

λ = maximum of
$0.85\, L/r_{vv}$ or $0.7\, L/r_{yy} + 30$

Member connected to gusset (parallel Y–Y) by two or more bolts

λ_c should not exceed 50. These requirements are given in clauses 4.7.13.1(d) and 4.7.9(c).

Slenderness for any strut should not exceed 180 for general members resisting loads other than wind loads, or 250 for members resisting self weight and wind load only (clause 4.7.3.2). For a member normally acting as a tie, but subject to reversal of stress due to wind, the slenderness should not exceed 350. In addition, very small sections should be avoided, so that damage during transport and erection does not occur, e.g. a minimum size of angle would be 50 × 50 × 6 generally.

6.4 COMPRESSION RESISTANCE

The compression resistance of struts is discussed also in Chapter 7. The compressive strength p_c depends on the slenderness λ and the design strength p_y. Tests on axially loaded, pin-ended struts show that their behaviour can be represented by a number of curves which relate to the type of section and the axis of buckling. These curves are dependent on material strength and the initial imperfections, which affect the inelastic behaviour and the inelastic buckling load. For design the value of p_c is obtained from one of four strut curves or tables (BS tables 27a to 27d). The appropriate table is chosen by reference to section type and thickness, and to the axis of buckling (BS table 25).

The compression resistance P_c is

either $P_c = A_g\, p_{cs}$ for slender sections (see Section 1.7)
or $P_c = A_g\, p_c$ for all other sections.

where A_g is the gross sectional area
p_{cs} is the compressive strength based on a reduced design strength (clause 3.6).

6.5 TENSION CAPACITY

The tension capacity P_t of a member is

$$P_t = A_e p_y$$

where A_e is the effective sectional area as defined in clause 3.3.3.

Where a member is connected eccentrically to its axis then allowance should be made for the resulting moment. Alternatively, such eccentric effects may be neglected by using a lower value of the effective area A_e. For a single angle connected through one leg:

$$A_e = a_1 + 3a_1\, a_2/(3a_1 + a_2)$$

where a_1 is the net sectional area of the connected leg
a_2 is the sectional area of the unconnected leg

Full details of these reduced effective areas are given in clause 4.6.3.

6.6 CONNECTIONS

Connections are required to join one member to another (internal joints), and to connect the truss to the rest of the building (external joints). Three main types of connection are used:

- bolting to a gusset plate
- welding to a gusset plate
- welding member to member

Examples of these are shown in Fig. 6.6. The choice of connection type is often made by the fabricator, and will depend on the available equipment, with welding becoming more economical the larger the number of truss and member repetitions.

Bolted to gusset plate　　　Welded member to member

Cap joint (bolted)

Fig. 6.6 Connection details

Ideally, members should be connected so that centroidal axes (or bolt centre lines in the case of angles or tees) meet at a point (Fig. 6.6). If this cannot be achieved, then both members and connection must be designed from the eccentricity. In many cases the gusset plate will not lie in the plane of the member centroidal axes, but stresses due to this eccentricity are ignored in construction using angles, channels and tees (clause 4.7.6c).

Design of the bolts or welds follows conventional methods[3,4]. Bolts (grade 4.6 or 8.8) must be designed for both shear stress and bearing stress (see Section 3.7g). Friction grip fasteners may be used[5], but are not usually economic unless bolt slip is unacceptable. Design stresses in the gusset may be checked as a short beam with a combined axial load. Such design is not realistic however, and it is sufficient to check the direct stress only at the end of the member (Fig. 6.7), on an area $b \times t$ as shown.

Minimum bolt size is usually 16 mm, and minimum gusset plate thickness is 6 mm (internal) or 8 mm (exposed).

Fig. 6.7 Gusset plate stress

6.7 EXAMPLE 11. ROOF TRUSS WITH SLOPING RAFTER

(a) Dimensions
(See Fig. 6.8.)

Span of truss	16.0 m
Rise of truss	3.2 m
Roof slope	21.8°
Truss spacing	4.0 m
Rafter length	8.62 m

Fig. 6.8 Roof truss

(b) Loading

Fig. 6.9

Cladding/insulation	0.12 kN/m²
Roof truss self weight (estimated)	8.0 kN
Snow/services	0.75 kN/m²
Wind pressures (q)	0.68 kN/m²

Using internal and external pressure coefficients given in reference (6), the worst case for wind loading on the roof slope is the case 'wind on end plus internal pressure' (Fig. 6.9). This gives an outward pressure of

$$-(0.7+0.2)0.68 = -0.61 \text{ kN/m}^2$$

Note must be taken of the effects of the slope on the values of each load as appropriate (see Section 2.7(d) and Fig. 6.10).

72 STRUCTURAL STEELWORK DESIGN TO BS 5950

Fig. 6.10

Dead load on each purlin
cladding $4.0 \times 2.0 \times 0.12$ = 0.96 kN
own weight of purlin (say, 0.11 kN/m) and truss
$0.11 \times 4.0 + 8.0/10$ = 1.24 kN
Total dead load on purlin = 2.20 kN

Dead load on rafter W_d = $2.20 \times 8.62/2.0 = 9.48$ kN
Imposed load on each purlin = $4.0 \times 2.0 \cos 21.8° \times 0.75 = 5.56$ kN
Imposed load on rafter W_i = $6.46 \times 8.62/2.0 = 24.0$ kN
Wind load on each purlin = $4.0 \times 2.0 \times 0.61 = 4.88$ kN (suction)
Wind load on rafter W_w = $4.88 \times 8.62/2.0 = 21.0$ kN (suction)

(c) Truss forces (dead)

Truss analysis is carried out placing concentrated loads at the nodes of the truss, i.e. dividing the rafter load proportional to the nodal centres (Fig. 6.11). Analysis by manual or computer techniques gives forces as in the table (owing to symmetry, half of the truss only is recorded).

$W_d = 9.48/4 = 2.37$ kN
$W_i = 24.0/4 = 6.00$ kN

Fig. 6.11

(d) Truss forces (imposed)

The forces are arranged as in (c) but have values of 6.00 kN, instead of 2.37 kN. Member forces are given in the table.

(e) Truss forces (wind)

Wind forces on the truss are as shown (Fig. 6.12) and member forces are again analysed and the results given in the table:

$W_w = 21.0/4 = 5.25$ kN

Fig. 6.12

Member	Dead load W_d	Imposed load W_i	Wind load W_w	$1.4W_d$ $+1.4W_i$	$1.0W_d$ $+1.4W_w$
1	22.4	56.6	−46.7	121.9	−43.0
2	21.4	54.3	−46.7	116.8	−44.0
3	20.6	52.1	−46.7	112.2	−44.8
4	19.7	49.9	−46.7	107.4	−45.7
5	−20.7	−52.5	42.8	−113.0	39.2
6	−17.8	−45.0	35.4	−96.9	31.8
7	−11.9	−30.0	20.7	−64.7	17.1
8	2.2	5.6	−5.3	12.0	−5.2
9	−3.0	−7.5	6.8	−16.2	6.5
10	4.4	11.1	−10.5	23.9	−10.3
11	−2.9	−7.4	6.8	−15.9	6.6
12	2.2	5.6	−5.3	12.0	−5.2
13	−5.9	−15.0	16.0	−32.3	16.5
14	−8.9	−22.5	22.8	−48.5	23.0

Compression force is positive.

As the wind load is suction, the load combinations $1.4W_d + 1.4W_w$ and $1.2(W_d + W_i + W_w)$ may be ignored.

(f) BM in rafter

Assuming purlin positions are not known

Purlin load $W = 1.4(0.96 + 0.44) + 1.6 \times 5.56 = 10.9$ kN

clause 4.10(c)   BM $= WL/6 = 10.9 \times 2.0 \cos 21.8°/6 = \mathbf{3.36\ kNm}$

(g) Rafter

Max. compression F = 121.9 kN
Max. tension = 43.0 kN
Nodal distance = 8.62/4 = 2.16 m

Use two $-80 \times 60 \times 7$ **unequal angles** spaced 8 mm apart to allow for gusset plates with spacing washers at quarter points, i.e. 0.54 m centres (Fig. 6.13). Note that the $y'-y'$ axis is for the single angle section and the $y-y$ axis is for the combined section.

Fig. 6.13

Fig. 6.4 and clause 4.10

$r_x = 25.0$ mm
Slenderness $\lambda_{xx} = 0.85 \times 2155/25.0 = 73$
Minimum $\lambda_{xx} = 50$

$r_{vv} = 12.7$ mm
$r_{yy} = 26.2$ mm

Slenderness $\lambda_m = 2155/26.2 = 82$
$\lambda_c = 2155/(4 \times 12.7) = 42$

(for connections at quarter points) where λ_c is based on the minimum r of the component (r_{vv} for single angle) as the angle can fail as a component between fasteners.

$$\lambda_b = \sqrt{(82^2 + 42^2)} = 92$$

clause 3.5.3   Section chosen is semi-compact with

$d/t = 80/8 = 10\ (< 15\varepsilon)$ and
$(b+d)/t = 17.5\ (< 23\varepsilon)$

Table 1.2   Design strength $p_y = 275$ N/mm^2

BS table 25   Strut table for angles is table 27c, and for $\lambda = 92$:

BS table 27c   Compressive strength p_c = 123 N/mm^2
Compression resistance $P_c = A_g p_c$
 $= 21.2 \times 123 \times 10^{-1} = 261$ kN

clause 4.2.5   Moment capacity M_{cx} $= p_y Z_{xx}$
 $= 275 \times 24.3 \times 10^{-3} = 6.7$ kNm

Simplified local capacity check (for further discussion see Section 7.5):

74 STRUCTURAL STEELWORK DESIGN TO BS 5950

clause 4.8.3.2 $\quad F/A_g\, p_y + M_x/M_{cx} \not> 1$

$121.9/(21.2 \times 275 \times 10^{-1}) + 3.36/6.7 = \mathbf{0.71}$

Section satisfactory.

Buckling resistance moment of the section must be checked using:

Slenderness $\lambda \quad = L_E/r_y$
$\quad = 2155/26.2 = 82$

clause 4.3.8 $\quad$ Buckling resistance $M_b = 0.8 p_b\, Z_x$
$\quad = 0.8 \times 275 \times 24.3 \times 10^{-3} = 5.35\,\text{kNm}$

Overall buckling check (simplified approach):

clause 4.8.3.3 $\quad F/A_g\, p_c + mM_x/M_b \not> 1$ where $m = 1.0$

$121.9/261 + 3.36/5.35 = \mathbf{1.00}$

(h) Bottom chord design

Max tension $\quad = 113.0\,\text{kN}$
Max compression $= 39.2\,\text{kN}$

Use two $- \mathbf{75 \times 50 \times 6}$ unequal angles spaced 8 mm apart (semi-compact)

clause 4.6.3.2 $\quad$ Effective area $A_e = a_1 + 5a_1 a_2/(5a_1 + a_2)$
$a_1 = (L - t/2)t - Dt$
$\quad = (75 - 3)6 - 22 \times 6 = 300\,\text{mm}^2$

allowing for one 22 mm diameter hole in calculating a_1:
$a_2 = (50 - 3)\, 6 = 282\,\text{mm}^2$
$A_e = 300 + 5 \times 300 \times 282/(5 \times 300 + 282)$
$\quad = 537\,\text{mm}^2/\text{angle}$

Note that A_e becomes net area if clause 4.6.3.3 is satisfied.

Tension capacity $P_t = A_e\, p_y = 2 \times 537 \times 0.275 = \mathbf{295\,kN}$

Compression resistance must be checked assuming lateral restraint to the bottom chord at the chord bracing connections, maximum spacing 4.64 m (Fig. 6.8).

$r_{yy} = 21.4\,\text{mm}$
$\lambda_m = 4640/21.4 = 217$
$\lambda_c = 580/10.8 = 54$

(for connections at quarter points, i.e. 0.58 m centre)

$\lambda_b = \sqrt{(217^2 + 54^2)} \quad = 224$
Maximum slenderness $\quad = 250$
clause 4.7.3.2b $\quad$ Compression strength $p_c = 34\,\text{N/mm}^2$
BS table 27c $\quad$ Compression resistance $P_c = A_g\, p_c$
$\quad = 14.4 \times 34 \times 10^{-1} = \mathbf{50.4\,kN}$

Section is satisfactory.

(i) Strut (member 10)

Max. compression $F = 23.9$ kN
Max. tension $= 10.3$ kN

Use **60 × 60 × 6 equal angle**
$r_{vv} = 11.7$ mm
$r_{yy} = 18.2$ mm
$L = 1.72$ m

Fig. 6.5 and
clause 4.7.10.2a
$\lambda = 0.85L/r_{vv} = 125$ or
$\lambda = 0.7L/r_{yy} + 30 = 96$

BS table 27c
Compressive strength $p_c = 91$ N/mm² for $\lambda = 125$
Compression resistance $P_c = A_g\, p_c$
$= 6.91 \times 91 \times 10^{-1} = \mathbf{63\ kN}$

Section is satisfactory.

(j) Connection

Check the strength of the connection joining members 6, 7, 10 and 13. Assume an 8 mm thick gusset plate and 20 mm bolts (grade 4.6), see Fig 6.14.

Max. force (member 13) = 32.3 kN

Force change between members 6 and 7 is 32.2 kN. If members 6 and 7 were to be joined at this connection (to change size or reduce fabrication lengths) then maximum force (member 6) would be 96.9 kN.

Fig. 6.14

clause 6.3.2
Shear capacity of bolts (double shear)
$P_s = p_s\, A_s = 160 \times 2 \times 0.245 = 78$ kN

clause 6.3.3.2
Bearing capacity of bolts $P_{bb} = dtp_{bb} = 2 \times 20 \times 6 \times 0.450 = 108$ kN

clause 6.3.3.3
Bearing capacity of angles $P_{bs} = dtp_{bs} = 2 \times 20 \times 6 \times 0.450 = 108$ kN
but $P_{bs} \not> etp_{bs}/2 = 40 \times 6 \times 0.450/2 = 54$ kN

Clearly a bolt size of as low as 16 mm would be possible.

Gusset plate stress can be based on an effective width (Fig. 6.7) of $60/\cos 30° - 22 = 47$ mm.

Plate stress $= 32.3 \times 10^3/(8 \times 47) = 86 \text{ N/mm}^2$
(Maximum value is $p_y = 275 \text{ N/mm}^2$)

The remaining truss members may be designed in the same way, with the compression force usually being the main design criterion. It should be noted that it is good practice to limit the number of different member sizes being used, probably not more than four in total. The deflection of a roof truss is not usually critical to the design, and the effects of sag under load may be offset by pre-cambering the truss during fabrication by, say, 50 mm or 100 mm. If the deflection is required then hand or graphical methods[7,8], or the virtual work method[9,10], or appropriate programs may be used. It should be noted that deflection due to bolt slip can be significant compared with dead load elastic deflections.

6.8 EXAMPLE 12. LATTICE GIRDER

(a) Dimensions

See Fig. 6.15: lattice girder fabricated from tubular sections, span 8.0 m; spacing 3.5 m.

Fig. 6.15 Lattice girder

(b) Loading

Timber joist floor	0.5 kN/m^2
Floor finishes	0.2 kN/m^2
Imposed load	4.0 kN/m^2

Dead load on timber joist:
    floor $0.5 \times 3.5 \times 0.5 = 0.88 \text{ kN}$
    finishes $0.2 \times 3.5 \times 0.5 = 0.35 \text{ kN}$
    weight of truss (estimated) $= \underline{0.37 \text{ kN}}$
    Total load W_d $= 1.60 \text{ kN}$

Imposed load on timber joist
$$W_i = 4.0 \times 3.5 \times 0.5 = 7.00 \text{ kN}$$

Load on girder from each joist $= 1.4W_d + 1.6W_i$
$$= 1.4 \times 1.60 \times 1.6 \times 7.00 = 13.4 \text{ kN}$$

Concentrated load at nodes $= \mathbf{26.8 \text{ kN}}$

TRUSSES

(c) Truss forces

Analysis of the truss[1,2] under the nodal loads gives the member forces (kN) in the table:

Top chord		Bottom chord		Diagonals	
Member	Force	Member	Force	Member	Force
1	188	5	−107	9	152
2	322	6	−268	10	−144
3	402	7	−375	11	114
4	429	8	−429	12	−76
				13	76
				14	−38
				15	38
				16	0

Fig. 6.16

Compression force is positive.

(d) BM in top chord

(See Fig. 6.16.)

$$\text{BM in continuous top chord} = WL/8$$
$$= 13.4 \times 1.0/8 = \mathbf{1.68\,kNm}$$

(e) Top chord

Max. compression $F = 429$ kN
Max. BM M_x $= 1.68$ kNm

Use **90 × 90 × 6.3 RHS**

$r_{xx} = 34.1$ mm

Fig. 6.4
BS table 25 Slenderness $\lambda = 0.85 \times 1000/34.1 = 25$
BS table 27a Strut table for RHS is table 27a
Compressive strength $p_c = 270\,\text{N/mm}^2$
Compression resistance $P_c = A_g\, p_c$
$= 20.9 \times 270 \times 10^{-1} = 564$ kN
Section chosen is plastic ($b/T = 14$)
Moment capacity M_c $= p_y\, S$
$= 275 \times 65.3 \times 10^{-3} = 17.96$ kNm

clause 4.8.3.2 Local capacity check

$$F/A_g\, p_y + m_x/M_{cx} \not> 1$$

$$429/(20.9 \times 275 \times 10^{-1}) + 1.68/17.96 = \mathbf{0.84}$$

Section is satisfactory.

Lateral torsional buckling need not be considered if compression flange is positively restrained connected to the timber joist floor, or for box sections (clause B.2.6.1).

(f) Bottom chord

Max. tension $F = 429$ kN

Use **$90 \times 90 \times 5$ RHS**

Tension capacity $P_t = A_e\, p_y$
$= 16.9 \times 275 \times 10^{-1} = \mathbf{465\,kN}$

Section is satisfactory.

(g) Diagonal

(See Fig. 6.17.)

Max. compression $F = 152$ kN
Max. tension $= 114$ kN

Use **$50 \times 50 \times 3.2$ RHS**

$r_x = 19.1$ mm

BS table 27a

Slenderness $\lambda = 0.7 L/r_x = 0.7 \times 500/19.1 \cos 45° = 26$
Compressive strength $p_c = 269$ N/mm²
Compression resistance $P_c = A_g\, p_c$
$= 5.94 \times 269 \times 10^{-1} = \mathbf{160\,kN}$
Tension resistance $P_t = 5.94 \times 275 \times 10^{-1} = \mathbf{163\,kN}$

Fig. 6.17

Section is satisfactory.

(h) Connections

All welded joints with continuous 4 mm welds.

clause 6.6.5.1

Weld length $= 2(50 + 50/\cos 45°) = 241$ mm
Design strength $p_w = 215$ N/mm²
Weld resistance $= 0.7 \times 4 \times 241 \times 215 = 145$ kN

The maximum forces which may be transmitted between hollow steel sections are more complex and detailed references may be consulted if required[11].

As in Section 6.7, the number of section sizes would be limited in practice, with the allowable variation in member sizes depending on the degree of repetition expected.

STUDY REFERENCES

Topic

1. Truss analysis

References

Marshall W.T. & Nelson H.M. (1990) Analysis of statically determinate structures, *Structures*, pp. 10–19. Longman.

2. Truss analysis	**Coates R.C., Coutie M.G. & Kong F.K.** (1988) Analysis of plane trusses, *Structural Analysis* pp. 41–51 Van Nostrand Reinhold
3. Bolt design	(1991) *Steelwork Design* vol. 4, pp. 10/1–10/10 Steel Construction Institute
4. Welding	(1991) *Steelwork Design* vol. 4, pp. 11/1–11/15 Steel Construction Institute
5. Friction grip bolts	(1991) *Steelwork Design* vol. 4, pp. 10/1–10/10 Steel Construction Institute
6. Wind pressure coefficients	BS 6399 *Design Loading for Buildings* Part 2: *Wind loads* (1995)
7. Truss deflection	**Marshall W.T. & Nelson H.M.** (1990) Deflection of Structures, *Structures*, pp. 273–98. Longman
8. Truss deflection	**Wang C.K.** (1983) The graphical method, *Intermediate Structural Analysis,* pp 93–99. Pergamon
9. Virtual work	**Marshall W.T. & Nelson H.M.** (1990) Virtual work and energy principles, *Structures*, pp 501–31. Longman
10. Virtual work	**Coates R.C., Coutie M.G. & Kong F.K.** (1988) Applications of the principle of virtual work, *Structural Analysis*, pp.134–75 Van Nostrand Reinhold
11. Connection of RHS	CIDECT (1985) Welded joints. *Construction with Hollow Steel Sections*, pp. 129–42, British Steel Corporation Tubes Division

7
SIMPLE AND COMPOUND COLUMNS

Columns, sometimes known as stanchions, are vertical steel members in both single and multi-storey frames. They are principally designed to carry axial loads in compression, but will also be subjected to moments due to eccentricities or lateral loads or as a result of being part of a rigid frame, i.e. continuity moments. In some structures, particularly single-storey frames and top lengths in multi-storey frames, the moments may have greater effect in the design than the axial compression; under some loading combinations axial tension may occur (see Section 2.7(d) and Table 12.7).

Where columns are principally compression members their behaviour and design are similar to those of struts (Section 6.4). The loads carried are usually larger than those in typical truss members, and the simplifications adopted in truss design (Section 6.2) should not be applied.

7.1 TYPES OF COLUMN

Typical cross-sections used in column design are shown in Fig. 7.1. While any section may be used, the problem of instability under axial compression results in preferred sections of circular or square types. Buckling is likely to occur about the axis of lower bending resistance, and the use of sections having one axis of very low bending resistance is usually uneconomic. However, as the range of hollow structural sections is limited, and making connections is more difficult, the H and I sections are frequently used.

Columns may also be built up from smaller sections or plates, being either welded or bolted together as in Fig. 7.2. In some cases, the column may take the form of a lattice girder, which will be particularly economic where large moments occur, such as tall building wind frames, masts and cranes. The laced column and the battened column shown in Fig. 7.3 are used in cases where overturning moments due to wind or eccentricity (crane gantry loading) are high.

Fig. 7.1 Column cross-sections

Preferred: UC, RHS, CHS
Others: UB, RSA

Fig. 7.2 Compound sections

RSCs and plate; RSAs and lattice; Welded box

Fig. 7.3 Column types

Laced; Battened

7.2 AXIAL COMPRESSION

Columns with idealized end connections may be considered as failing in an Euler-type buckling mode[1,2]. However, practical columns usually fail by inelastic bending and do not conform to the Euler theory assumptions[3,4], particularly with respect to elastic behaviour. Only extremely slender columns remain linearly elastic up to failure. Local buckling of thin flanges rarely occurs in practice when normal rolled sections are used, as their flange thicknesses usually conform to clause 3.5 of BS 5950.

The behaviour of columns and their ultimate strength are assessed by their slenderness λ and the material design strength p_y, which are described briefly in Section 6.4.

7.3 SLENDERNESS

Slenderness is given by:

$$\lambda = L_E/r$$

where L_E = effective length
    r = radius of gyration

The effective length of a column is dependent on the restraint conditions at each end. If perfect pin connections exist then L_E equals the actual length. In practice, beam type members connected to the end of a column, as well as giving positional restraint, provide varying degrees of end restraint from virtually full fixity to nearly pinned.

BS table 24 indicates in broad terms the nominal effective length of a column member, provided the designer can define the amount of positional and rotational restraint acting on the column in question. The various classes of end fixity alluded to in BS table 24 represent a crude classification, as indicated in Fig. 7.4. The nominal effective lengths tend to be longer than those obtained using the Euler values as they take cognizance of practical conditions: that is, there is no such condition as 'pinned' (unless a pin is deliberately manufactured, which would be costly) or 'fixed'. Therefore, taking the case having both ends 'fixed', in reality the ends would have some flexibility; hence a value of 0.7, rather than the idealized Euler value of 0.5, is used. Although these values in BS table 24 are adequate for column members in multi-storey buildings designed by the 'simple design' method, a more accurate assessment is provided by section 5.7 and appendix E of BS 5950: Part 1 for columns in 'rigid' frames (continuous construction).

The restraint provided may be different about the two column axes, and in practice in steelwork frames this will generally be the case. The effective lengths in the two planes will therefore generally be different, and so will be the slenderness (λ_x and λ_y). For axially loaded columns the compressive strength p_c is selected from BS tables 27a to 27d for both values of slenderness, and the lower strength is used in the design, irrespective of axis.

Fig. 7.4 End fixity/effective length

The effective length of a column in a single-storey frame is difficult to assess from considerations of fixity as in Fig. 7.4 and hence BS 5950: Part 1 gives these special consideration. Clause 4.7.2c and appendix D1 should be used for these cases.

It is essential to have a realistic assessment of effective length, but it should be realized that this assessment is not precise and is open to debate, and consequently it is not reasonable to expect highly accurate strength predictions in column design. Interpolation between the values in BS table 24 might produce more accurate estimates, but such interpolation must reflect the actual restraint conditions.

7.4 BENDING AND ECCENTRICITY

In addition to axial compression, columns will usually be subjected to moments due to horizontal loading and eccentricity of connections carrying vertical loads. The types of load that can occur are summarized in Fig. 7.5 in which:

W_A is the vertical load from a roof truss, taken as applied concentrically (clause 4.7.6a(2))
W_B are horizontal loads due to wind, applied by side rails
W_C and W_{HC} are the crane gantry loads, applied through a bracket at a known eccentricity
W_D is the self weight of the column/sheeting
W_E is the resultant force carried through the truss bottom chord

In single-storey column and truss structures load W_E occurs whenever the columns carry unequal horizontal loads or unequal moments. Values of the resultant force for different arrangements of horizontal loading are shown in Fig. 7.6.

In cases such as most beam/column connections where eccentricity is not known precisely, clause 4.7.6 states that a value of 100 mm from the column face (flange or web as appropriate) should be used.

For the design of a column under a load system such as that shown in Fig. 7.5, load factors γ_f must be included in the calculations. Each load may comprise one or more load types, e.g. load W_B will be wind load only, while W_A may consist of dead load, imposed and wind load. When loads are applied in combination, different load factors may be required for each load group (see Chapter 2). Unlike simple load cases, it will no longer be clear which combination produces the highest axial forces and moments. Complex load cases, as shown in Fig. 7.5, may require all possible combinations to be examined, although with experience three or four worst arrangements may be selected. The best way of examining the effects of all or some combinations is the use of matrices. Matrix manipulation should be arranged in the following manner:

Unfactored load matrix $= [W]$
Load factor/combination matrix $= [\gamma_f]$
Factored load matrix $[W_\gamma]$ $= [W] \times [\gamma_f]$
Axial force and moment coefficient matrix $[\alpha]$
Axial force and moment matrix $[F,M] = [W_\gamma]^T \times [\alpha]$

This matrix method is demonstrated in Section 7.7.

Fig. 7.5 Column loads

Fig. 7.6 Resultant force roof truss bottom chord

7.5 LOCAL CAPACITY

At any section in a column the sum of effects of axial load F and moments M_x and M_y must not exceed the local capacity. This is considered to be satisfactory if the combination relationship satisfies the following interaction equation:

(a) for **plastic** and **compact** sections:
UB, UC and joist sections

$$(M_x/M_{rx})^2 + M_y/M_{ry} \not> 1$$

RHS, CHS and solid sections

$$(M_x/M_{rx})^{5/3} + (M_y/M_{ry})^{5/3} \not> 1$$

Channel, angle and all other sections

$$M_x/M_{rx} + M_y/M_{ry} \not> 1$$

where M_{rx} and M_{ry} are the reduced plastic moments in the presence of axial load, as defined in reference (5).

(b) For **semi-compact** and **slender** sections, and as a simplified method for compact sections in (a):

$$F/A_g p_y + M_x/M_{cx} + M_y/M_{cy} \not> 1$$

where A_g is the gross sectional area and M_{cx} and M_{cy} are the moment capacities defined in Section 3.4.

SIMPLE AND COMPOUND COLUMNS

7.6 OVERALL BUCKLING

The failure of columns, whether carrying moments combined with an axial load or not, will commonly involve member buckling. The assessment of overall buckling resistance involves the same factors and procedures as given in Section 3.2, but with an additional term for axial load. Using a simplified approach (clause 4.8.3.3.1) the overall buckling check is considered to be satisfactory if the combination relationship is satisfied.

$$F/A_g\, p_c + m_x M_x/M_b + m_y M_y/p_y\, Z_y \not> 1$$

where p_c is the compressive strength (Section 6.4)
m_x, m_y are the equivalent uniform moment factors (Section 3.2)
M_b is the buckling resistance (Section 3.5)
Z_y is the minor axis elastic section modulus (compression face)

A more exact approach (clause 4.8.3.3.2) may be carried out using M_{ax} and M_{ay}, the maximum buckling moments about major and minor axes in the presence of axial load. In this approach:

$$m_x M_x/M_{ax} + m_y M_y/M_{ay} \not> 1$$

7.7 EXAMPLE 13. COLUMN FOR INDUSTRIAL BUILDING

(a) Dimensions

(See Fig. 7.7.)

Overall height	12.5 m
Height of crane rail	10.0 m
Free standing brick wall	2.5 m
Crane rail eccentricity	0.5 m
Cladding eccentricity	0.25 m

(b) Loading (unfactored)

Roof truss reactions W_A:
Dead load	50 kN
Imposed load	78 kN
Wind load (suction)	90 kN

Crane girder reactions (vertical) W_C:
Dead load (girder self weight)	20 kN
Crane load incl. impact (nearside)	220 kN
Crane load (far side)	150 kN

Crane girder reaction (horizontal) W_{HC}:
Crane surge load	6 kN

Wind on side of building W_B:
Wind load (u.d.l. above brickwork) 55 kN

Self weight (column) W_D
Dead load   10 kN

Cladding W_{DC}
Dead load   16 kN

The resultant force (W_E for wind) depends on the difference between the wind loading on the column being designed and that on the similar column on the far side of the building (Fig. 7.6). If the wind load on the far side is zero:

$$W_E = 3(55+0)/16 \qquad = 10.3 \text{ kN}$$

The resultant force (W_E for crane) depends on the difference in the crane (vertical) loading on the near and far side columns (Fig. 7.6):

$$W_E = 3(220+150)0.5/(4 \times 12.5) \qquad = 10.2 \text{ kN}$$

The resultant force (W_E for dead load) depends on the crane girder dead load producing moments in opposite directions:

$$W_E = 3(20+20)0.5/(4 \times 12.5) \qquad = 1.2 \text{ kN}$$

Note that the crane surge loading (horizontal) produces no resultant force in the truss as it induces equal loads in both columns, and in the same direction.

All the unfactored loads may be shown in the load matrix [W] kN:

	Dead load W_d	Imposed load W_i	Crane load vertical W_{cv}	Crane load horizontal W_{ch}	Wind load W_w
W_A	50	78	0	0	−90
W_B	0	0	0	0	55
W_C	20	0	220	0	0
W_{HC}	0	0	0	6	0
W_D	10	0	0	0	0
W_{DC}	16	0	0	0	0
W_E	1.2	0	10.2	0	10.3

Fig. 7.7

(c) Load factors and combinations

clause 2.4.1.1   Possible combinations of the different loads are:

$1.4\ W_d + 1.6\ W_i$
$1.4\ W_d + 1.4\ W_w$
$1.4\ W_d + 1.6\ W_{cv}$
$1.2\ W_d + 1.6\ W_i + 1.4\ W_{cv} + 1.4\ W_{ch}$ case (i)
$1.2\ W_d + 1.6\ W_i + 1.6\ W_{cv}$ case (ii)
$1.2\ W_d + 1.2\ W_i + 1.2\ W_w$
$1.2\ W_d + 1.2\ W_i + 1.2\ W_{cv} + 1.2\ W_{ch} + 1.2\ W_w$ case (iii)

Of these combinations, three cases are selected as shown and are given in matrix form $[\gamma_f]$:

	Case (i)	Case (ii)	Case (iii)
W_d	1.2	1.2	1.2
W_i	1.6	1.6	1.2
W_{cv}	1.4	1.6	1.2
W_{ch}	1.4	0	1.2
W_w	0	0	1.2

(d) **Factored loading**

As shown in Section 7.4, the factored loading matrix is
$[W_\gamma] = [W] \times [\gamma_f]\,\text{kN}$

	Case (i)	Case (ii)	Case (iii)
W_A	185	185	46
W_B	0	0	66
W_C	332	376	288
W_{HC}	8.4	0	7.2
W_D	12	12	12
W_{DC}	19	19	19
W_E	15.7	17.8	26.0

The loads for each loading case can be shown on diagrams as in Figs. 7.8, 7.9 and 7.10.

Fig. 7.8   Fig. 7.9   Fig. 7.10

(e) Coefficients

Bending moments and axial forces are produced by the factored loads. To calculate axial force all vertical loads above the cross-section being checked should be added together. To calculate bending moments the products of force and distance should be added together. This is again usefully shown in a matrix $[\alpha]$, for axial force F and bending moment M. The BM values are to be calculated at three positions as shown in Fig. 7.7. The coefficients for BM are in metres.

	F_1	M_1	M_2	M_3
W_A	1	0.0	0.0	0.0
W_B	0	7.5	2.81	0.13
W_C	1	0.5	0.5	0.5
W_{HC}	0	10.0	5.0	0.0
W_D	1	0.0	0.0	0.0
W_{DC}	1	−0.25	−0.19	−0.06
W_E	0	−12.5	−7.5	−2.5

The coefficients for BM in relation to the load W_B allow for a reduced value of W_B as well as a reduced distance in calculating α for positions 2 and 3. For example:

$$M_2 = W_B \ (7.5/10) \times (7.5/2) = 2.81 W_B$$

hence $\alpha = 2.81$ m.

Similarly the value of W_{DC} decreases for points 2 and 3 and this is reflected in the value of α:

$$M_2 = W_{DC} \ (7.5/10) \times (-0.25) = -0.19 \ W_{DC}$$

(f) BM and axial force

As shown in Section 7.4, the BM and axial force matrix is

$$[F, M] = [W_\gamma]^T \times [\alpha]$$

	F_1	M_1	M_2	M_3
Case (i)	548	49	87	125
Case (ii)	592	−40	50	142
Case (iii)	365	381	167	86

The BM diagrams for each load case may be drawn as in Fig. 7.11.

Fig. 7.11

(g) Local capacity

BS table 6

Use a **305 × 305 × 137 UC** section (grade 43). This is a plastic section ($b/T = 7.8$, $d/t = 17.9$) and has a design strength $p_y = 265 \text{ N/mm}^2$

Moment capacity $M_{cx} = p_y S_x$
$= 265 \times 2298 \times 10^{-3} = 609 \text{ kNm}$
$A_g p_y = 174.6 \times 265 \times 10^{-1} = 4630 \text{ kN}$

CASE (i)

clause 4.8.3.2

Local capacity check (simplified):

$F/A_g p_y + M_x/M_{cx} \not> 1$

$548/4630 + 125/609 = \mathbf{0.32}$

CASE (ii)

Local capacity check:

$592/4630 + 142/609 = \mathbf{0.36}$

CASE (iii)

Local capacity check:

$365/4630 + 381/609 = \mathbf{0.70}$

(h) Overall buckling

For compressive strength, the slenderness λ is based on an effective length L_E:

BS appendix D (fig. 19)

$L_{EX} = 1.5 \times 12.5 = 18.75 \text{ m}$, or
$L_{EY} = 0.85 \times 10.0 = 8.5 \text{ m}$
$\lambda_x = L_{EX}/r_x = 18750/137 = 137$

BS table 27b

$p_c = 86 \text{ N/mm}^2$
$\lambda_y = 8500/78.2 = 109$

BS table 27c

$p_c = 110 \text{ N/mm}^2$

Use lower value of p_c to obtain compression resistance:

$$A_g\, p_{cy} = 174.6 \times 86 \times 10^{-1} = 1500\,\text{kN}$$

For moment resistance, the buckling strength is always based on the minor axis slenderness:

$$\lambda_y = L_E/r_y = 109$$
$$x = 14.1\ (\text{torsional index})$$
$$\lambda/x = 7.7$$
$$N = 0.5$$
BS table 14
$$v = 0.71$$
$$u = 0.851$$

CASE (i)

BS table 13 $n = 1.0$, $m \leq 1.0$, as the column is not loaded along its length for this combination.

BS table 18
$$\beta = 49/125 = 0.39$$
$$m = 0.72$$
$$\lambda_{LT} = nuv\lambda$$
$$= 1.0 \times 0.851 \times 0.71 \times 109 = 66$$
BS table 11 Buckling strength $p_b = 194\,\text{N/mm}^2$
Buckling resistance $M_b = p_b\, S_x$
$$= 194 \times 2298 \times 10^{-3} = 446\,\text{kNm}$$

Simplified overall buckling check:

$$F/A_g\, p_c + m_x M_x/M_b \not> 1$$

$$548/1500 + 0.72 \times 125/446 = \mathbf{0.57}$$

CASE (ii)

As for case (i), $n = 1.0$, $m \leq 1.0$.

BS table 18
$$\beta = -40/142 = -0.28$$
$$m = 0.48$$

As for case (i), buckling resistance $= 446\,\text{kNm}$

Simplified overall buckling check:

$$592/1500 + 0.48 \times 142/446 = \mathbf{0.55}$$

CASE (iii)

BS table 13 $m = 1.0$, $n \leq 1.0$, as the column is loaded along its length for this combination.

BS tables 16, 17

BS tables 11

$\beta = 86/381 = 0.23$
$\gamma = -381/67 = -5.7$
$n = 0.71$
$\lambda_{LT} = 0.71 \times 0.851 \times 0.71 \times 109 = 47$
Buckling strength $p_b = 238$ N/mm^2
Buckling resistance $M_b = 238 \times 2298 \times 10^{-3} = 547$ kNm

Simplified overall buckling check:

$$365/1500 + 1.0 \times 381/547 = \mathbf{0.94}$$

Section is satisfactory. Using a smaller section (305 × 305 × 118 UC) results in the simplified buckling check > 1.

7.8 EXAMPLE 14. LACED COLUMN FOR INDUSTRIAL BUILDING

(a) Dimensions

Principal dimensions are as Section 7.7(a) but the cross-section is revised as shown (Fig. 7.12).

(b) Loading

As Section 7.7(b).

(c) Combined force and BM

Combinations of loads may be considered as before (including γ_f factors) using the same values of W_A to W_E inclusive. Maximum values of BM and force will vary slightly owing to the revised cross-section.
Consider case (iii) only, i.e. 1.2 (dead + crane + wind + imposed):

Max. BM $= 288.0 \times 0.4 + 7.2 \times 10.0 + 66.0 \times 7.5$
$\qquad - 26.0 \times 12.5 - 19.0 \times 0.65 - 46.0 \times 0.4 = 326$ kNm
Axial force $= 46.0 + 288.0 + 12.0 + 19.0 = \mathbf{365}$ **kN**

Fig. 7.12

(d) Local capacity

clause 4.7.8

The design of a laced column may be carried out assuming it to be a single integral member, and checking the design of the lacings.
Use two **−305 × 127 × 37 UB** sections. These are semi-compact sections ($b/T = 14.2$) with a design strength $p_y = 275$ N/mm^2.

Moment of inertia of combined section
$$I_a = 2\,(337 + 47.5 \times 40^2) = 152\,670\,\text{cm}^4$$
$$Z_a = 152\,670/55 = 2776\,\text{cm}^3$$

clause 4.2.5
$$M_{cx} = p_y\,Z_a = 275 \times 2776 \times 10^{-3} = 763\,\text{kNm}$$

Local capacity check
$$F/A_g\,p_y + M_x/M_{cx} \not> 1$$
$$365/(2 \times 47.5 \times 275 \times 10^{-1}) + 326/909 = \mathbf{0.50}$$

(e) Overall buckling

For compressive strength the overall slenderness λ is based on an effective length:

BS appendix D (fig. 20)
$$L_{Ea} = 1.5 \times 10.0 = 15.0\,\text{m, or}$$
$$L_{Eb} = 0.85 \times 10.0 = 8.5\,\text{m}$$
$$r_a = \sqrt{(I_a/A)}$$
$$\quad = \sqrt{[152\,670/(2 \times 47.5)]} = 40.1\,\text{cm}$$
$$\lambda_a = L_{Ea}/r_a$$
$$\quad = 1500/40.1 = 37$$
$$\lambda_b = L_{Eb}/r_b$$
$$\quad = 850/12.3 = 69$$
Local slenderness $\lambda_c = L/r_y$
$$\quad = 1000/26.7 = 38$$

clause 4.7.8g
Limiting value for $\lambda_c = 50$
Overall slenderness $\lambda_b \not> 1.4\lambda$
$$\quad = 1.4 \times 38$$
$$\quad = 53$$

BS table 27a Compressive strength p_c based on the highest value of λ, i.e. 69:
$$p_c = 225\,\text{N/mm}^2$$

Bending about the A–A axis (Fig. 7.12) can be assumed to produce axial forces in the UB sections.

Axial force = moment/centroidal distance between UBs
$$\quad = 326/0.8 = 408\,\text{kN (tension and compression)}$$
Maximum compression in one UB
$$\quad = 408 + 365/2 = 591\,\text{kN}$$
Compression resistance $P_c = A_g\,p_c$
$$\quad = 47.5 \times 225 \times 10^{-1} = \mathbf{1069\,\text{kN}}$$

Section is satisfactory.

(f) Lacings

$$\text{Max. force in lowest diagonal} = (66.0 + 7.2 - 26.0)/\cos 51.3°$$
$$= 75.5 \text{ kN (compression)}$$

In addition, the lacings should carry a transverse force equal to 2.5% of the axial column force, i.e. $0.025 \times 365 = 9.1$ kN. Note that the original value of 1% in clause 4.7.8i was considered too small.

$$\text{Lacing force} = 9.1/\cos 51.3° = 13.3 \text{ kN}$$
$$\text{Total force} = 75.5 + 13.3 = 88.8 \text{ kN}$$

i.e. 44.4 kN on each side of column.

Use **60 × 60 × 6 equal angle**:

clause 4.7.8h
$$\text{Effective length of lacing} = \sqrt{[0.80^2 + 1.00^2]} = 1.28 \text{ m}$$
$$\lambda = 1280/11.7$$
$$= 109$$

BS table 27c
$$\text{Compressive strength } p_c = 111 \text{ N/mm}^2$$
$$\text{Compression resistance } P_c = A_g \, p_c$$
$$= 6.91 \times 111 \times 10^{-1}$$
$$= \mathbf{76.7 \text{ kN}}$$

Section is satisfactory.

STUDY REFERENCES

Topic	References
1. Euler load	**Marshall W.T. & Nelson H.M.** (1990) Elastic stability analysis, *Structures*, pp. 420–52. Longman
2. Euler load	**Coates R.C., Coutie M.G. & Kong F.K.** (1988) Instability of struts and frameworks, *Structural Analysis*, pp. 58–71. Van Nostrand Reinhold
3. Column behaviour	**Dowling P.J., Knowles P. & Owens G.W.** (1988) *Structural Steel Design*. Steel Construction Institute.
4. Column behaviour	**Kirby P.A. & Nethercot D.A.** (1979) In-plane instability of columns, *Design for Structural Stability*. Collins
5. Reduced plastic moment	(1987) *Steelwork Design*, vol. 1, Section properties member capacities. Steel Construction Institute

8

COLUMN BASES & BRACKETS

The transfer of force between one element of a structure and the next requires particular care by the designer. A good detail may result from long experience of the use of structural steelwork, and many examples are available for students to copy or adapt[1,2]. However, equally good details may be developed with less experience providing the following basic principles are adhered to:

- The forces to be carried must be set out and transferred between different elements of the connection, i.e. a realistic load path must be assumed within the connection.
- Simplicity of detail usually produces the most effective and robust engineering solution, provided strength and stiffness requirements are satisfied.
- Any detail must be practicable and cost-effective, both from the point of view of the steelwork fabricator, and from that of the site erector.

Column bases and brackets are connections which carry forces and reactions to a column element. Bases transfer reactions from the foundation to the column, while brackets may be used to transfer loads from crane girders or similar members. In addition, columns may receive loads via beam connections (see Section 3.7g) or cap plates (see Section 6.6).

8.1 COLUMN BASES

Two main types of column base are used and these are shown in Fig. 8.1. Welded or bolted construction can be used, or a combination of both, the decision being dependent on whether or not the base is attached to the column during fabrication, or later during site erection. In general, the simpler slab base is used in small and medium construction when axial load dominates. The gusseted base is used in heavy construction with larger column loads and where a certain amount of fixity is required.

Construction requirements and details are given in reference (3).

COLUMN BASES & BRACKETS **95**

Fig. 8.1 Column bases — Slab base — Gusseted base

8.2 DESIGN OF COLUMN BASES

Column bases transmit forces and moments from the column to its foundation. The forces will be axial loads, shear forces and moments about either axis or any combination of them. Shear force is in reality probably transmitted by friction between the base plate and the foundation concrete, but it is common in design for this shear force to be resisted totally by the holding down bolts.

The common design case deals with axial load and moment about one axis. If the ratio of moment/axial load is less than $L/6$ where L is the base length, then a positive bearing pressure exists over the whole base and may be calculated from equilibrium alone. Nominal holding down bolts are provided in this case, and at least two are in fact provided to locate the base plate accurately. If the ratio exceeds base length $L/6$, holding down bolts are required to provide a tensile force. Both arrangements are shown in Fig. 8.2.

Fig. 8.2 Bearing pressure

Where tension does occur in the holding down bolts, a number of methods of design are possible:

1. It is assumed that the bearing pressure has a linear distribution to a maximum value of $0.4 f_{cu}$, where f_{cu} is the concrete cube strength. This basis is suggested in clause 4.13.1 and analysis of the bearing pressure and bolt stresses may follow reinforced concrete theory (Fig. 8.3).

Fig. 8.3 Reinforced concrete theory

f_c = bearing pressure
f_t = tensile stress in bolts
m = modular ratio
A_s = total bolt cross section
$d = L - n$
$d_1 = 0.5(d - n) + M/F$
$A_1 = 6md_1A_s/B$

y is solution of $y^3 - 3(d - d_1)y^2 + A_1y - A_1d = 0$
$f_c = 6d_1F/[By(3d - y)]$
$f_t = mf_c(d/y - 1)$

2. An alternative approximate analysis is sometimes used which assumes that the permissible stresses for steel tension and concrete bearing are reached together. This method is shown in Fig. 8.4.

Fig. 8.4 Approximate analysis

Resultant compression C
Resultant tension T
Permissible compressive stress p_c
Permissible tensile stress p_t

$d = L - n$
$y = mp_cd/(mp_c + p_t)$
$C = \dfrac{M + F(L/2 - n)}{d - y/3}$
$T = C - F$
$f_c = 2C/By$

3. A rectangular pressure distribution may also be used, which leads to a slightly different plate thickness and bolt sizes. The analysis is based on reinforced concrete theory for ultimate limit state.

In general, the first method is used to obtain bearing pressures and bolt stresses. The thickness of the base plate is obtained using the steel strength p_{yp} from Table 1.2 of Chapter 1, but not more than 270 N/mm².

Maximum moment in plate $\not> 1.2p_{yp} Z$ (clause 4.13.2.3) where Z is the elastic modulus of the plate section.

For the case of concentric forces only, the base plate thickness may be obtained from clause 4.13.2.2.

8.3 BRACKETS

Brackets are used as an alternative to cleated connections (Section 3.7g) only where the latter are unsuitable. A common case of this situation is the crane girder support on a column (Section 5.3k). The eccentric connection by the bracket is necessary for the chosen structural arrangement. It does, however, generate large moments in the column (Section 7.7c) and is therefore used only where it is essential to the steelwork layout.

Brackets may be connected either to the web or to the flange of the column using bolts, welds or a combination of the two. Examples are shown in Fig. 8.5. In Fig. 8.5(a), the moment acts out of plane producing tension in the bolts, while in Fig. 8.5(b) the moment is in the plane of the connection resulting in a shear effect in the bolts. The bracket may be fabricated from offcuts of rolled sections, or from plates appropriately shaped and welded together. Connection to the column may be made during fabrication, or the brackets may be attached during site erection.

(a) Face connected moment (out of plane)

(b) Lapped torsional moment (in plane)

Fig. 8.5 Brackets

8.4 DESIGN OF BRACKETS

Brackets are subjected both to a vertical shear load and to a moment due to the eccentricity of the vertical load. Note that the moment will vary with the point in the bracket under consideration. A bracket may also be subjected to horizontal loads, but these are usually of a secondary nature, or may be covered by a special detail (Fig. 5.11).

Vertical loads are supported by welds or by bolts acting in shear. Ideally, the moments also should be carried by bolts in shear (or by welds), but some bracket arrangements shown in Fig. 8.5(a) will give rise to bolt tension.

Vertical load W is divided between the bolts or weld group uniformly so that:

Bolt shear $= W/N$ kN

or

Weld shear $= W/L_w$ kN/mm

where N is the number of bolts
L_w is the total weld length (mm)

For a weld design strength of 215 N/mm² (BS table 36), weld capacities (kN/mm) may be calculated for each weld size. Note that the weld size is in fact defined by the leg length, and the design dimension is the throat distance, i.e. throat size = leg length/$\sqrt{2}$. Values of weld capacity are given in reference (4).

Where the moment produces bolt shear only (Fig. 8.5) then the shear on each bolt is given approximately by:

$$\text{Bolt shear} = We d_{max}/\Sigma d^2$$

where d is the bolt distance from the bolt group centroid.

For a weld group the approximation is:

$$\text{Weld shear} = We d_{max}/I'_p$$

where I'_p is the $I_x + I_y$ for the weld group (Fig. 8.11).

In some cases the moment produces bolt tension and in these cases bolt force is considered proportional to distance from a neutral axis. The neutral axis may be taken as $d_c/7$ in depth[2], and as a result:

$$\text{Bolt tension} = We d_{max}/\Sigma d^2$$

where d is the bolt distance from the neutral axis.

For a weld group:

$$\text{Weld shear} = We d_{max}/I_x$$

where I'_x is the equivalent second moment of area of the weld about the weld group centroid.

The effects of vertical load and moment due to eccentricity must be added either for individual bolts, or for points in a weld run. Clearly, those points of maximum force or stress need to be checked, which occur at positions furthest from the group centroid or neutral axis.

Where the moment produces shear in a bolt, vectorial addition may be used. In cases where the moment produces tension a combined check may be used (clause 6.3.6.3):

$$F_s/P_s + F_t/P_t \not> 1.4 \text{ for ordinary bolts}$$

where F_s is the applied shear
P_s is shear capacity
F_t is applied tension
P_t is tension capacity

For a weld group all combinations of vertical load and moment produce shear in the weld, and vectorial addition is used as necessary.

8.5 EXAMPLE 15. DESIGN OF SLAB BASE

(a) Dimensions

305 × 305 × 137 UC column

(b) Loading

All loads include appropriate values of γ_f
Case (i)   Maximum vertical load 1400 kN
Case (ii)   Largest moment under maximum load conditions:
moment 60 kNm and 850 kN
Case (iii)   Largest moment under minimum load conditions:
moment 85 kNm and 450 kN

(c) Bearing pressure

(See Fig. 8.6.) Assume base **520 × 520 plate** and four bolts (grade 4.6) 20 mm diameter.
$$\text{Tension bolt area } A_s = 2 \times 245 = 490 \text{ mm}^2$$
$$d = 520 - 50 = 470 \text{ mm}$$

Fig. 8.6

CASE (i) LOADING

clause 4.13.1
Assuming a concrete cube strength $f_{cu} = 30 \text{ N/mm}^2$
Permissible pressure $= 0.4 \times 30 = 12.0 \text{ N/mm}^2$
Pressure $= 1400 \times 10^3/(520 \times 520) =$ **5.2 N/mm²**

CASE (ii) LOADING

$M/F = 60/850 = 71$ mm
$L/6 = 520/6 = 87$ mm
$M/F < L/6$
Base area $A = 520 \times 520 \text{ mm}^2 = 2700 \text{ cm}^2$
Base modulus $Z = 520 \times 520^2/6 \text{ mm}^3 = 23\,400 \text{ cm}^3$
Pressure $= F/A + M/Z =$ **6.11 N/mm²**

CASE (iii) LOADING

Fig. 8.3
$M/F = 85/450 = 189$ mm
$d_1 = 0.5(470-50) + 85 \times 10^3/450 = 339$ mm
$A_1 = 6 \times 399 \times 15 \times 490/520 = 33.8 \times 10^3 \text{ mm}^2$

The distance y (Fig. 8.7) is the solution of:
$$y^3 - 3(d-d_1)y^2 + A_1 y - A_1 d = 0$$
$$y^3 - 3(470-399)y^2 + 33.8 \times 10^3 y - 33.8 \times 10^3 \times 470 = 0$$
hence $y = 288$ mm
Pressure $f_c = 6d_1 F/[By(3d-y)]$
$= 6 \times 339 \times 450 \times 10^3/[520 \times 288(3 \times 470 - 288)]$
$=$ **6.41 N/mm²**

Bearing pressure satisfactory ($< 12 \text{ N/mm}^2$)

Fig. 8.7

100 STRUCTURAL STEELWORK DESIGN TO BS 5950

(d) Bolt capacity

$$\begin{aligned}
\text{Bolt stress } f_t &= mf_c(d/y - 1) \\
&= 15 \times 6.41\,(470/288 - 1) = 61\,\text{N/mm}^2 \\
\text{Force/bolt} &= 61 \times 245 \times 10^{-3} = 14.9\,\text{kN}
\end{aligned}$$

clause 6.3.6.1

$$\begin{aligned}
\text{Bolt capacity } P_t &= p_t\,A_t \\
&= 195 \times 245 \times 10^{-3} = \mathbf{47.8\,kN}
\end{aligned}$$

Bolts are satisfactory.

(e) Plate thickness

(See Fig. 8.8.)

Fig. 8.8

Maximum bearing pressure from case (ii) loading $= 6.41\,\text{N/mm}^2$
Maximum BM (assuming constant pressure)
$= 6.41 \times 520 \times 100^2/2 = 16.7\,\text{kNm}$
Some reduction of BM may be found by using the trapezium pressure distribution.
Try 25 mm thick plate:

Table 1.2

$$p_{yp} = 265\,\text{N/mm}^2$$
$$\text{Plate modulus } Z = 520 \times 25^2/6 = 54.2 \times 10^3\,\text{mm}^3$$

clause 4.13.2.3

$$\begin{aligned}
\text{Moment capacity} &= 1.2 p_{yp}\,Z \\
&= 1.2 \times 265 \times 54.2 \times 10^{-3} = \mathbf{17.2\,kNm}
\end{aligned}$$

Plate is satisfactory.

For larger loads and/or moments, a gusseted base may be required, particularly if the thickness of a slab base would otherwise exceed 50 mm. The design is the same as given above, but in Section 8.5(e) the plate modulus Z is based on the combined effect of plate and gussets. At thicknesses greater than 25 mm, steel grades other than 43A may be needed to avoid the possibility of brittle fracture (BS table 4).

(f) Column/base plate weld

Fig. 8.9

The weld is commonly designed to carry the maximum moment, ignoring the effect of vertical load. All compression is taken in direct bearing (Fig. 8.9).

Maximum tension in flange	$= M/(D-T)$
	$= 85 \times 10^3/300 = 283\,\text{kN}$
For one flange weld length	$= 2 \times 308 = 616\,\text{mm}$
Weld shear $= 283/616$	$= 0.459\,\text{kN/mm}$

clause 6.6.5

Use **6 mm fillet weld**, capacity[4] $= \mathbf{0.903\,kN/mm}$

Weld is satisfactory.

8.6 EXAMPLE 16. DESIGN OF CRANE GIRDER BRACKET (FACE)

(a) Dimensions

(See Fig. 8.10.)
Column 610 × 229 × 140 UB
Crane girder eccentricity 550 mm

(b) Loading

Maximum rail reaction 462 kN
(including appropriate values of γ_f)
Crane surge load carried by diaphragm restraint

(c) Bracket

Use offcut of **457 × 191 × 89 UB** (grade 43A)
Maximum BM in bracket

$M_x = 462 \times 0.2215 = 102$ kNm
Shear capacity $P_v = 0.6 p_y A_v$
$= 0.6 \times 275 \times 10.6 \times 463.6 \times 10^{-3}$
$= 811$ kN
Shear force $F_v = 462$ kN
$F_v / P_v = 0.57 < 0.60$
Moment capacity $M_{cx} = p_y S_x$
$= 265 \times 2010 \times 10^{-3} = \mathbf{533}$ **kNm**

Bracket is satisfactory.

Fig. 8.10

(d) End plate weld

(See Fig. 8.11.)

Shear force = 462 kN
Moment = 102 kNm

clause 6.6.5.3 Use **6 mm fillet weld**

Weld length = $4 \times 190 + 2 \times 420 = 1600$ mm
Weld force (vertical load) = $462/1600 = 0.289$ kN/mm
Weld second moment $I'_{xx} = 2 \times 420^3/12 + 4 \times 190 \times 220^2$
$= 49 \times 10^6$ mm^3
Weld shear (moment) $= My/I'_{xx}$
$= 102 \times 10^{-3} \times 220/(49 \times 10^6)$
$= 0.458$ kN/mm

Fig. 8.11

Note that in this case, the vertical shear and the shear due to moment act perpendicular to each other, and the resultant shear is obtained by vectorial addition.

$$\text{Resultant} = \sqrt{[0.289^2 + 0.458^2]}$$
$$= \mathbf{0.542\ N/mm}$$
Weld capacity = 0.903 N/mm

Weld is satisfactory.

(e) Connection bolts

Shear force = 462 kN
Out-of-plane moment = 462 × 0.2415 = 112 kNm

Use **eight no. 22 mm diameter bolts** (grade 8.8).

	Shear/bolt F_s	= 462/8	= 57.8 kN
clause 6.3.2	Shear capacity P_s	= $p_s A_s$	
		= 0.375 × 303	= 114 kN
clause 6.3.3	Bearing capacity of plate $P_{bs} = dtp_{bs}$		
		= 22 × 20 × 0.450	= 198 kN
clause 6.3.6	Tensile force (Fig. 8.12)		

$$d_c/7 = 457/7 = 65\ mm$$
$$F_t = Md_{max}/\Sigma d^2$$
$$= 112 \times 10^3 \times 315/[(15^2 + 115^2 + 215^2 + 315^2)2] = 111\ kN$$
Tension capacity $P_t = p_t A_t$
$$= 0.450 \times 303 = 136\ kN$$

Combined check

$$F_s/P_s + F_t/P_t \not> 1.4$$
$$57.8/114 + 111/136 = 0.51 + 0.82 = \mathbf{1.33}$$

Bolts are satisfactory.

Fig. 8.12

8.7 EXAMPLE 17. DESIGN OF CRANE GIRDER BRACKET (LAPPED)

(a) Dimensions

(See Fig. 8.13.)
Column 305 × 305 × 158 UC
Crane girder eccentricity 550 mm

(b) Loading

As Section 8.6(b)
Maximum reaction 462 kN

Fig. 8.13

8 no.
29 dia. holes
for M27 bolts

(c) Bracket

Use **two 20 mm thick plates** (grade 43A) shaped as Fig. 8.13.
Maximum BM in bracket:

$$M_x = 462 \times 0.550 = 254 \text{ kNm}$$

clause 4.2.3(c) Shear area $A_v = 0.9(450 - 4 \times 29)20 \times 2 = 12\,020 \text{ mm}^2$
Shear capacity $P_v = 0.6 p_y A_v$
$$= 0.6 \times 265 \times 12\,020 \times 10^{-3} = 1910 \text{ kN}$$
hence $F_v/P_v = 0.24$

Second moment of area of plate (cm units)
$$= 2 \times 20 \times 450^3/12 = 30\,380 \text{ cm}^4$$
Minus bolt holes:
$$4 \times 20 \times 26 \times 50^2 = 580 \text{ cm}^4$$
$$4 \times 20 \times 26 \times 150^2 = 5220 \text{ cm}^4$$
Net $I = 24\,580 \text{ cm}^4$
Modulus $Z = 24\,580/22.5 = 1090 \text{ cm}^3$

For brackets of this type it may be assumed that the bolts or welds provide lateral restraint to the compression zones. The moment capacity should be taken as:

$$M_{cx} = p_y Z_x = 265 \times 1090 \times 10^{-3} = \mathbf{289 \text{ kNm}}$$

Bracket is satisfactory.

(d) Column bolts

Shear force $= 462 \text{ kN}$
Moment $= 462 \times 0.550 = 254 \text{ kNm}$
Use **eight no. 27 mm diameter bolts** (grade 8.8) on each face.
Shear/bolt due to vertical load $= 462/8 \times 2 = 28.9 \text{ kN}$
Shear/bolt due to moment $= Md_{max}/\Sigma d^2$
$$= 254 \times 10^{-3} \times 168/8\,(90^2 + 168^2)$$
$$= 147 \text{ kN}$$
Vector sum of shear $= \mathbf{162 \text{ kN/bolt}}$
Shear capacity/bolt $P_s = p_s A_s$
$$= 375 \times 459 = \mathbf{172 \text{ kN}}$$
Bearing capacity of plate $P_{bs} = dt p_{bs}$
$$= 27 \times 20 \times 0.450 = 243 \text{ kN}$$
Bolts are satisfactory.
Note that the lapped bracket requires twice the number of bolts of a larger size compared with the face bracket.

Fig. 8.14

STUDY REFERENCES

Topic	References
1. Connections	(1993) *Joints in Simple Construction*, vol. 1. Steel Construction Institute
2. Connections	(1987) Bolt & weld capacities, *Steelwork Design* vol. 1, Section properties, member capacities, pp. 22–4, Steel Construction Institute

3. Column bases   (1980) *Holding Down Systems for Steel Stanchions.* Concrete Society/BCSA/Steel Construction Institute

4. Weld capacity   (1987) Strength of fillet welds, *Steelwork Design* vol. 1, Section properties, member capacities, p. 205. Steel Construction Institute

9
COMPOSITE BEAMS & SLABS

The term 'composite' can be used of any structural medium in which two or more materials interact to provide the required strength and stiffness. In steelwork construction the term refers to cross-sections which combine steel sections with concrete in such a way that the two act together. Typical cross-sections of beams and slabs are shown in Fig. 9.1.

Fig. 9.1 Composite sections

The performance of composite beams is similar to that of reinforced concrete beams[1], but there are two main differences. Firstly, the steel section has a significant depth and its second moment of area may not be ignored, unlike that of the steel bar reinforcement. Secondly, the concrete to reinforcement bond, which is essential for reinforced concrete action, is absent in composite beams generally and must be provided by shear connection. Design methods for composite beams therefore follow those methods for reinforced concrete with modifications as indicated. Owing to the presence of the concrete slab, problems of steel compression flange instability and local buckling of the steel member are not usually relevant in simply supported members except during erection.

Recommendations for design in composite construction are not included in Part 1 of BS 5950 but are included in:

Part 3.1: *Design of composite beams* (1990)
Part 4: *Design of floors with profiled steel sheeting* (1982)

The basis of design used in this chapter is given in Section 9.7.

9.1 COMPOSITE BEAMS

The advantages of composite beams compared with normal steelwork beams are the increased moment capacity and stiffness, or alternatively the reduced steel sizes for the same moment capacity. Apart from a saving in material, the reduced construction depth can be worthwhile in multi-storey frames. The main disadvantage of composite construction is the need to provide shear connectors to ensure interaction of the parts.

As in all beam design, shear capacity and moment capacity of a composite section must be shown to be adequate. But in addition, the strength of the shear connection must be shown to be satisfactory, with regard to both connector failure and also local shear failure of the surrounding concrete (see Section 9.4). For full interaction of the steel and concrete, sufficient shear connection must be provided to ensure that the ultimate moment capacity of the section can be reached. Lower levels of connection will result in partial interaction which is not covered in this chapter[2].

Composite beams are essentially T beams with wide concrete flanges. The non-uniform distribution of longitudinal bending stress must be allowed for and this is usually achieved by use of an effective breadth for the concrete flange. For buildings the effective breadth B_e may be taken as one-quarter of the span (simply supported). Continuous beams and cantilevers are treated differently (see BS 5950: Part 3.1).

9.2 SHEAR AND MOMENT CAPACITY OF COMPOSITE BEAMS

The shear capacity of a composite beam is based on the resistance of the web of the steel section alone. Calculation of the shear capacity P_v is given in Section 3.7(d):

$$P_v = 0.6 p_y A_v$$

Moment capacity is based on assumed ultimate stress conditions shown in Figs. 9.2 and 9.3. When the neutral axis lies in the concrete slab (Fig. 9.2) the value of x_p may be found by equilibrium of the tension and compression forces. The moment capacity M_c is given by:

$$M_c = A p_y (D_s + D/2 - x_p/2)$$

When the neutral axis lies in the steel section (Fig. 9.3) the value of A_{sc} may be found by equilibrium. The centroid of the compression steel A_{sc} must be located, and moment capacity M_c is given by:

$$M_c = A p_y (D/2 + D_s/2) - 2 A_{sc} p_y (d_{sc} - D_s/2)$$

Alternatively, formulae given in BS 5950: Part 3.1 may be used.

Fig. 9.2 Moment capacity (NA in slab)

Fig. 9.3 Moment capacity (NA in steel beam)

9.3 SHEAR CONNECTORS

Many forms of shear connector have been used, of which two are shown in Fig. 9.4, but the preferred type is the headed stud. This combines ease of fixing with economy. Shear connectors must perform the primary function of transferring shear at the steel/concrete interface (equivalent to bond) and hence control slip between the two parts. In addition, they have the secondary function of carrying tension between the parts and controlling separation.

The relationship between shear force and slip for a given connector is important in design where partial interaction is expected. For the design in this section, where full interaction is assumed, a knowledge of only the maximum shear force which the connector can sustain is required. The strengths of standard headed studs embedded in different normal weight concretes are given in Table 9.1.

The strength of alternative shear connectors can be found by use of a standard push-out test (BS 5400: Part 5). The performance of all shear connectors is affected by lateral restraint of the surrounding concrete, the

Table 9.1 – Shear strength of headed studs

Diameter (mm)	Height (mm)	Shear strength Q_k (kN) for concrete f_{cu} (N/mm²)			
		25	30	35	40
22	100	119	126	132	139
19	100	95	100	104	109
16	75	70	74	78	82

Fig. 9.4 Shear connectors   Headed studs   RSA offcuts

presence of tension in the concrete, and the type of concrete used, i.e. normal concrete or lightweight. For design of composite beams in these cases further references[2] should be consulted.

The shear connection in buildings may be designed on the assumption that at the ultimate limit state the shear force transmitted across the interface is distributed evenly between the connectors. The shear force is based on the moment capacity of the section and connector force Q_p is shown in Fig. 9.5.

$$Q_p = R_c/N_p$$

where $R_c = 0.45 f_{cu} B_e x_p$ (when NA in concrete)
or $R_c = 0.45 f_{cu} B_e D_s$ (when NA in steel)

The connector force Q_p must be checked:

$$Q_p \not> 0.8 Q_k$$

Fig. 9.5 Connector force

9.4 LOCAL SHEAR IN CONCRETE

The total shear connection depends not only on the shear connector (headed stud, etc.) but also on the ability of the surrounding concrete to transmit the shear stresses. Longitudinal shear failure is possible on the planes shown in Fig. 9.6. Transverse reinforcement combined with the concrete should give a strength greater than the applied shear per unit length v, such that:

$$v \not> 0.8 L_s \sqrt{f_{cu}} + v_p \quad \text{and}$$
$$v \not> 0.03 L_s f_{cu} + 0.7 A_{sv} f_y + v_p$$

where A_{sv} is either $(A_{rt}+A_{rb})$ or $2A_{rb}$, depending on the shear path
f_y is design strength of the reinforcement
f_{cu} is the concrete cube strength
L_s is either (twice slab depth)
or (connector width + twice stud height)
v_p is the contribution of the profiled steel sheeting, if present

Fig. 9.6 Shear in concrete

9.5 DEFLECTIONS

As in steel beam design, deflection must be calculated at the serviceability limit state, i.e. with unfactored loads. The presence of concrete in the section means that the two different elastic moduli (steel and concrete) must be included, which is usually achieved by use of the transformed (or equivalent) cross-section[3,4]. The elastic modulus for concrete is usually modified to allow for creep. Under sustained loading the elastic modulus is about one-third that under short term loading. The modular ratio α (= E_s/E_c) is taken as 6 for short term loading, and 18 for long term loading. An equivalent ratio α_e may be used, based on the proportion of loading considered to be long term, and is a linear interpolation between these values.

The values of neutral axis depth x_e and equivalent second moment of area I_g are shown in Fig. 9.7. This allows deflections to be calculated using normal elastic formulae with a value for E_s for 205 kN/mm².

Fig. 9.7 Transformed section

$x_e = [D_s/2 + \alpha r(D/2 + D_s)]/(1 + \alpha r)$
$I_g = I_x + A(D + D_s)^2/4(1 + \alpha r) + B_e D_s^3/12\alpha$

9.6 COMPOSITE SLABS

Composite slabs are constructed from profiled steel sheeting with two typical sections, as shown in Fig. 9.8. The sheeting alone resists the moments due to the wet concrete and other construction loads. When the concrete has hardened the composite section resists moments due to finishes and imposed loads. Composite action is achieved by bond as well as web indentations, and in some cases by end anchorage where the connectors for composite beams are welded through the sheeting.

Fig. 9.8 Profiled sheeting

In most cases design is controlled by the construction condition rather than by the performance as a composite section. In general, the failure of the slab as a composite section takes place owing to incomplete interaction, i.e. slip on the steel/concrete interface. For these reasons, design of composite slabs with profiled sheeting has evolved from testing. Details of the test information are available from manufacturers and SCI[4]. The effects of the sheeting profile on connector performance and on beam behaviour are also given in the SCI publication[4].

9.7 EXAMPLE 18. COMPOSITE BEAM IN BUILDING

The design follows that given in Section 3.7 for a non-composite beam. The notation follows that of BS 5950: Part 3.1.

(a) Dimensions

(See Fig. 3.2.)

Span 7.5 m simply supported
Beams at 6.0 m centres
Concrete slab 250 mm thick ($f_{cu} = 30$ N/mm^2) spanning in two
   directions
Finishing screed 40 mm thick

(b) Loading

As Section 3.7b allowing the same self weight of beam.

Dead load W_d = 180 kN
Imposed load W_i = 135 kN

(c) BM and SF

Ultimate moment $M_x = 592$ kNm
Ultimate shear force $F_x = 242$ kN

(d) Shear capacity

Assume the beam to be **406 × 140 × 46 UB**.

Shear capacity $P_v = 0.6 p_y A_v$
$= 0.6 \times 0.275 \times 402.3 \times 6.9 = 458$ kN
Shear force $F_x / P_v = 0.52$

(e) Moment capacity

Use effective breadth B_e as $L/4$, i.e. 1.85 m.
For neutral axis in the concrete slab, see Fig. 9.2.

$x_p = A p_y / (0.45 B_e f_{cu})$
$= 5900 \times 275/(0.45 \times 1850 \times 30) = 65$ mm

In slab 250 mm thick, see Fig. 9.9.

Moment capacity $M_c = A p_y /(D_s + D/2 - x_p/2)$
$= 5900 \times 275/(250 + 402.3/2 - 65/2) 10^{-6}$
$= \mathbf{679\ kNm}$
$M_x / M_c = 0.84$

Section is satisfactory.

Fig. 9.9

(f) Shear connectors

Force in concrete at mid-span:

$R_c = 0.45 f_{cu} B_e x_p$
$= 0.45 \times 30 \times 1850 \times 65 \times 10^{-3} = 1623$ kN

Use **19 mm diameter by 100 mm high headed stud connectors**.

Table 9.1 $Q_k = 100$ kN
$N_p = 1620/(100 \times 0.8) = \mathbf{21\ studs}$

These are distributed evenly in each half span.

Spacing $= 3700/21 = \mathbf{175\ mm}$

(See Figs. 9.6 and 9.10.)

19 dia. stud 100 high at 175 mm spacing

10 dia. HT bars at 200 crs.

$A_{sv} = 0.785$ mm^2/mm
$f_y = 410$ N/mm^2

Fig. 9.10

Length of shear path L_s = 40 + 2 × 100 = 240 mm
Shear per unit length v = $R_c/(L/2)$
= 1620/3700 = 438 N/mm
Longitudinal shear capacity $\not> 0.8 L_s \sqrt{f_{cu}}$
= 0.8 × 240 × $\sqrt{30}$ = 1050 N/mm
and $\not> 0.03 L_s f_{cu} + 0.7 A_{sv} f_y$
= 0.03 × 240 × 30 + 0.7 × 0.785 × 410
= 441 N/mm

Local shear is satisfactory.

(g) Deflection

Using unfactored imposed loads as in Section 3.7f, $W = 132$ kN. The properties of the transformed sections[4] are:

Fig. 9.7

Section 9.7

$r = A/(B_e D_s)$
= 5900/(1850 × 250) = 0.0128
$\alpha_e = 10$
$x_e = [250/2 + 10 × 0.0128(201 + 250)]/(1 + 10 × 0.0128)$
= 176 mm
$I_g = 79\,700$ cm^4
Deflection = $WL^3/60EI_g$
= 132 × 7400^3/(60 × 205 × 79 700 × 10^4) = **5.5 mm**
Deflection limit = 700/360 = 20.6 mm

Comparing the section used (406 × 140 × 46 UB) with that required in non-composite (533 × 210 × 92 UB) gives a clear indication of the weight saving achieved in composite construction. However, as discussed in Section 9.1, some other costs must be taken into account in any cost comparison.

STUDY REFERENCES

Topic	Reference
1. Reinforced concrete	**Kong F.K. & Evans R.H.** (1987) Reinforced concrete beams – the ultimate limit state, *Reinforced and Prestressed Concrete*, pp. 85–155. Van Nostrand Reinhold
2. Composite construction	**Johnson R.P.** (1982) Simply supported composite beams and slab, *Composite Structures of Steel and Concrete*, pp. 40–100. Granada Publishing
3. Transformed cross-section	**Kong F.K. & Evans R.H.** (1987) Elastic theory, *Reinforced and Prestressed Concrete*, pp. 157–67. Van Nostrand Reinhold
4. Composite slabs	**Lawson R.M.** (1989) *Design of Composite Slabs and Beams with Steel Decking*. Steel Construction Institute

10

BRACING

10.1 LOADING RESISTED BY BRACING

Bracing members, or braced bay frames, consist usually of simple steel sections such as flats, angles, channels or hollow sections arranged to form a truss (Section 6.1). The members are often arranged, using cross-bracing, so that design may be on a tension only basis.

A bracing will carry loading which is usually horizontal, derived from a number of sources:

- wind, crane and machinery loads acting horizontally on a structure;
- earthquake loads derived as an equivalent static horizontal load;
- notional loads to ensure sway stability;
- beam or column bracing forces as a proportion of the longitudinal force;
- loads present during the temporary construction stage.

In addition, bracing, whether permanent or temporary, is usually necessary for steelwork erectors to line and level properly the steel framework during construction.

10.2 SWAY STABILITY

It is important that all structures should have adequate stiffness against sway. Such stiffness is generally present where the frame is designed to resist horizontal forces due to the wind loading. To ensure a minimum sway provision, notational forces are suggested in clause 2.4.2.3 applied horizontally:

1.0% of $\gamma_f W_d$ or
0.5% of $\gamma_f (W_d + W_i)$ if greater

acting in conjunction with $1.4W_d + 1.6W_i$ vertically.

This requirement is in place of the horizontal wind or other loads and in practice forms a minimum provision.

10.3 MULTI-STOREY BRACING

In multi-storey frames horizontal forces may be resisted by:

- rigidly jointing the framework with connections capable of resisting the applied moments and analysing the frame accordingly;
- providing stiff shear concrete walls usually at stair and lift wells, and designing these to absorb all the horizontal loads;
- arranging braced bay frames of steel members forming trusses as shown in Fig. 6.3.

In all but the first case the steel beams and columns may be designed as simply supported.

The arrangement of steel bracing or wind towers of concrete walls requires care to ensure economy and simplicity. Alternative arrangements are shown in Fig. 10.1. Symmetrical arrangements are preferred as they avoid torsion in plan of the braced frames.

The vertical bracing must be used in conjunction with suitable horizontal framing. Wind loads are transmitted by the cladding of the building on to the floors, and then to the vertical braced bays or towers. Design should ensure that adequate horizontal frames exist at floor levels to carry these loads to vertical bracing. Where concrete floors are provided no further provision may be required but in open frame industrial buildings horizontal bracing is also needed (Fig. 10.1).

Braced bay frames may take a number of different forms as shown in Fig. 10.2. Cross-bracing, while it allows a tension only design, creates

Fig. 10.1 Wind towers and bracing

Fig. 10.2 Braced bay frames

difficulty where door or window openings are required. The alternatives shown may be used to accommodate openings, but will involve compression in the bracing members. In the design of such members slenderness must be kept as low as possible by use of tubes or hollow sections, and by reducing lengths as far as practicable.

10.4 SINGLE-STOREY BRACING

The principal loading which requires the provision of bracing in a single-storey building is that due to wind. In addition the longitudinal crane forces will require braced bay support. The horizontal (wind and crane surge) loads transverse to the building are supported by portal frame action, or column cantilever action, and no further bracing is needed in this direction.

Longitudinal forces do, however, require support by a braced bay frame as shown in Fig. 6.3. The wind forces arise from pressures or suctions on the gable end and frictional drag on the cladding of both the roof and sides of a building (see Section 12.4.3). Gable wind girders are needed therefore at each end of the building, and may be provided at the level of the rafters (low-pitch) or at the level of the eaves, as shown in Fig. 10.3. The gable wind girders are supported by vertical side bracing as shown, which is also used to support the longitudinal crane forces. The gable posts themselves are designed to span vertically carrying the wind load between the base and the gable wind girder.

Fig. 10.3 Gable wind girder

In addition some bracing may be required by the truss lower chord members. This is a restraint against buckling and is needed in cases where reversal of stress in the bottom chord can occur. Lightweight roof structures often have this design condition, when wind suction on the roof causes compression in the lower chord of the truss.

10.5 BEAM TRUSS AND COLUMN BRACING

Both flexural and compression members may require lateral bracing or restraint to improve their buckling resistance. This provision has been discussed in the appropriate chapters:

> Beams in buildings – Chapter 3
> Crane girders – Chapter 5
> Trusses – Chapter 6
> Columns – Chapter 7

In each case, the effective length of the portion of the member in compression may be reduced by providing single members or frameworks capable of resisting the lateral buckling forces. The values of these lateral forces have been assessed from test data and given in the appropriate clauses of BS 5950. In some cases, e.g. crane girders, the buckling force is combined with other lateral forces in the design of the bracing.

The designer should always be aware of the need of bracing in unusual positions, and should examine all compression members, and compression flanges, to ensure that adequate lateral restraint exists and is satisfactory. Examples of restraints needed in lattice frameworks and portal frames (plastic design) are given in Fig. 10.4.

Fig. 10.4 Special restraints

10.6 EXAMPLE 19. GABLE WIND GIRDER AND SIDE BRACING

(a) Dimensions

(See Fig. 10.5.)

Gable end panel widths (6 no.)	5.0 m each
Depth of girder (in plan)	3.0 m
Side bay width	6.0 m
Eaves height	12.5 m

Fig. 10.5 Wind girder dimensions

(b) Loading

Reactions (excluding γ_f) from gable stanchions (spanning vertically) vary as shown in Fig. 10.6.
Wind pressure or suction (in brackets) results in two sets of reactions. These values are derived knowing C_{pe} and C_{pi} and are given in Section 2.3.
Longitudinal load from crane (Section 5.1) = 12.0 kN.

Fig. 10.6 Wind girder loading

(c) Member forces

Member forces may be obtained by any of the methods of analysis (Section 6.2a) and the pressure and suction cases are shown in the table; the loads incorporate the factor $\gamma_f = 1.4$.

Member	Factored member force (kN)	
	Pressure	Suction
1	184	−147
2	299	−239
3	338	−270
4	−215	172
5	111	−89
6	−133	106
7	69	−55
8	−46	37
9	48	−38
10	−184	147
11	−299	239

Compression is positive.

118 STRUCTURAL STEELWORK DESIGN TO BS 5950

(d) External chord

Maximum force (compression) 338 kN

Use **254 × 146 × 37 UB** (grade 43)
Slenderness $\lambda = 0.85 L/r_x$ or
$ = 1.0\ L/r_y$
(see Fig. 6.4 and Section 12.7.2)
hence max. $\lambda = 1.0 \times 5000/34.7 = 144$

BS table 27b   Compressive strength $p_c = 80\ \text{N/mm}^2$
Compression resistance $P_c = A_g\, p_c$
$ = 47.5 \times 80/10 = \mathbf{380\ kN}$

Section is satisfactory.

(e) Internal chord

Maximum compression 239 kN

Use **203 × 133 × 30 UB** (grade 43)
Slenderness $\lambda = 1.0 \times 5000/31.8 = 157$

BS table 27b   Compressive strength $p_c = 68\ \text{N/mm}^2$
Compression resistance $P_c = 38.0 \times 68/10 = \mathbf{258\ kN}$

Maximum tension 299 kN

clause 3.3.3   Effective area $A_e = 1.2\ A_{net}$ but $\not> A_g$
Allowing for two 26 mm diameter holes
$A_e = 1.2\ (32.3 - 2 \times 2.6 \times 0.58) = 35.1\ \text{cm}^2$
but $ A_e \not> 32.3\ \text{cm}_2$

Tension capacity $P_t = A_e\, p_y$
$ = 32.3 \times 275/10 = \mathbf{888\ kN}$

Section is satisfactory.

(f) Diagonals/struts

Maximum compression (diagonal) 172 kN.

Use **203 × 133 × 30 UB** (grade 43)
Slenderness $\lambda = 1.0 \times 5830/31.8 = 183$

BS table 27b $p_c = 52\ \text{N/mm}^2$
Compression resistance $P_c = 38.0 \times 52/10 = \mathbf{198\ kN}$
Maximum compression (strut member 5) $= 111$ kN
$\lambda = 1.0 \times 3000/31.8 = 94$
Use same section.

(g) Side bracing

Reaction from wind girder 93 kN
Crane load 12 kN

BS table 2

Maximum design load $= 1.4 \times 93 = 130$ kN or
$= 1.2 \times 93 + 1.2 \times 12 = 126$ kN

With reference to Fig. 10.7, the factored member forces (kN) are:

1	-130
2	188
3	-130
4	188
5	135
6	-135
7	270

Fig. 10.7

Maximum tension in diagonals 2 and 4 (assuming cross-bracing to avoid compression) $= 188$ kN.

Use **100 × 65 × 7 Angle**

clause 4.6.3.1

Effective area $A_e = a_1 + 3a_1\,a_2/(3a_1+a_2)$
$a_1 = (100-7/2)7 - 22 \times 7 = 522$ mm^2
$a_2 = (65-7/2)7 \qquad = 431$ mm^2

allowing for one 22 mm diameter hole in connected leg (100 mm)
$A_e = 522 + 3 \times 522 \times 431/(3 \times 522 + 431)$
$\qquad = 860$ mm^2
Tension capacity $P_t = A_e\,p_y$
$\qquad = 860 \times 275 \times 10^{-3} = 237$ kN

Maximum compression in strut $3 = 130$ kN

Use **203 × 133 × 25 UB** (grade 43)

BS table 27b

$\lambda = 1.0 \times 6000/31.0 = 194$
$p_c = 46$ N/mm^2
Compression resistance $P_c = 32.3 \times 46/10 = \mathbf{149\ kN}$

Forces in the eaves girder 1, and main frame members 5, 6 and 7 should be considered in the design of these members when appropriate. The values of these forces will need to be adjusted for γ_f used in the combination of forces for each member.

10.7 MULTI-STOREY WIND BRACING

(a) Dimensions

(See Fig. 10.8.)
7 storeys at 3.5 m high
Bay width 4.0 m
Cross-bracing with K-bracing on alternative floors to allow door openings

(b) Loading

Wind loading transmitted to bracing by concrete floor slabs at each level.

(c) Member forces

Member forces (kN) excluding γ_f are given for the lower two storeys only:

1	993
2	716
3	−1304
4	−716
5	472
6	−197
7	319
8	−319
9	−317
10	0

Note that wind loading can act in the reverse direction, which will generally reverse the force direction in each member. Member 10, which has zero load, however, will carry 317 kN in this wind reversal case.

Fig. 10.8

(d) Cross bracing

$$\text{Maximum tension} = 1.4 \times 472 = 661 \text{ kN}$$

Use **203 × 133 × 30 UB** (grade 43)

Section 10.6e Tension capacity = **888 kN**

(e) K-bracing

$$\text{Maximum compression} = 1.4 \times 319 = 447 \text{ kN}$$

Use **203 × 133 × 37 UB** (grade 43)

$$\lambda = 0.85 \times 4030/34.7 = 116$$
$$p_c = 114 \text{ N/mm}^2$$

BS table 27b

$$\text{Compression resistance } P_c = 47.5 \times 114/10 = \mathbf{541 \text{ kN}}$$

As in Section 10.6g the forces in columns 1, 2, 3 and 4 and beams 6, 9 and 10 must be taken into account in the overall design of these members which will include dead and imposed loading from floors, etc. The value of γ_f appropriate to each combination of loads must be used (Section 2.7).

STUDY REFERENCES

Topic   *Reference*

1. Frame stability   (1988) *Stability of Buildings*. The Institution of Structural Engineers

11

PLATE GIRDERS

11.1 INTRODUCTION

Occasionally, the required bending resistance of a member cannot be provided by the largest available universal beam (914 × 419 × 388 UB) and therefore the designer has to resort to using a plate girder. Basically, a plate girder is built up from three plates (one web and two flanges) fastened together to form an I-shape, see Fig 11.1. There are many examples of plate girders, e.g. crane girders in heavy mill buildings, road and rail bridges, roof construction of stadia, balcony girders in concert halls.

Examination of older forms of plate girders would show that the connection between the flanges and web plates is made by rivets or bolts, via angle sections, as shown in Fig 11.1(b). These plate girders were relatively more prevalent as the depth of rolled sections, prior to the 1950s, was limited to about 0.6 m, whereas today universal beams are rolled up to 0.92 m deep. Splices for this type of plate girder, required because of transportation considerations and/or maximum available length of plate, were provided by means of riveted or bolted cover plates.

The advent of welding allowed the designer the freedom to tailor-make a member to suit any design requirement. As the choice of plate is restricted

Fig. 11.1 Plate girder sections

Table 11.1(a) Maximum rolled lengths (m) for selected range of wide flats[1]

Wide flats	Thickness (mm)												
Width (mm)	12	15	20	25	30	35	40	45	50	55	60	65	70	75
200	12	13	14	15	15	15	15	15	15	16	16	16	16	16
220	12	13	14	15	15	15	15	15	15	16	16	16	16	16
250	12	13	14	15	15	16	16	17	17	17	18	18	18	18
275	12	14	14	15	15	16	17	17	17	18	18	18	18	18
300	12	14	15	16	18	18	18	18	18	19	19	19	19	19
325	12	12	15	16	18	18	18	18	18	19	19	19	19	19
350		12	15	16	18	18	18	18	20	20	20	20	20	20
375		12	16	16	20	20	20	20	20	20	21	21	21	21
400		12	16	17	21	21	22	22	22	22	23	23	23	23
425		12	16	18	21	23	23	23	23	23	23	23	23	23
450			16	18	21	23	23	23	23	23	23	23	23	23
475			15	18	21	23	23	23	23	23	23	23	23	23
500			15	18	21	23	23	23	23	23	23	23	23	23
525				18	21	23	23	23	23	23	23	23	23	22
550				18	21	23	23	23	23	23	23	23	22	20
575				18	21	23	23	23	23	23	23	23	21	19
600				18	21	23	23	23	23	23	23	22	21	19
625				18	21	23	23	23	23	23	23	21	20	19
650				18	21	23	23	23	23	23	22	21	19	18

Table 11.1(b) Maximum rolled lengths (m) for selected range of plates

Plates	Thickness (mm)											
Width (mm)	5	6	7	8	9	10	12.5	15	20	25	30	35	40	45
$1220 < d \leq 1250$	12	13	13	13	18	18	19	19	19	19	19	19	19	17
$1250 < d \leq 1300$	12	13	13	13	18	18	19	19	19	19	19	19	19	17
$1300 < d \leq 1500$	12	13	13	13	18	18	19	19	19	19	19	19	19	17
$1500 < d \leq 1600$	12	13	13	13	18	18	19	19	19	19	19	19	19	17
$1600 < d \leq 1750$	12	13	13	13	18	18	19	19	19	19	19	19	19	17
$1750 < d \leq 1800$	12	13	13	13	18	18	19	19	19	19	19	19	19	17
$1800 < d \leq 2000$	12	13	13	13	18	18	19	19	19	19	19	19	19	17
$2000 < d \leq 2100$	12	13	13	13	18	18	19	19	19	19	19	19	19	17
$2100 < d \leq 2250$	12	13	13	13	18	18	19	19	19	19	19	19	19	17
$2250 < d \leq 2500$	12	13	13	13	18	18	19	19	19	19	19	19	19	17
$2500 < d \leq 2750$		12	13	13	18	18	19	19	19	19	19	19	19	17
$2750 < d \leq 3000$		12	13	13	18	18	19	19	19	19	19	19	19	17

only by the discrete sizes of rolled plate (see Table 11.1) then this form of construction can be economic in terms of material. The flanges and web are normally connected together by fillet welds, using semi- or fully automatic welding procedures. The designer must assume that the load transfer is entirely through the welds as there is no guarantee there is a perfect bearing between flange and web. Where a splice is required, owing to maximum length of plate rolled or because there is a change in plate thickness, a full strength butt weld is required. As there are a number of different types of butt weld, the selection of the appropriate type should be discussed with the fabricator in order to produce an economic solution.

Generally, a plate girder is made doubly symmetrical, i.e. both flange plates are identical, like the universal sections, see Figs. 11.1(a) and 11.1(b). However, should design conditions dictate that a single axis of symmetry is necessary, then a judicious choice of different plates for the flanges can

Fig. 11.2 Typical plate girder

readily accommodate this requirement, see Figs. 11.1(c) and 11.1(d). The design strength of *each* plate is dependent only on its own thickness and grade of steel, unlike the universal sections where the design strength is based on the thickest part of the I-section, i.e. the flanges. Also, different grades of steel can be used for the plates within one girder, e.g. the use of a notch ductile steel, say 50C, for the tension flange of a road bridge in order to eliminate the possibility of low cycle brittle fracture.

Plate girders usually require load carrying stiffeners and intermediate stiffeners (non-load carrying), see Fig. 11.2, dividing the web into panels. These have the following functions:

- **Load carrying stiffeners** are used to diffuse any concentrated load locally, into the web. This load can result from axial load in columns connected to a flange or an end reaction from an intersecting beam member, which can be connected to either the flange or the stiffener itself.
- The sole function of **intermediate stiffeners** is to control the shear buckling resistance of any web area/panel bounded by the flanges and an adjacent pair of intermediate stiffeners.
- The elastic critical shear buckling of a **web panel** is a function of a/d (known as the **aspect ratio**) and d/t, where a is the distance between the two stiffeners bounding the web panel being considered, d is the actual depth of the web plate and t is the web plate thickness. Note that for universal sections, the overall depth D is allowed for calculating the shear area.

clause 4.4.5.1

When the web is relatively thin, the presence of intermediate stiffeners is useful in maintaining the I-shape, particularly during transportation and erection. In very deep plate girders, additional horizontal longitudinal stiffening may be necessary in the compression zone, in order to maintain an economic web thickness. This particular design variation lies outside the scope of BS 5950: Part 1, but is covered in BS 5400 and therefore will not be discussed here.

The two main forces that a plate girder has to resist are bending moment and shear force, though axial force, if present, would need to be taken into account. Though in reality, the bending moment (plus any axial load) and shear force would be resisted by the whole section, the usual assumption made for small and medium plate girders is that the flanges resist the bending moment and the web carries the shear force.

The further apart the two flanges of a beam member are positioned from the member's centroidal axis, the better is the member bending capacity. The depth of plate girders usually lies within the range of 1/12 to 1/8 of the span. Occasionally, the depth of a girder might be limited by minimum headroom considerations. Furthermore, there might be a constraint on the overall deflection of the girder, to which the self weight of the girder could make a significant contribution.

Clause 4.4 of the steel code BS 5950: Part 1 allows the designer three different ways of proportioning web plates:

- The web is made deliberately thick, removing the necessity for intermediate stiffeners; the disadvantage is that the weight of the girder would be relatively heavy compared with that obtained by using the other methods and could increase the cost of the foundations. However, the overall fabrication costs would be lower (no stiffeners), which may more than offset the extra material and foundation costs.
- The web is made thin enough to require intermediate stiffeners to control the shear buckling action within any web panel.
- Finally, the web thickness is minimized by taking into account **tension field action**, thereby maximizing the effectiveness of the web and flanges. Tension field action is discussed in Section 11.7. *This method cannot be applied to gantry girder design.*

The design of the plate girder must also comply with the guidance given in clause 4.3, BS 5950 (lateral torsional buckling of beam members). The design of load carrying and intermediate stiffeners is covered by clause 4.5. The different methods of designing plate girders are illustrated by Examples 21–24 inclusive, which include further design information where appropriate.

11.2 DESIGN OF UNSTIFFENED PLATE GIRDER

An unstiffened plate girder is similar to a universal beam section, where the web is generally thick enough not to necessitate shear stiffening/intermediate stiffeners. According to BS 5950: Part 1, the moment capacity of an unstiffened plate girder depends on the value of d/t, i.e.

clause 4.4.4.1
- If $d/t < 63\varepsilon$ (**thick web**), then the moment capacity of the plate girder can be determined as for universal beams, i.e. according to clause 4.2.5 or 4.2.6, BS 5950.

clause 4.4.4.2
- If $d/t \geq 63\varepsilon$ (**thin web**), then the moment capacity can be calculated by one of two methods or a combination of the two methods (clause 4.4.4.2). The two methods are:
  - moment plus any axial load resisted by flanges only, and the web is designed for shear only (clause 4.4.5). It is assumed that each flange is subject to a uniform stress p_y.
  - moment and axial load resisted by whole section with the web resisting the combined shear and longitudinal stresses (see clause H.3).

Similar to universal beams, a plate girder has to comply with clauses 3.5 (local buckling), 4.2 (members in bending) and 4.3 (lateral torsional buckling), BS 5950. Also, a plate girder may require end bearing stiffeners in order to transfer the end shear into the supports, and load carrying stiffeners where large concentrated loads have to be supported within the span of a member.

11.3 EXAMPLE 21. DESIGN OF UNSTIFFENED PLATE GIRDER – THICK WEBS

A plate girder, simply supported over a span of 22 m, is required to carry the loads indicated in Fig 11.3; the uniformly distributed dead load includes the self weight of the girder. The concentrated applied load is the axial load transferred from a column member ($203 \times 203 \times 46$ UC) and the ends of the girder and those of adjacent girders are supported by $254 \times 254 \times 73$ UC columns. The depth of the plate girder is to be limited to 1.5 m owing to minimum headroom requirements.

For the purpose of this example, the top (compression) flange is assumed to be restrained laterally and prevented from rotating. The plate girder is to be fabricated from grade 43C steel.

Fig. 11.3 Details of girder

Fig. 11.4 BM & SF diagrams (factored loads)

(a) Factored loading

column load	– dead	1.4×240	$= 336\,\text{kN}$
	– imposed	1.6×450	$= 720\,\text{kN}$
u.d.l.	– dead (including SW)	1.4×20	$= 28\,\text{kN/m}$
	– imposed	1.6×25	$= 40\,\text{kN/m}$

(b) Moment and shear force

The distributions of bending moment and shear force in the simply supported plate girder can readily be determined by conventional *elastic* methods and are shown in Fig. 11.4.

(c) Shear capacity

There are two design requirements regarding the minimum web thickness for the condition of no intermediate stiffeners, i.e.

clause 4.4.2.2 for serviceability: $d/t \leq 250$
clause 4.4.2.3 to avoid flange buckling: $d/t \leq 250\,(345/p_{yf})$

where p_{yf} is the design strength of the compression flange. As the web is to be deliberately made thick, i.e. $d/t < 63\varepsilon$, these requirements are automatically complied with. A quick estimate of the minimum web thickness can be derived using the overall depth of the section (D), i.e.

$$t \geq 1500/63 = 23.8\,\text{mm}$$

From the relevant table for *plates* in Table 11.1b, the nearest appropriate plate thickness for the web is 25 mm.

When the web is thick, then the moment capacity for a plate girder is calculated according to clause 4.2.5 or 4.2.6 depending on the magnitude of the shear load coexistent with maximum moment, i.e. 1166 kN. The shear capacity of a web of a built-up section is defined as:

clause 4.2.3 $P_v = 0.6\,p_y\,A_v$

where $A_v = td$. A good estimate of the shear capacity of the web can be obtained by substituting the overall depth of the girder (D) for d, which is unknown at this stage:

$$P_v = 0.6 \times 0.265 \times 25 \times 1500 = 5962\,\text{kN}$$

and as

$$F_v \leq 0.6\,P_v, \quad \text{i.e. } 1166\,\text{kN} < 3577\,\text{kN}$$

clause 4.2.5 Then the member has a 'low shear load'. Note that the use of D instead of d does not affect the outcome of this design check.

(d) Moment capacity

In order to maximize its moment capacity, the cross-section of the plate girder should be proportioned so as satisfy the requirements for a *compact* section. The moment capacity for a compact plate girder with a thick web is given by:

clause 4.2.5 $M_{cx} = p_y S_x$ but $1.2 p_y Z_x$ or $\gamma p_y Z_x$ if $S > 1.2 Z_x$

BS table 7   The b/T ratio for the outstand of the compression flange for a compact built-up section should not exceed 8.5ε, and assuming that the design strength p_y is 265 N/mm², then $B \leq (17.3T + t)$. The plastic moment capacity of a plate girder is:

$$M_{cx} = p_y [BD^2 - (B-t)(D-2T)^2]/4$$

hence

$$7442 \leq 0.265[(17.3T+25)1500^2 - (17.3T)(1500-2T)^2]/(4 \times 10^3)$$

Solving this equation gives $T = 23.4$ mm and hence $B = 430$ mm. Note that the assumption regarding p_y is correct. Select 450 mm × 25 mm from the range of **wide flats** given in Table 11.1(a) for the flanges, from which it follows that the web size must be 1450 mm × 25 mm. The actual plastic moment capacity of the plate girder is

$$\begin{aligned}M_{cx} &= 0.265[(450)1500^2 - (425)1450^2]/(4 \times 10^3) \\ &= 0.265 \times 29736 \\ &= 7880 \text{ kNm} > 7442 \text{ kNm}\end{aligned}$$

and $1.2 p_y Z_x = 1.2 \times 0.265[(450)1500^3 - (425)1450^3]/(12 \times 750 \times 10^3]$
$ = 1.2 \times 0.265[18.59 \times 10^6]/750$
$ = 7882 \text{ kNm} > 7880 \text{ kNm}$

Therefore, the moment capacity of the design plate girder (7880 kNm) is adequate.

Use **two 450 mm×25 mm wide flats**
1450 mm×25 mm plate

(i) WELD AT WEB/FLANGE JUNCTION

Next, the weld size required for the connection between the flanges and web is determined from the magnitude of the horizontal shear/mm at the web/flange interface, assuming a fillet weld on each side of the web,

$$q_w = \frac{FA_f \bar{y}_f}{2I_x}$$
$$= 1540(450 \times 25)737.5/(2 \times 18.59 \times 10^9) = 0.35 \text{ kN/mm}$$

Use **6 mm FW**

An examination of Table 11.1 reveals that both plates would need to be spliced, as the appropriate maximum lengths available from the rolling mills (flanges – 18 m ; web – 19 m) are less than the overall length of the plate girder. If the girder can be transported as one unit, then make a welded shop splice about 18 m from right-hand end, the welds being full strength butts. On the other hand, if the girder has to be delivered in two parts owing to transport considerations, then make a bolted site splice near the centre of the girder. The use of numerical controlled cutting machines in the modern fabrication shops would minimize any wastage, by utilizing the plate offcuts to provide stiffeners and other plate components for this girder and other projects.

(e) Lateral torsional buckling

According to the design brief, the compression flange of the girder is restrained laterally and therefore there is no need for a lateral torsional buckling check to be undertaken. If the flange had not been restrained, then the recommendations of section 4.3, BS 5950 must be satisfied.

(f) Check bearing capacity and buckling resistance of web

At points of concentrated applied load and support reactions the web of a plate girder must be checked for local web bearing and web buckling. If necessary, load carrying stiffeners must be introduced to prevent these forms of local failure. The design checks are similar to those applied to universal beams, as outlined in Section 5.3, Example 9.

(i) AT POSITION OF APPLIED COLUMN LOAD

$$F_x = 1056 \text{ kN}$$

acting in the plane of the web of the plate girder, i.e. no moment generated in stiffeners. Assuming that the column base (supported by the compression flange) provides a minimum stiff bearing of 203 mm, then the bearing capacity of the unstiffened web at the junction of the web/flange is:

clause 4.5.3

$$P_{crip} = (b_1 + n_2) t p_{yw}$$
$$= [203 + 2.5(2 \times 25)] \times 25 \times 0.265 = 2170 \text{ kN}$$

The associated buckling resistance (P_w) is dependent on the slenderness of the unstiffened web and a design strength of 265 N/mm^2;

clause 4.5.2.1

$$\lambda = 2.5d/t = 2.5 \times 1450/25 \quad = 145$$
$$p_c = 71 \text{ N/mm}^2$$
$$P_w = (b_1 + n_1) t p_c$$
$$= [(203 + 2 \times 750) \times 25] \, 0.071 \quad = 3020 \text{ kN}$$

The web is adequate and therefore requires no load carrying stiffeners under the concentrated load.

(ii) AT THE SUPPORTS

The web at the right-hand support, at which the greater reaction occurs, i.e. 1540 kN, also needs to be checked for bearing and buckling. It is assumed that a minimum stiff bearing provided by support is 254/2 mm:

clause 4.5.3
$$P_{crip} = [254/2 + 2.5(25)] \times 25 \times 0.265 = 2095 \text{ kN}$$
$$p_c = 71 \text{ N/mm}^2 \quad \text{(as above)}$$
clause 4.5.2.1
$$P_w = [(254/2 + 750) \times 25] \, 0.071 = 1555 \text{ kN}$$

The web is adequate at both supports and therefore requires no load bearing stiffeners. In this example, the use of a thick web has eliminated the use of load bearing stiffeners, thereby minimizing fabrication.

See Figs. 11.21 and 11.25 for the construction details of this girder.

11.4 EXAMPLE 22. DESIGN OF UNSTIFFENED PLATE GIRDER – THIN WEBS

In the previous example, the bending moment was resisted by the whole section, while the shear capacity of the thick web is clearly unutilized; this is similar to the design of universal beams. The difference between the universal sections and plate girders is that the designer can select the web thickness.

The plate girder in Example 21 is redesigned using a *thin* web plate, in order to make the web work more efficiently. This is achieved by comparing the design shear load with the web shear resistance, based on the critical shear strength (q_{cr}) and not on the design strength (p_y) as is the case for thick webs. The normal design practice for plate girders with thin webs will be employed, by which the bending moment and axial load are assumed to be resisted wholly by the flanges and the shear load by the web.

(a) Moment capacity of flanges

clause 4.4.4.2a The design assumption that the moment is carried only by the flanges means that a good estimate of flange area can be determined.

$$A_f \approx M_x / (p_{yf} \times D)$$

Anticipating that the flange thickness lies within the range 16–40 mm, then $p_y = 265$ N/mm^2 and hence:

$$A_f = 7442 \times 10^3 / (0.265 \times 1500) = 18\,720 \text{ mm}^2$$

BS table 6 The limiting b/T ratio of 8.5ε (compact sections) is still applicable, hence the approximate flange thickness is given by:

$$T = \sqrt{[18720/17.3]} = 32.9 \text{ mm, say 35 mm}$$

hence

$$B = 18720/35 = 535 \text{ mm, say 550 mm}$$

and
$$M_{cx} = 0.265 \times 550[1500^2 - 1430^2]/4 = 7473 \text{ kNm} > 7442 \text{ kNm}$$

(b) Shear resistance of web

The design requirements regarding the minimum web thickness for the condition of no intermediate stiffeners are the same as in Example 21, i.e.

clause 4.4.2.2   for serviceability: $d/t \leq 250$
clause 4.4.2.3   to avoid flange buckling: $d/t \leq 250 \; (345/p_{yf}) = 325$

For plate girders which are not designed for tension action, the elastic critical shear strength (q_{cr}) for thin webs can be determined from table 21, BS 5950, knowing the d/t and the a/d values. In the special case of *unstiffened* thin webs, $a/d = \infty$, therefore the appropriate shear strength for a given d/t is noted in the *last* column of table 21.

The web thickness (t) is unknown at this stage, hence it is suggested that a thickness approximately half that of the thick web (Example 21) is selected, i.e. 12.5 (=25/2) mm, which in fact is a rolled thickness, see Table 11.1(b). This results in a d/t ratio of 1430/12.5 = 114 (< 250) and as t < 16 mm, then an inspection of table 21(b) indicates that,

$$q_{cr} = 77.4 \text{ N/mm}^2$$

hence

$$V_{cr} = 0.0774 \times 1430 \times 12.5 = 1384 \text{ kN}$$

The web thickness of 12.5 mm is adequate for the part-length from the left-hand end to just right of the applied point load, i.e. > 1012 kN, but not for the remainder of the girder, i.e. < 1540 kN. Recalculate the shear resistance of the length between the applied load and the right-hand end, using a 15mm plate, which has a shear buckling resistance of:

$$V_{cr} = 0.110 \times 1430 \times 15.0 = 2360 \text{ kN} < 1540 \text{ kN}$$

based on

$$d/t = 95 \text{ and } q_{cr} = 110 \text{ N/mm}^2$$

Note that a 20% increase in web thickness produces a disproportionate increase (70%) in shear buckling resistance.

Again from Table 11.1(b), the maximum available length of plate, i.e. flanges – 23 m and web – 19 m, means that the web plate has to be spliced and therefore it is economic to change the web thickness at the splice position, say at 6 m from the right-hand end.

Use **two 550 mm ×35 mm wide flats**
1430 mm ×15 mm plate and
1430 mm ×12.5 mm plate

PLATE GIRDERS

(i) WELD AT WEB/FLANGE JUNCTION

The determination of the weld connecting the flanges to the web follows the same method as that outlined in Example 21.

Use **6 mm FW**

(c) Lateral torsional buckling

As in Example 21, there is no need for a lateral torsional buckling check as the compression flange is restrained.

(d) Design of load carrying stiffeners

Examination of the detailed calculations for web bearing and buckling in Example 21 would readily indicate that a web thickness of 15 mm would require load carrying stiffeners at both the concentrated load and end reaction positions.

The code recommends, for the condition where the outer edge of these stiffeners is not stiffened (normal practice), that the outstand of the stiffener should not exceed $19t_s\varepsilon$. However, where the outstand is between $13t_s\varepsilon$ and $19t_s\varepsilon$, then the stiffener design must be based on a core area of the stiffeners having an outstand of $13t_s\varepsilon$. In deriving the compressive strength p_c of stiffeners for *welded* plate girders, the design strength (p_y) is the lesser value for the web or stiffener, **less 20 N/mm²**.

clause 4.5.1.2

clause 4.5.1.5

(i) AT POSITION OF APPLIED COLUMN LOAD

The applied load, 1056 kN, acting in the plane of the web of the plate girder, i.e. no moment generated in stiffeners. Similar to the calculations outlined in Example 21, the bearing capacity of the *unstiffened web* is:

clause 4.5.3

$$P_{crip} = [(203 + 2 \times 35) \times 15]\, 0.275 = 1126\,\text{kN} > 1056\,\text{kN}$$

The associated buckling resistance (P_w) is dependent on the slenderness of the *unstiffened web* and a design strength of 275 N/mm².

clause 4.5.2.1

$$\lambda = 2.5d/t = 2.5 \times 1430/15 = 238$$
$$p_c = 30.4\,\text{N/mm}^2$$
$$P_w = [(203 + 2 \times 750) \times 15]\, 0.0304 = 777\,\text{kN} < 1056\,\text{kN}$$

Clearly, stiffeners are necessary to prevent the local buckling failure of the web at the point load position.

Try a pair of 160 mm × 12 mm flats.

clause 4.5.4.2

The bearing check for load carrying web stiffeners consists of checking that:

$$F_x < A\, p_{ys}/0.8$$

where A is the area of the stiffener **in contact** with the flange and p_{ys} is the design strength of the stiffeners. In this example, the stiffeners are subject to external load and therefore must extend to the flanges, though may not be necessarily connected to them, *unless an external load induces tension in the stiffener*. Normally the flanges and web would be welded together, using semi- or fully automatic welding, before the stiffeners are 'fitted'. This means that the inside corners of the stiffeners need to be coped/chamfered, say 15 mm, at the junction of the web and flange, so that they do not foul the web/flange weld. Hence,

$$A = 2 \times (160-15) \times 12 = 3480 \text{ mm}^2$$
$$A\, p_{ys}/0.8 = 3480 \times 0.275/0.8 = 1196 \text{ kN} > 1056 \text{ kN}$$

i.e. the stiffeners are adequate in bearing.

As the outstand of the stiffeners is slightly greater than $13 t_s \varepsilon$ (156), then the local buckling resistance of a stiffened web is based on the stiffener core area of 156×12 mm^2, together with an effective web area limited to $2 \times 20t$.

$$A_s = 12(2 \times 156) + (2 \times 20 \times 15)15 = 12\,744 \text{ mm}^2$$

and the corresponding radius of gyration, about an axis parallel to the web, is

$$r = \sqrt{\frac{12(2 \times 156 + 15)^3 + (2 \times 20 \times 15)15^3}{12 \times 12\,744}}$$
$$= 52.7 \text{ mm}$$

clause 4.5.1.5

It is noted that the flange is restrained against lateral movement and rotation. As a result of this flange restraint, it can be assumed that the column base is restrained laterally. Therefore, the effective length (L_E) of the load carrying stiffeners can be taken as $0.7L$ and with a reduced design strength of $(p_y - 20)$, where p_y is the lesser strength of web or stiffener, hence the reduced strength is 255 N/mm^2.

$$\lambda = 0.7\ L/r = 0.7 \times 1430/52.7 = 19.1$$
$$p_c = 252 \text{ N/mm}^2$$
$$P_x = A_s\, p_c = 12744 \times 0.252 = 3210 \text{ kN} > 1056 \text{ kN}$$

The buckling resistance of the stiffener is more than satisfactory.

Use **two 160 mm × 12 mm flats**

(ii) **WELD FOR LOAD CARRYING STIFFENERS**

The minimum weld size required for connecting the stiffeners to the web, assuming a weld on each side of the stiffener, is determined as follows:

clause 4.4.6.7

$$q_1 = t^2/(2 \times 5b_s) = 15^2/(2 \times 5 \times 160) \qquad = 0.14 \text{ kN/mm}$$

Fig. 11.5 Load carrying stiffener

When the stiffener is also subject to an external load, then the shear load/mm must be added to the above shear load. The load resisted by the stiffeners is the difference between the applied load and the minimum load that can be carried safely by the unstiffened web.

$$q_2 = (1056 - 777)/[2 \times (1430 - 2 \times 15)] = 0.10 \text{ kN/mm}$$
$$q_w = q_1 + q_2 = 0.14 + 0.10 = 0.24 \text{ kN/mm}$$

Use **6 mm FW**

clause 4.5.1.5b   Note that if the rotation of the flange had not been restrained, then $L_E = L$. Also, had the column base (compression member) not been laterally restrained, then the stiffeners would needed to be designed as part of the compression member and the interfacing connection checked for any effects from strut action.

(iii) **AT THE SUPPORTS**

The reaction (1540 kN) at the right-hand support of the member is greater than the applied point load, and therefore stiffeners are necessary. The restraint conditions with respect to the flange apply also to this location. The design of stiffeners at both supports follows a similar pattern as the previous design calculations, except that the effective web area is limited to only $20t$.

Try a 450 mm × 15 mm wide flat.

The outstand of the stiffeners is equivalent to $14.5 t_s \varepsilon$, which means the core area of the stiffeners is reduced to 2×195 mm × 15 mm. The local buckling
clause 4.5.1.2   resistance of the stiffened web is based on this core area of the stiffeners, plus effective web area of 20 mm × 15 mm.

$$A_2 = 15(390) + (20 \times 15)15 = 10\,350 \text{ mm}^2$$

$$r = \sqrt{\frac{15(390)^3 + (20 \times 15)15^3}{12 \times 10350}}$$

$$= 84.7 \text{ mm}$$

Again, as the flange is restrained against lateral movement and rotation, the effective length (L_E) of the load carrying stiffeners is $0.7L$ and with a design strength of 255 N/mm², the slenderness is:.

clause 4.5.2.1 $$\lambda = 0.7 \times 1430/84.7 = 11.8$$

hence

$$p_c = 255 \text{ N/mm}^2$$
$$P_x = 10350 \times 0.255 = 2639 \text{ kN} > 1540 \text{ kN}$$

The buckling resistance of the stiffener is satisfactory. Make the load carrying stiffener for the left-hand end of girder the same size. Now check the bearing capacity of the end stiffener, note that as the stiffener is welded to end of girder there is no coping, i.e. full stiffener area can be used.

$$P_{crip} = 450 \times 15 \times 0.275 = 1855 \text{ kN} > 1470 \text{ kN}$$

Use **450 mm × 15 mm wide flat**

However, it might be deemed necessary that the ends of the plate girder be torsionally restrained during transportation and erection. This can be accomplished by checking the second moment of area of the end-bearing stiffeners at the supports against the guidance given in BS 5950:

clause 4.5.8

$$I_s \geq 0.34\, \alpha_s D^3 T_c$$

where $\alpha_s = 0.006$ for $\lambda \leq 50$
$ = 0.3/\lambda$ for $50 < \lambda \leq 100$
$ = 30/\lambda^2$ for $100 < \lambda$

Assuming that there is no restraint to the compression flange for the noted situations, i.e. L_E and taking the lower value of r_y for the girder, then:

$$\lambda = 22\,500/225 = 100$$

hence

$$I_s \geq 0.34 \times (0.3/100) \times 1500^3 \times 35 = 12\,050\ \text{cm}^4$$

now

$$I_s = 15 \times 450^3/12 = 11\,400\ \text{cm}^4$$

i.e. change end stiffeners to 450 mm × 20 mm flats ($I_s = 15200\ \text{cm}^4$).

(iv) **WELD FOR END STIFFENERS**

Finally, calculate the size of the welds connecting the stiffeners to the web, noting that the web is connected on one side of the stiffener. The minimum weld size required for connecting the stiffeners to the web, assuming a weld on each side of stiffener, is determined as follows:

$$q_1 = 15^2/[2 \times 5 \times (450-15)/2] = 0.10\ \text{kN/mm}$$

In addition, the stiffener has to resist the difference in load between the reaction and the minimum load carried by the unstiffened web when buckling or bearing is taken into consideration. Assume a stiff bearing $254/2 = 127$ mm and with $\lambda = 238$:

$$p_c = 30.4\ \text{N/mm}^2$$
$$P_w = [(127+750)15]\,0.0304 = 400\ \text{kN}$$

The support reaction induces an external load to the stiffener, and is the difference between the load carried by an unstiffened web and the reaction, therefore the additional shear load/mm is

$$q_2 = (1540-400)/[2 \times (1430-2 \times 15)] = 0.41\ \text{kN/mm}$$
$$q_w = q_1 + q_2 = 0.10 + 0.41 = 0.51\ \text{kN/mm}$$

Use **6 mm FW**

Fig. 11.6 End bearing stiffener

See Figs. 11.22 and 11.25 for construction arrangement of this girder.

11.5 DESIGN OF STIFFENED PLATE GIRDER

It is the normal practice in the United Kingdom to use intermediate stiffeners, particularly for large span or heavily loaded plate girders. The use of the intermediate transverse stiffeners improves the shear capacity of a *thin* web, and generally leads to a reduction in the web thickness, compared with the corresponding unstiffened web thickness, as Example 23 will demonstrate.

The spacing of the intermediate stiffeners controls the critical shear strength of the web. The designer should attempt to minimize web thickness without the use of too many stiffeners, otherwise the fabrication cost could become uneconomic. Usually a/d ratios of about 1.4 will lead to economic construction; however, practical considerations may dictate a wider spacing.

The outstand of intermediate stiffeners should comply with the same requirements as load bearing stiffeners, as outlined in Section 11.4(d).

clause 4.4.6.4 Intermediate stiffeners are required to have a minimum stiffness about the centre of the web, i.e. when

$$a/d < 1.41, \text{ then } I_s \geq 1.5 \, d^3 t^3/a^2$$
$$a/d \geq 1.41, \text{ then } I_s \geq 0.75 \, dt^3$$

where t is the minimum thickness required for the actual spacing a, *using the tension field action*, see Example 24.

No increase in the minimum stiffness I_s is required when the stiffener is subjected only to transverse loads in plane of the web. However, if the stiffener is subject to lateral forces or a net moment arising from transverse load(s) acting eccentric to the plane of the web, then the minimum value needs to be increased to satisfy the guidance outlined in clause 4.4.6.5, BS 5950.

clause 4.4.6.6 Intermediate stiffeners not subject to external loads or moments should be checked for a stiffener force (F_q):

$$F_q = V - V_s \leq P_q$$

where V is the maximum shear in the web adjacent to the stiffener.
V_s is the shear buckling resistance of the web panel (q_{cr}), designed without using tension field action
P_q is the buckling resistance of an intermediate stiffener

When intermediate stiffeners, or *load carrying stiffeners which also act as intermediate stiffeners*, are subject to external loads and moments, they must satisfy the following interaction expression:

$$\frac{F_q - F_x}{P_q} + \frac{F_x}{P_x} + \frac{M_s}{M_{ys}} \leq 1$$

where F_q is the stiffener force previously defined
F_x is the external load or reaction
P_x is the buckling resistance of load carrying stiffener
M_s is the moment on stiffener due to eccentric applied load
M_{ys} is the *elastic* moment capacity of the stiffener

and if $F_q < F_x$, then ($F_q - F_x$) is made zero.

clause 4.5.2.2 If loads or reactions are applied direct or through a flange, in between web stiffeners, then the guidance given in clause 4.5.2.2, BS 5950 must be satisfied with respect to the stress, f_{ed}, along the compression edge of the appropriate web panel. This stress is a combination of any applied uniform load (kN/m) acting along the flange plus any point loads or distributed loads shorter than the smaller panel dimension, divided by this smaller dimension, all divided by the appropriate web thickness. The stress, f_{ed}, must not exceed the compression strength, p_{ed}, which is dependent on whether or not the compression flange is restrained against rotation relative to flange:

rotationally restrained

$$p_{ed} = \left[2.75 + \frac{2}{(a/d^2)}\right] \frac{E}{(d/t)^2}$$

not rotationally restrained

$$p_{ed} = \left[1.0 + \frac{2}{(a/d)^2}\right] \frac{E}{(d/t)^2}$$

11.6 EXAMPLE 23. DESIGN OF STIFFENED PLATE GIRDER – EXCLUDING TENSION FIELD ACTION

Example 22 is redesigned here to illustrate the effect on the web thickness when intermediate stiffeners are used. This particular plate girder design does not use tension field action. Any additional benefit that accrues from utilizing this structural action is demonstrated in Example 24. Reference to Example 22 is advisable.

(a) Moment capacity of flanges

Again, assuming that the flanges resist only the moment, then the design of the flanges is identical to that for Example 22, i.e. **550 mm × 35 mm.**

(b) Shear resistance of web

The design requirements regarding the minimum thickness for webs using intermediate stiffeners are different from those used in previous examples, i.e.

clause 4.4.2.2 for serviceability: $a/d > 1.0$ $d/t \leq 250$
 $a/d \leq 1.0$ $d/t \leq 250/(a/d)^{0.5}$
clause 4.4.2.3 to avoid flange buckling: $a/d > 1.5$ $d/t \leq 250\,(345/p_{yf})$
 $a/d \leq 1.5$ $d/t \leq 250\,(455/p_{yf})^{0.5}$

Bearing in mind the overall girder dimensions and the location of the load carrying stiffener at the ends and under the applied load (see Fig. 11.3), preliminary decisions must be made with respect to the spacing of the intermediate stiffeners. Applying the suggested $a/d = 1.4$ produces a stiffener spacing of 2.0 m. The part of the girder subjected to the largest shear loading, i.e. between the right-hand end and at the applied load position, is 5.5 m long. The spacing of 2 m would not therefore be practical for that length. It may be more appropriate to place an intermediate stiffener 2.5 m from the right-hand end, i.e. 3 m from the applied load. For the stiffened girder to be economic, compared with the unstiffened girder, the web thickness must be reduced, i.e. try 12.5 mm. This results in a value of $d/t = 1430/12.5 = 115$, which satisfies the minimum serviceability and flange buckling criterion of $d/t \leq 250$.

Check the web panel adjacent to right-hand end:

$$a/d = 2500/1430 = 1.75$$
$$q_{cr} = 94 \text{ N/mm}^2$$
$$V_{cr} = 0.094 \times 1430 \times 12.5 = 1680 \text{ kN} > 1540 \text{ kN}$$

Check the web panel adjacent to applied load:

$$a/d = 3000/1430 = 2.1$$
$$q_{cr} = 89 \text{ N/mm}^2$$
$$V_{cr} = 0.089 \times 1430 \times 12.5 = 1590 \text{ kN} > 1370 \text{ kN}$$

Fig. 11.7 RH portion of girder (stiffener spacing)

A thinner web plate (say 10 mm) can be shown to be inadequate for this part-length.

The maximum shear in the web panel adjacent to left-hand end is 1012 kN. From the calculations in Example 22, it can be seen that the shear resistance of an unstiffened 12.5 mm plate (condition $a/d = \infty$) is 1384 kN. That is, if a uniform web of thickness 12.5 mm is used along the length of the girder, then the part-length from the left-hand end to the applied load need not be stiffened. Alternatively, as a splice is required near the point load (6 m from right-hand end, see Example 2), then a stiffened 10 mm web plate could prove to be more economic, when the net cost of stiffener fabrication against that of material saving is evaluated. In order to determine the number of intermediate stiffeners required in the left-hand part-length, estimate the minimum spacing for the maximum shear load within that length:

BS table 21b

Minimum shear strength required for end panel
$$= 1012 \times 10^3/(1430 \times 10)$$
$$= 71 \text{ N/mm}^2$$

Using the appropriate d/t value (143), the maximum a/d to produce this strength can be obtained from table 21(b), BS 5950, i.e. about 1.3, which represents a maximum spacing of 1.86 m. Therefore, place the first intermediate stiffener 1.8 m from left-hand end, and then calculate the maximum shear in the adjacent panel and apply the same procedure to establish the spacing of the next stiffener, etc. Based on these calculations, it is proposed that the spacing of the stiffeners is 1.8 m, 2.3 m, 4.0 m and 8.4 m. Check that these spacings are acceptable:

Left-hand end panel:
$$a/d = 1800/1430 = 1.26,$$
$$q_{cr} = 74.5 \text{ N/mm}^2$$
$$V_{cr} = 1065 \text{ kN} > 1012 \text{ kN}$$

Fig. 11.8 LH end of girder (stiffener spacing)

Second panel from left-hand end:

$a/d = 2300/1430 = 1.61$
$q_{cr} = 62.9 \text{ N/mm}^2$
$V_{cr} = 899 \text{ kN} > 890 \text{ kN}$

Third panel from left-hand end:

$a/d = 4000/1430 = 2.8$
$q_{cr} = 53.8 \text{ N/mm}^2$
$V_{cr} = 769 \text{ kN} > 733 \text{ kN}$

The last panel, 8.4 m long, immediately left of the applied point load, has to resist a maximum shear of 618 kN, which generates a shear stress of 43.2 N/mm². This shear stress is less than the value of q_{cr} for an *unstiffened* 10 mm web, i.e. 49.2 N/mm² (see last column of table 21(b), BS 5950). Structurally, there is no need for an additional stiffener to reduce the panel length.

However, it is probably advisable to introduce another intermediate stiffener, so as to subdivide the panel into two equal panels. This additional stiffener will help to reduce any flange and web distortion during transportation and erection owing to the thin web.

Note that when considering the shear buckling resistance of those web panels bounded on one side by a load carrying stiffener, the implication is that these stiffeners also act as intermediate stiffeners and must be designed accordingly.

clause 4.5.2.2

Finally, as the uniformly distributed load (68 kN/m) is applied directly to the flange, then a check on the web between the stiffeners is necessary. The maximum value of the compression stress acting on the edge of the web for this example is obtained by using the thinner plate size, i.e.

$f_{ed} = 68/10 = 6.8 \text{ N/mm}^2$

On other hand, the minimum value of compression strength is obtained by taking the largest stiffener spacing. Taking into account that in this example, the flange is rotationally restrained:

$$p_{ed} = \left[2.75 + \frac{2}{(4200/1430)^2} \right] \frac{205 \times 10^3}{(1430/10)^2}$$
$$= 29.9 \text{ N/mm}^2$$

The edge of the web on the compression side is satisfactory.

Use **two 550 mm × 35 mm wide flats**
1430 mm × 12.5 mm plate and
1430 mm × 10 mm plate

(i) WELD AT WEB/FLANGE JUNCTION

Calculation for the weld connecting the flanges to the web is as per Example 21, and results in $q_w = 0.46$ kN/mm.

Use **6 mm FW**

(c) **Lateral torsional buckling**

As in Example 21, there is no need for a lateral torsional buckling check as the design brief states that the compression flange is restrained.

(d) **Design of intermediate stiffeners**

Examining the a/d ratios for the different panels shows that only the left-hand end panel has a value less than 1.41 ($=\sqrt{2}$). Therefore, both criteria for the minimum stiffness apply. Using the calculated web thicknesses produces safe estimates of the minimum stiffness (t should be based on tension field action as determined in Example 24):

clause 4.4.6.4

| LH end panel | $I_s > 1.5\,(1430 \times 10)^3/(1.800)^2 = 135$ cm^4 |
| remaining panels | $I_s > 0.75 \times 1430 \times 12.5^3 = 210$ cm^4 |

Try 75 mm × 10 mm flats.

The outstand of the stiffener (75 mm) is less than $13 t_s \varepsilon$ (130 mm).

$$I_s = 10\,(2 \times 75 + 10)^3/(12 \times 10^4) = 341 \text{ cm}^4 > 210 \text{ cm}^4$$

clause 4.4.6.6 Check the stiffener force (F_q) does not exceed the buckling resistance of the stiffener (P_q) for the intermediate stiffener located in the right-hand part-length ($p_y = 255$ N/mm^2):

$$r = \sqrt{\frac{10(2 \times 75 + 12.5)^3 + (20 \times 12.5)12.5^3}{12[(10(2 \times 75)) + (20 \times 12.5)12.5]}}$$

$= 28.0$ mm
$\lambda = 0.7 \times 1430/28.0 = 36$
$p_c = 218$ N/mm^2
$P_q = 4625 \times 0.218 = 1008$ kN
$V = 1370$ kN (shear at stiffener position, see Fig. 11.4)
$V_s = q_{cr}\,dt$ (choose lower q_{cr} of the adjacent panels)
$= 0.089 \times 1430 \times 12.5 = 1591$ kN
$F_q = V - V_s = (1370 - 1591) \rightarrow 0 \leq P_q$

Fig. 11.9 Intermediate stiffener

clause 4.4.6.7 Intermediate stiffeners should extend to the compression flange, but not necessarily be connected to it. Stiffeners not subject to external load or moment can be terminated at a distance of about $4t$ from the tension flange. For this example, taking the smallest web thickness (10 mm), then stiffener can end within 40 mm of flange. Note that intermediate stiffeners at any specific position can consist of a pair of stiffeners placed symmetrically about the plane of the web or a single stiffener placed on one side of the web. The latter is effective for the outer girders of a bridge when single stiffeners are welded to the inner face of the web to give the girder an appearance of being unstiffened when viewed by the public.

The arrangement of the intermediate stiffeners is given in Fig. 11.23.

(i) **WELD FOR INTERMEDIATE STIFFENERS**

For intermediate stiffeners subject to no external loading, the minimum weld size required for connecting the stiffeners to the web, assuming a weld on each side of the stiffener, is determined as follows:

$$q_1 = 12.5^2/(2 \times 5 \times 75) = 0.21 \text{ kN/mm}$$

Use **6 mm FW**

(e) **Design of load carrying stiffeners**

Examination of the detailed calculations for web bearing and buckling in Example 21 would readily indicate that a web thickness of 12.5 mm requires load carrying stiffeners at both the concentrated load and end reaction positions.

(i) **AT POSITION OF APPLIED COLUMN LOAD**

The applied load, 1056 kN, acts in the plane of the web of the plate girder, i.e. no moment is generated in stiffeners. Similar to the calculations outlined in Example 21, the bearing capacity of the unstiffened web is:

clause 4.5.3
$$P_{crip} = [(203 + 2 \times 35) \times 12.5] \, 0.275 = 938 \text{ kN}$$

The associated buckling resistance (P_w) is dependent on the slenderness of the *unstiffened* web and a design strength of 275 N/mm².

clause 4.5.2.1
$$\lambda = 2.5 d/t = 2.5 \times 1430/12.5 = 286$$
$$p_c = 22.0 \text{ N/mm}^2$$
$$P_w = [(203 + 2 \times 750) \times 12.5] \, 0.022 = 468 \text{ kN}$$

Clearly, stiffeners are necessary to prevent the local bearing and buckling failure of the web at the position of the point load.

Try a pair of 160 mm × 12 mm flats.

clause 4.5.1.2   As the outstand of the stiffeners is slightly greater than $13t_s\varepsilon$ (156), the local buckling resistance of a stiffened web is based on a stiffener core area of 156×12 mm², together with an effective web area limited to $2 \times 20t$.

$$A_s = (2 \times 156 \times 12) + (2 \times 20 \times 12.5) = 4244 \text{ mm}^2$$

$$r = \sqrt{\frac{12(2 \times 156 + 12.5)^3 + (2 \times 20 \times 12.5)12.5^3}{12 \times 4244}}$$

$$= 89.8 \text{ mm}$$

clause 4.5.1.5   As the flange is restrained against lateral movement and rotation and by implication the column base, the effective length of the stiffeners can be taken as $0.7L$. With a reduced design strength of 255 N/mm², i.e. $(p_y - 20)$, then:

$$\lambda = 0.7L/r = 0.7 \times 1430/89.8 = 11.1$$
$$p_c = 255 \text{ N/mm}^2$$
$$P_x = A_s p_c = 4244 \times 0.255 = 1082 \text{ kN}$$

When calculating the shear resistance of panel adjacent to the load carrying stiffener, there is the implied assumption that the stiffener acts as an intermediate stiffener. Therefore, an additional check must be made using the interaction expression previously noted. Using the lesser critical shear

clause 4.4.6.6   resistance of the two adjacent panels and assuming that the panel has a uniform 10 mm plate, i.e. ignore the splice, then:

$$V_s = 0.0492 \times 1430 \times 10 = 703 \text{ kN}$$
$$V = 1166 \text{ (see Fig. 11.4)}$$
$$F_q = V - V_s = 1166 - 703 = 463 \text{ kN}$$
$$F_q - F_x = 463 - 1056 \rightarrow 0$$

As there is no moment acting on the stiffener, $M_s = 0$ then the interaction expression reduces to:

$$F_x/P_x = 1056/1082 = 0.98 < 1$$

clause 4.5.4.2   The bearing capacity for load carrying web stiffeners is obtained, based on the stiffeners being coped/chamfered (15 mm) at the inside corner:

$$F_x < Ap_{ys}/0.8$$
$$A = 2 \times (160 - 15) \times 12 = 3480 \text{ mm}^2$$
$$Ap_{ys}/0.8 = 3480 \times 0.275/0.8 = 1196 \text{ kN} > 1056 \text{ kN}$$

The buckling resistance and bearing capacity of the stiffened web is satisfactory.

Use **two 160 mm × 12 mm flats**

clause 4.5.1.5b   Note that if the rotation of the flange had not been restrained, then $L_E = L$. Also, had the column base (compression member) not been laterally restrained, then the stiffeners would need to be designed as part of the compression member and the interfacing connection checked for any effects from strut action.

(ii) WELD FOR LOAD CARRYING STIFFENER

The minimum weld size required for connecting the stiffeners to the web, assuming a weld on each side of the stiffener, is determined as follows:

clause 4.4.6.7

$$q_1 = 12.5^2/(2 \times 5 \times 160) = 0.10 \text{ kN/mm}$$

In addition, there is the external load to be taken into account:

$$q_2 = (1166 - 703)/[2 \times (1430 - 2 \times 15)] = 0.17 \text{ kN/mm}$$
$$q_w = q_1 + q_2 = 0.10 + 0.17 = 0.27 \text{ kN/mm}$$

Use 6 mm FW

(iii) AT THE SUPPORTS

Load carrying stiffeners are required at both ends of the girder. Their design is based on the worse situation, i.e. the reaction (1540 kN) at the right-hand support. The restraint conditions with respect to the flange apply also to this location. The design of these stiffeners at both supports follows a similar pattern as the design calculations noted in Example 22.

Try a 450 mm × 15 mm wide flat.

The outstand of the stiffeners is equivalent to $14.5t_s\varepsilon$, which means the core area of the stiffeners is reduced to 2 × 195 mm × 15 mm. The local buckling resistance of the stiffened web is based on this core area of the stiffeners,

clause 4.5.1.2 plus an effective web area of 20 mm × 12.5 mm.

$$A_s = (2 \times 195 \times 15) + (20 \times 12.5) = 6100 \text{ mm}^2$$

$$r = \sqrt{\frac{15(390 + 12.5)^3 + (20 \times 12.5)12.5^3}{12 \times 6100}}$$

$$= 116 \text{ mm}$$

Again, as the flange is restrained against lateral movement and rotation, the effective length (L_E) of the load carrying stiffeners is 0.7L and with a design strength of 255 N/mm², the slenderness is:

clause 4.5.2.1

$$\lambda = 0.7 \times 1430/116 = 8.6$$

hence

$$p_c = 255 \text{ N/mm}^2$$
$$P_x = 6100 \times 0.255 = 1550 \text{ kN}$$
$$F_q = 1540 - 1680 \to 0$$
$$M_s = 0$$

Therefore, the interaction formula becomes:

$$F_x/P_x = 1540/1550 = 0.994 < 1$$

The buckling resistance of the stiffener is satisfactory. Make the load bearing stiffener for the left-hand end of girder the same size. Now check the bearing capacity of the end stiffener; note that as the stiffener is welded to end of girder there is no coping, i.e. the full stiffener area can be used.

clause 4.5.3 $P_{crip} = 450 \times 15 \times 0.275/0.8 = 2320 \text{ kN} > 1540 \text{ kN}$

Use **450 mm × 15 mm wide flat**

Should the ends of the girder be required to be torsionally restrained during transportation and erection, then the end stiffeners would need to be changed to 450 mm × 20 mm flats, see last paragraph, Section 11.4(d) for details.

(iv) WELD FOR END STIFFENERS

Minimum weld is given by:

clause 4.4.6.7 $q_1 = 12.5^2/[5 \times (425 - 12.5)/2] = 0.15 \text{ kN/mm}$

Assuming a stiff bearing of 127 mm (=254/2) and with $\lambda = 286$ then the minimum load capacity for an *unstiffened web*:

$p_c = 22 \text{ N/mm}^2$
$P_w = [(127 + 750)12.5] \, 0.022 = 241 \text{ kN}$

In addition, the support reaction induces a load into the stiffener, and is the difference between the load carried by an unstiffened web and the reaction, therefore the additional shear load/mm is:

$q_2 = (1540 - 241)/[2 \times (1430 - 2 \times 15)] = 0.46 \text{ kN/mm}$
$q = q_1 + q_2 = 0.15 + 0.46 \qquad\qquad = 0.61 \text{ kN/mm}$

Use **6 mm FW**

See Figs. 11.23 and 11.25 for the construction details of this girder.

11.7 DESIGN OF GIRDER INCLUDING TENSION FIELD ACTION

Note that this method of design cannot be applied to gantry girders. The main advance in plate girder design has been the introduction of tension field action, whereby the benefit accrued from the post-buckled strength of the web can be utilized. Generally, any plate element subject to a dominant shearing action, such as a web of a plate girder, is deemed to have 'failed', when the magnitude of the shear causes it to buckle out of plane owing to the compression component within the shear field. However, if the edges of the plate element are reinforced, say like a web panel in a stiffened plate girder, then that portion of the panel, which is parallel to the tension component, continues to resist additional shear load. The web no longer has any strength in the direction parallel to the compression direction, as it has buckled. In effect, the plate girder behaves like an N-type lattice girder, with the flanges acting as the top and bottom chords and the 'tension components' of the web acting as pseudo diagonal members, as shown in Fig. 11.10.

Fig. 11.10 Tension field action

clause 4.4.5.4.1

Designers are allowed to take advantage of this extra web strength, and thereby reduce the web thickness, i.e. the basic shear resistance of a *web panel* utilizing the tension field action is defined as:

$$V_b = q_b \, dt$$

where q_b is the basic shear strength, which is a function of d/t and a/d, and is obtained from the appropriate table 22, BS 5950

The mean longitudinal stress f is the stress in the smaller flange (associated with the web panel being considered) due to the moment and/or axial load. If this flange plate is not fully stressed ($f < p_y$), then additional shear resistance can be generated from this reserve of flange strength ($p_y - f$), giving a combined shear resistance for the web panel of:

clause 4.4.5.4.1

$$V_b = (q_b + q_f \sqrt{K_f})dt \quad \text{but} \quad \leq 0.6 p_y dt$$

$$K_f = \frac{M_{pf}}{4 M_{pw}} \left(1 - \frac{f}{p_{yf}}\right)$$

where q_f is the flange dependent shear strength being a function of d/t and a/d, and is obtained from the appropriate table 23, BS 5950

$$f = M/[(D-T)BT] + f_a \text{ (axial load stress)}$$
$$M_{pf} = 0.25 BT^2 p_{yf}$$
$$M_{pw} = 0.25 td^2 p_{yw}$$

Note that the parameter K_f is dependent on the maximum moment that exists in the web panel being considered.

The disadvantage of **tension field action** is that in order to develop the post-buckled strength of the web, the top corners at the ends of the girder have to be prevented from being pulled inwards, under the action of the diagonal tension forces. This is achieved by designing the 'ends' to act as anchors. The **anchor force** (H_q) required to produce the necessary rigidity at the ends is generated as a direct result of the tension field action, and is defined as:

clause 4.4.5.4.4

$$H_q = 0.75 dt p_y \sqrt{1 - \frac{q_{cr}}{0.6 p_y}}$$

If ($f_v < q_b$), then this force H_q can be multiplied by $(f_v - q_{cr})/(q_b - q_{cr})$, where:

f_v is the applied shear stress *
q_b is the basic shear strength *
q_{cr} is the critical shear strength *

* the values of f_v, q_b and q_{cr} should be based on the conditions that appertain to *nearest* panel to the end which utilizes tension field action, i.e. represented by the shaded areas in the Figs. 11.11–13 inclusive.

The anchor force, H_q, induces a longitudinal shear force, R_{tf}, and moment, M_{tf}, which have to be resisted by the end-post/stiffeners and are defined as:

$$R_{tf} = H_q/2$$
$$M_{tf} = H_q \, d/10$$

clause 4.4.5.4.2
clause 4.4.5.4.3

There are three different methods by which the ends of the girder can be designed to resist the additional forces induced by the anchor force:

Fig. 11.11 End panel resists tension field action

- The end panels are designed *without* tension field action, though the remainder of the panels are designed for tension field action, see Fig. 11.11. In addition, these end panels have to be designed as a beam spanning between the flanges for a shear force R_{tf} and a moment M_{tf}. The end stiffener must be designed to resist the end reaction plus the compression force due to the moment M_{tf}. This method has the advantage of relatively stiff end panels, but at the expense of smaller end panels or thicker webs in the end panels, depending on practical and fabrication considerations.
- The end panels are designed *to utilize* tension field action. There are two alternatives for the design of the end post(s):

Fig. 11.12 End stiffener resists tension field action

  – the end post comprises a **single stiffener**, see Fig. 11.12: the end post must be designed to resist the end reaction plus a moment equal to M_{tf}. This results in substantial end stiffeners, but the width and thickness of the end post must not exceed those of the flange, otherwise the designer must resort to using the first method. The top of the end post must be connected by full strength welds to the flange.

  – the end post comprises a **double stiffener**, see Fig. 11.13: both stiffeners of the end post must be checked as part of a beam spanning between the flanges resisting a shear force R_{tf} and a moment M_{tf}. In addition, the inner stiffener (over the support) must be designed for the compression due to bearing (reaction). This method requires sufficient space beyond the centre of the support member to extend the girder to accommodate the double stiffener.

Fig. 11.13 End post resists tension field action

clause 4.4.5.5

The design of any web panel with an opening must satisfy the recommendations given in clause 4.15, BS 5950. In addition, for any panel in which there is an opening with a dimension greater than 10% of the minimum panel dimension, that panel must be designed without utilizing tension field action. However, the adjacent panels can be designed with or without the utilisation of tension field action, as appropriate.

11.8 EXAMPLE 24. DESIGN OF STIFFENED PLATE GIRDER – UTILIZING TENSION FIELD ACTION

The design specification is the same as for Example 21, with the plate girder being designed utilizing tension field action.

(a) Moment capacity of flanges

clause 4.4.4.2a   Again, assuming that only the flanges resist the moment, then the design of the flanges is identical to that for Example 22, i.e. **550 mm × 35 mm**.

(b) Shear resistance of web

The design requirements regarding the minimum thickness for webs using intermediate stiffeners with respect to serviceability and local flange buckling are outlined in Section 11.6(b), Example 23.

First consider the web in the right-hand end panel, which carries a shear of 1540 kN and a moment. The magnitude of the moment acting on this panel can be defined only after the stiffener spacing has been decided. In order to make preliminary decisions regarding the spacing of the intermediate stiffeners, use the same spacing for the girder designed in Example 23 as a guide. Place the first intermediate stiffener 1.7 m from the end support; therefore, the appropriate moment is 2520 kNm. Bearing in mind that the web plate in the previous example (no tension field action) is 12.5 mm thick, check the adequacy of a 10 mm plate, which gives $d/t = 143$:

clause 4.4.5.4.1

$a/d = 1700/1430 = 1.19$
$q_b = 108 \text{ N/mm}^2$ (table 22b)
$q_f = 323 \text{ N/mm}^2$ (table 23b)

$f = (2520 \times 10^6)/[(1500-35)550 \times 35] + 0 = 89 \text{ N/mm}^2$
$M_{pf} = 0.25 \times 550 \times 35^2 \times 0.265 \times 10^{-3} = 44.6 \text{ kNm}$
$M_{pw} = 0.25 \times 10 \times 1430^2 \times 0.275 \times 10^{-3} = 1405 \text{ kNm}$

$$K_f = \frac{44.6}{4 \times 1405}\left(1 - \frac{89}{265}\right) = 0.005\,27$$

$V_b = (108 + 323\sqrt{0.005\,27})1430 \times 10 \times 10^{-3} = 1880 \text{ kN}$
$< 0.6 \times 0.275 \times 1430 \times 10 \times 10^{-3} = 2360 \text{ kN}$

Fig. 11.14 Stiffener positions at RH end

Place another intermediate stiffener 3.6 m from the support.

Check second panel from right-hand end:

$M = 5103 \text{ kNm}$ $F_v = 1424 \text{ kN}$
$a/d = 1.33$

$$K_f = \frac{44.6}{4 \times 1405}\left(1 - \frac{181}{265}\right) = 0.002\,52$$

$V_b = (102 + 302\sqrt{0.002\,52})1430 \times 10 \times 10^{-3} = 1675 \text{ kN}$

Check third panel from right-hand end

$M = 7442 \text{ kNm}$ $F_v = 1295 \text{ kN}$
$a/d = 1.33$

$$K_f = \frac{44.6}{4 \times 1405}\left(1 - \frac{264}{265}\right) = 0.000\,03$$

Fig. 11.15 BM & SF diagrams

$V_b = (102 + 302\sqrt{0.000\,03})1430 \times 10 \times 10^{-3} = 1482 \text{ kN}$

Using the next plate thickness (9 mm) would show that the third panel fails to satisfy the shear resistance criterion. Note that there is virtually no shear contribution from the flange in the third panel, as the flange is almost fully stressed, i.e. 264 N/mm².

Now considering the left-hand part-length, again assume a change in plate thickness to the left of the point load and make the web 7 mm thick. Though the stiffener spacing of the previous example was used as a basis, it is proposed that the final intermediate stiffener spacing from the left-hand end is 1.9 m, 2.2 m, 2.8 m and then three equal panels of 3.2 m. The web panels need to be checked with $d/t = 204$, which still satisfies the serviceability and flange buckling criteria of $d/t \leq 250$.

Fig. 11.16 Stiffener positions at LH end

Check left-hand end panel:

$$M = 1800 \text{ kNm} \qquad F_v = 1012 \text{ kN}$$
$$a/d = 1900/1430 = 1.33$$

$$K_f = \frac{44.6}{4 \times 983}\left(1 - \frac{64}{265}\right) = 0.008\,60$$

$$V_b = (77.1 + 350\sqrt{0.008\,60})1430 \times 7 \times 10^{-3} = 1071 \text{ kN}$$

Check second panel from left-hand end:

$$M = 3578 \text{ kNm} \qquad F_v = 883 \text{ kN}$$
$$a/d = 2200/1430 = 1.54$$

$$K_f = \frac{44.6}{4 \times 983}\left(1 - \frac{127}{265}\right) = 0.00591$$

$$V_b = (68.0 + 306\sqrt{0.005\,91})1430 \times 7 \times 10^{-3} = 916 \text{ kN}$$

Fig. 11.17 BM & SF diagrams

Check third panel from left-hand end:

$$M = 5364 \text{ kNm} \qquad F_v = 733 \text{ kN}$$
$$a/d = 2800/1430 = 1.96$$

$$K_f = \frac{44.6}{4 \times 983}\left(1 - \frac{190}{265}\right) = 0.00321$$

$$V_b = (59.0 + 256\sqrt{0.003\,21})1430 \times 7 \times 10^{-3} = 735 \text{ kN}$$

Ignoring the flange contribution in the fourth panel, then $q_b = 59.0$ N/mm² ($\equiv 590$ kN) which is more than adequate for a shear of 543 kN.

Note that when considering the shear buckling resistance of those web panels bounded on one side by a load carrying stiffener, the implication is that these stiffeners also act as intermediate stiffeners and must be designed accordingly.

The uniformly distributed load (68 kN/m) is applied directly to the flange, and therefore a check on the web between the stiffeners becomes necessary. The maximum value of the compression stress acting on the edge of the web for this example is obtained by using the thinner plate size, i.e.

clause 4.5.2.2

$$f_{ed} = 68/7 = 9.7 \text{ N/mm}^2$$

On other hand, the minimum value of compression strength is obtained by taking the largest stiffener spacing. Taking into account that, in this example, the flange is rotationally restrained:

$$p_{ed} = \left[2.75 + \frac{2}{(3300/1430)^2}\right] \frac{205 \times 10^3}{(1430/7)^2}$$
$$= 42.2 \, \text{N/mm}^2$$

The edge of the web on the compression side is satisfactory. Smaller spacing and/or thicker webs would give more conservative results.

Use **two 550 mm × 35 mm wide flats**
1430 mm × 10 mm plate and
1430 mm × 7 mm plate

(i) WELD AT WEB/FLANGE JUNCTION

Calculation for the weld connecting the flanges to the web is as per Example 21, resulting in $q_w = 0.47$ kN/mm.

Use **6 mm FW**

(c) Lateral torsional buckling

As in Example 21, there is no need for a lateral torsional buckling check as the design brief states that the compression flange is restrained.

(d) Design of intermediate stiffeners

clause 4.4.6.4

Examining the a/d ratios for the different panels shows that only the left-hand end panel has a value less than $1.41 (= \sqrt{2})$. Therefore, both criteria for the minimum stiffness apply. Using the calculated web thicknesses (t must be based on tension field action):

For 10 mm web and $a/d < 1.41$
$$I_s > 1.5 \, (1430 \times 10)^3/(1700)^2 = 152 \, \text{cm}^4$$
For 7 mm web and $a/d < 1.41$
$$I_s > 1.5 \, (1430 \times 7)^3/(1800)^2 = 46 \, \text{cm}^4$$
For 7 mm web and $a/d \geq 1.41$
$$I_s > 0.75 \times 1430 \times 7^3 = 37 \, \text{cm}^4$$

Try 75 mm × 10 mm flats.

The outstand of the stiffener (75 mm) is less than $13 t_s \varepsilon$ (130 mm).

$$I_s = 10 \, (2 \times 75 + 10)^3/(12 \times 10^4) = 341 \, \text{cm}^4 > 152 \, \text{cm}^4$$

clause 4.4.6.6

Check the stiffener force (F_q) does not exceed the buckling resistance of the stiffener (P_q) for the first intermediate stiffener from the right-hand end ($p_y = 255$ N/mm^2):

PLATE GIRDERS 149

$$r = \sqrt{\frac{10(2 \times 75 + 10)^3 + (20 \times 10)10^3}{12[(10(2 \times 75) + (20 \times 10)10]}}$$

$= 31.4\,\text{mm}$
$\lambda = 0.7 \times 1430/31.4 = 32$
$p_c = 234\,\text{N/mm}^2$
$P_q = 3500 \times 0.234 = 819\,\text{kN}$
$V = 1424\,\text{kN}$ (shear at stiffener position, see Fig. 11.15)
$V_s = q_{cr}\, dt$ (choose lower q_{cr} of the adjacent panels)
$= 0.065 \times 1430 \times 10 = 930\,\text{kN}$
$F_q = V - V_s$
$= 1424 - 930 = 494 \leq P_q$

Also, check the first intermediate stiffener from left-hand end:

$$r = \sqrt{\frac{10(2 \times 75 + 7)^3 + (20 \times 7)7^3}{12[(10(2 \times 75) + (20 \times 7)7]}}$$

$= 36.1\,\text{mm}$
$\lambda = 0.7 \times 1430/36.1 = 28$
$p_c = 240\,\text{N/mm}^2$
$P_q = 2480 \times 0.240 = 595\,\text{kN}$
$V = 883\,\text{kN}$ (shear at stiffener position, see Fig. 11.17)
$V_s = q_{cr}\, dt$ (choose lower q_{cr} of the adjacent panels)
$= 0.032 \times 1430 \times 7 = 320\,\text{kN}$
$F_q = V - V_s$
$= 883 - 320 = 563 \leq P_q$

clause 4.5.11

The remaining intermediate stiffeners can be shown to be adequate. When tension field action is utilized, all stiffeners, including intermediate stiffeners, constitute the 'compression components' of the N-type lattice girder model, see Fig. 11.10. Therefore, the intermediate stiffeners should be fitted between or connected with continuous weld to both flanges. See additional notes given in Section 11.6(d) in Example 23.

The arrangement of the intermediate stiffeners is given in Fig. 11.24.

(i)   WELD FOR WEB/INTERMEDIATE STIFFENERS

For intermediate stiffeners subject to no external loading, the minimum shear per mm length required is the same as for Example 23.

Use **6 mm FW**

(e) **Load carrying stiffeners**

Examination of the detailed calculations for web bearing and buckling in Example 23 would readily indicate that load carrying stiffeners are required at both the concentrated load and end reaction positions. The load carrying stiffeners at the supports in effect become end posts/stiffeners.

(f) Load carrying stiffener at position of applied column load

The applied load, 1056 kN, acts in the plane of the web of the plate girder, i.e. no moment is generated in the stiffeners.

Try a pair of 160 mm × 12 mm flats.

clause 4.5.1.2 The outstand of the stiffeners is slightly greater than $13t_s \varepsilon$ (156), therefore the local buckling resistance of a stiffened web is based on a stiffener core area of $156 \times 12 \text{ mm}^2$, together with an effective web area limited to $2 \times 20t$:

$$A_s = (2 \times 156 \times 12) + (2 \times 20 \times 10) = 4144 \text{ mm}^2$$

$$r = \sqrt{\frac{12(2 \times 156 + 10)^3 + (2 \times 20 \times 10)10^3}{12 \times 4144}}$$

$$= 89.8 \text{ mm}$$

The effective length of the stiffeners is taken as $0.7L$, as the flange is restrained against lateral movement and rotation and, by implication, the column base. With a reduced design strength of 255 N/mm², i.e. $(p_y - 20)$, then:

$$\lambda = 0.7 \ L/r = 0.7 \times 1430/89.8 = 11.1$$
$$p_c = 255 \text{ N/mm}^2$$
$$P_x = A_s p_c = 4144 \times 0.255 = 1056 \text{ kN}$$

clause 4.4.6.6 When calculating the shear resistance of panels adjacent to the load carrying stiffener, there is the implied assumption that the stiffener acts as an intermediate stiffener. Therefore, an additional check must be made using the interaction expression previously noted. Using the lesser critical shear resistance of the two adjacent panels and assuming that the panel has a uniform 10 mm plate, i.e. ignoring the splice, then:

$$V_s = 0.028 \times 1430 \times 7 = 280 \text{ kN}$$
$$V = 1166 \text{ (see Fig. 11.15)}$$
$$F_q = V - V_s = 1166 - 280 = 886 \text{ kN}$$
$$F_q - F_x = 886 - 1056 \to 0$$

As there is no moment acting on the stiffener, $M_s = 0$ then the interaction expression reduces to:

$$F_x/P_x = 1056/1056 = 1.00 \leq 1$$

clause 4.5.4.2 The bearing capacity for load carrying web stiffeners is obtained, based on the stiffeners being coped/chamfered (15 mm) at the inside corner:

$$F_x < Ap_{ys}/0.8$$
$$A = 2 \times (160 - 15) \times 12 = 3480 \text{ mm}^2$$
$$Ap_{ys}/0.8 = 3480 \times 0.275/0.8 = 1196 \text{ kN} > 1056 \text{ kN}$$

The buckling resistance and bearing capacity of the stiffened web is satisfactory.

Use two 160 mm × 12 mm flats

PLATE GIRDERS

clause 4.5.1.5b   Note that if the rotation of the flange had not been restrained, then $L_E = L$. Also, had the column base (compression member) not been laterally restrained, then the stiffeners would need to be designed as part of the compression member and the interfacing connection checked for any effects from strut action.

(i) WELD FOR LOAD CARRYING STIFFENER

The minimum weld size required for connecting the stiffeners to the web, assuming a weld on each side of stiffener, is determined as follows:

clause 4.4.6.7 $q_1 = 12.5^2/(2 \times 5 \times 160) = 0.10 \text{ kN/mm}$

In addition, there is the external load to be taken into account.

$$q_2 = (1166 - 280)/[2 \times (1430 - 2 \times 15)] = 0.32 \text{ kN/mm}$$
$$q_w = q_1 + q_2 = 0.10 + 0.32 \qquad\qquad = 0.41 \text{ kN/mm}$$

Use **6 mm FW**

(g) Design of end-post/stiffeners at the supports

End-posts/stiffeners are required at both ends of the girder. In plate girders where tension field action is utilized, the 'end stiffeners' play an important structural role, in that they have to resist the anchor force without which tension field action would not be generated. Section 11.7 outlines three methods of providing the necessary resistance to the anchor force. In the example being considered, the design specification states that the ends of the girder must not project beyond the centre lines of the supporting columns. Therefore, the choice is reduced to one of two methods:

clause 4.4.5.4.2   (a) Design the panel immediately adjacent to the end-post/stiffener without tension field action and to resist additional forces due to anchor force; design end post/stiffener to withstand reaction and force due to moment.

clause 4.4.5.4.3   (b) Design an end-post/stiffener which has to provide the total resistance to the anchor force.

The restraint conditions with respect to the flange apply also to the ends.

(i) METHOD (a)

clause 4.4.5.4.2   (See Fig. 11.11.) Redesign the right-hand end panel without tension field action. This can be accomplished in two ways:

(a/1) Retain the web thicknesses obtained in Section 11.8(b), but reduce the width of the end panels, so that the actual shear stress is less than the corresponding shear buckling strength. The positioning of other intermediate stiffeners may have to be amended.

(a/2) Retain the spacing of the intermediate stiffeners as determined in Section 11.8(d), but increase the web thicknesses until the actual shear stress is less than the shear buckling strength of the panel.

Applying method (a/1), first calculate the actual shear stress in the end panel:

$$f = (1540 \times 10^3)/(1430 \times 10) = 108 \text{ N/mm}^2$$

and knowing that $d/t = 143$, determine the a/d ratio from table 21(b), BS 5950 which would give a value of q_{cr} at least equal to 108 N/mm², i.e. 0.83, which would result in a panel width of about 1.18 m. Place the two intermediate stiffeners in the right-hand part of the girder at 1.1 m and 3.1 m from the end and check the adequacy of the modified panels.

Check first panel from right-hand end – without tension field action:

$$d/t = 143 \qquad a/d = 0.77$$
$$q_{cr} = 120 \text{ N/mm}^2$$
$$V_{cr} = 0.120 \times 1430 \times 10 = 1715 \text{ kN} > 1540 \text{ kN}$$

Fig. 11.18 Revised stiffener spacing

Check second panel from right-hand end – utilising tension field action:

$$M = 4447 \text{ kNm} \qquad F_v = 1465 \text{ kN}$$
$$a/d = 1.40$$

$$K_f = \frac{44.6}{4 \times 1405}\left(1 - \frac{158}{265}\right) = 0.00320$$

$$V_b = (99.2 + 295\sqrt{0.00320})1430 \times 10 \times 10^{-3} = 1660 \text{ kN}$$

Check third panel from right-hand end – utilizing tension field action:

$$M = 7442 \text{ kNm} \qquad F_v = 1329 \text{ kN}$$
$$a/d = 1.68$$

$$K_f = \frac{44.6}{4 \times 1405}\left(1 - \frac{264}{265}\right) = 0.000\,03$$

$$V_b = (92.5 + 267\sqrt{0.000\,03})1430 \times 10 \times 10^{-3} = 1344 \text{ kN}$$

Fig. 11.19 MB & SF diagrams

In addition, the end panel bounded by the end post and an intermediate stiffener has to be checked as a beam spanning between the flanges of the girder. This means that the end stiffener has to be capable of resisting the end reaction and the compression force arising from M_{tf} and the intermediate stiffener has to resist an additional force from M_{tf}.

Trial section for right-hand end-post/bearing stiffener – 450 mm × 20 mm

BS table 6 $\qquad b/T = (450 - 10)/(2 \times 20) = 11 > 8.5\varepsilon$ but $< 13\varepsilon$
$\qquad\qquad\qquad d/t = 1100/10 \qquad\qquad = 110 > 98\varepsilon$ but $< 120\varepsilon$

This means that the 'beam' is semi-compact and its moment capacity, irrespective of its shear load, can be determined from $M_{cx} = p_y Z_x$. Check the shear capacity of the web in the 'beam'.

clause 4.2.3 $\qquad P_v = 0.6 \times 0.275 \times 1100 \times 10 = 1815 \text{ kN} > 1120 \text{ kN}$

Making the conservative assumption that only the 'flanges' of the 'beam' resist the moment, and that the adjacent intermediate stiffener has the same section size,

$$M_{cx} = 0.275 \times 450[1110^3 - 1070^3]/(12 \times 555 \times 10^3)$$
$$= 2650 \text{ kNm}$$

which indicates that the moment capacity of the 'beam' is more than adequate, compared with 320 kNm.

Check the buckling resistance of the end-post/bearing stiffener:

clause 4.5.1.2 The full section of the end stiffener can be used in the buckling check, as $b/T < 13\varepsilon$.

$$A_s = (450 \times 20) + (20 \times 10) = 9200 \text{ mm}^2$$

$$r = \sqrt{\frac{20(450)^3 + (20 \times 10)10^3}{12 \times 9200}}$$
$$= 128 \text{ mm}$$

clause 4.4.6.6 The flange is restrained against lateral movement and rotation and, using a design strength of 245 N/mm², the slenderness becomes:

$$\lambda = 0.7 \times 1430/128 = 7.8$$

hence

$$p_c = 245 \text{ N/mm}^2$$
$$P_x = 9200 \times 0.245 \quad = 2255 \text{ kN}$$
$$F_x = 1540 + 320/1.09 = 1835 \text{ kN}$$
$$F_q = 1540 - 1715 \rightarrow 0$$
$$M_s = 0$$

Therefore, the interaction formula becomes:

$$F_x/P_x = 1835/2255 = 0.814 < 1$$

The buckling resistance of the end-post/bearing stiffener is satisfactory.

clause 4.4.5.4.2 Now check the bearing capacity of the end stiffener, noting that as the stiffener is welded to the end of the girder, there is no coping, i.e. the full stiffener area can be used:

$$P_{crip} = 450 \times 20 \times 0.265/0.8 = 2980 \text{ kN} > 1835 \text{ kN}$$

If the end stiffener had been fitted between the girder flanges, then the stiffener would have been coped and a check on the reduced bearing capacity at the bottom of the stiffener would have needed to be undertaken.

Use **450 mm × 20 mm wide flat**

As the first intermediate stiffener from the right-hand end forms part of the 'beam', use a pair of 225 mm × 20 mm wide flats.

The calculations given in the last paragraph, Section 11.4(d) indicate that this stiffener provides sufficient torsional restraint to the ends of the girder, should this be a design requirement.

(ii) **WELD FOR RH END POST/BEARING STIFFENERS**

Minimum weld is given by:

$$q_1 = 10^2/[5 \times (450-10)/2] = 0.09 \text{ kN/mm}$$

Assuming a stiff bearing of 127 mm ($=254/2$) and with $\lambda = 357$ then the minimum load capacity for an *unstiffened web*:

$$p_c = 15 \text{ N/mm}^2$$
$$P_w = [(127+750)10]\, 0.015 = 132 \text{ kN}$$

In addition, the support reaction induces a load into the stiffener, and is the difference between the load carried by an unstiffened web and the reaction; therefore the additional shear load/mm is:

$$q_2 = (1835-132)/(2 \times 1430) = 0.60 \text{ kN/mm}$$
$$q = q_1 + q_2 = 0.09 + 0.60 \quad = 0.68 \text{ kN/mm}$$

Use 6 mm FW

Now consider the left-hand portion of the girder. In order to keep the number of different web plate thicknesses to a minimum, make the web of the left-hand end panel 10 mm thick, while retaining the stiffener spacing determined in Section 11.8(b). Check the shear buckling resistance of the end panel without tension field action:

$$d/t = 143 \quad a/d = 1.33$$
$$q_{cr} = 70.2 \text{ N/mm}^2$$
$$V_{cr} = 0.0702 \times 1430 \times 10 = 1004 \text{ kN} \approx 1012 \text{ kN}$$

It appears that the 10 mm plate just fails (< 1%); however, a panel dimension of 1900 mm was used for ease of calculation. Examination of the details of the panel would indicate that the clear dimension of the web should be 1870 mm ($=1900-20-10$). Recalculation would show that the 10 mm plate is just adequate.

Make the end-post/bearing stiffeners at the left-hand end the same size as those for the right-hand end. Checking these sizes would reveal them to be more than satisfactory as the design conditions are less severe.

Examination of Table 11.1(b) indicates that the maximum length for a 7 mm plate is 13 m and for a 10 mm plate the length is 18 m. The design solution indicates that the web plate for both the right-hand portion (from the point load to the right-hand end) and the left-hand end needs to be 10 mm thick, with the remaining plate being 7 mm thick. By making the splice joints at 3 m and 16 m from left-hand end, it is possible to utilize the maximum length of 7 mm thick plate and a total length of 9 m for the 10 mm plate which is exactly half the maximum length for that thickness of plate. The actual maximum length for 10 mm plate rolled by British Steel is 18.3 m, which means that two web plates can be cut from this length. The extra 0.3 m would allow for the cutting and necessary edge preparation.

If the second method (a/2) is applied, i.e. retaining the original spacing of the intermediate stiffener of 1.7 m, 1.9 m and 1.9 m within the right-hand part of the girder, then a 15 mm web plate would be required for the end panel. This would introduce another plate thickness. The design solution from the method (a/1) would be more economic.

See Figs. 11.24 and 11.25 for the construction arrangement of the girder, which is based on the design solution determined by method (a/1).

(iii) **METHOD (b)**

clause 4.4.5.4.3

clause 4.4.5.4.4

(See Fig. 11.12.) Design the right-hand end stiffener for the condition of the adjoining panel designed utilizing tension field action, as designed in the previous section. Calculate the anchor force, and the corresponding longitudinal shear load and moment, using $t = 10$ mm:

$$H_q = 0.75 \times 1430 \times 10 \times 0.275 \sqrt{1 - \frac{70}{0.6 \times 275}} = 2240 \text{ kN}$$

$$R_{tf} = 2240/2 = 1120 \text{ kN}$$

$$M_{tf} = 224 \times 1.430/10 = 320 \text{ kNm}$$

The stiffener has to act as an end bearing stiffener and an end post, and has to resist both the end reaction (1540 kN) and a moment equal to 213 kNm ($=\frac{2}{3} \times 320$). Also, the width and thickness of the stiffener must not exceed the width and thickness of the flange. Assuming that the end stiffener has the dimensions of the flange, i.e. 550 mm × 35 mm, then its plastic moment capacity about its own axis is $275 \times 550 \times 35^2/(4 \times 10^6) = 46.3$ kNm which is considerably less than the 213 kNm required. It would require a stiffener significantly larger than 550 mm × 35 mm, thereby rendering this method invalid for this example, as the limitation on maximum size would be violated.

(h) **Alternative design solution for end posts**

(See Fig. 11.13.) If the design brief had allowed the girder to be supported across the whole width of the supporting columns, i.e. the situation when there are no adjacent girders, see Fig. 11.13, then the method outlined in clause 4.4.5.4.3(b) could be applied. The method is similar to method (a/1), i.e. the web between the two end posts, together with the end posts would be designed as the 'beam', while the web of the adjacent end panel would be designed for tension field action, as outlined in Section 11.8(b).

11.9 OTHER CONSIDERATIONS

The examples have dealt with the design of the common form for plate girders, i.e. constant depth and doubly symmetric cross-section, as recommended by BS 5950: Part 1. An examination of the different designs for the design example, i.e. Examples 21–24 inclusive, would reveal that as the overall weight of the girder (and hence material costs) decrease, there is a corresponding increase in fabrication cost, e.g. splices, stiffeners. The actual least cost solution will depend on the combined material and fabricator costs. Experience has shown that the costs between different fabricators can vary considerably, partly owing to the their expertise and the fabrication facilities. Therefore, it is essential that the design engineer has a good working relationship with the nominated fabricator, as the latter's knowledge of fabrication details could lead to reduced costs, e.g. what is the most cost effective edge preparation for butt welds? Is the best solution for joining two flanges of different thicknesses obtained by tapering the thicker plate down to the thickness of the thinner plate at a splice position? Is intermittent fillet welding more economic than continuous welding when the calculated weld size for a continuous weld is much less than the minimum size of 6 mm?

With large span girders, the designer can take advantage of shop fabrication splices necessary owing to limiting lengths of rolled plate by varying the web and/or flange thicknesses. Indeed, in Examples 22, 23 and 24, the web thickness was varied along the length of the girder according to the shear requirements. Likewise, the size of the flanges can be made to reflect the moment distribution along the girder length, *provided an improvement in the overall economy of the structure, including foundations, can be clearly demonstrated.* Normally, changes in flange size along the girder are not economic for buildings, because of the modern practice of automatic/semi-automatic welding. However, the emergence of the welded tapered portal frame (with varying flanges and web depth) produced by dedicated equipment has seen an increasing use of the tapered/haunched beams, in multi-storey buildings as they can readily accommodate services within their depth.

Other practical situations can arise in which varying the web depth/flanges could prove economic. For example, in a heavy industrial building, a crane girder may be required to span twice the normal distance because the 'central column' has to be removed to allow a railway siding to enter the side of the building. If the columns are standardized for economy reasons, then the ends of all crane girders must have the same depth at the column supports. For the double-length girder, this means that its depth has to be varied downwards in order to maintain the crane rail level at a specified height, as illustrated by Fig. 11.20. The reverse situation can arise for through road/rail bridges where the soffit of the bridge needs to be horizontal, i.e. the height of the girder would be varied along the girder length.

Fig. 11.20 Typical variable depth girders

Fig. 11.21 Details of unstiffened plate girder – thick web

Fig. 11.22 Details of unstiffened plate girder – thin web

Fig. 11.23 Details of unstiffened plate girder – excluding tension field action

Fig. 11.24 Details of stiffened plate girder – utilizing tension field action

Fig. 11.25 Stiffener details of the plate girder examples

As the reader becomes familiar with BS 5950: Part 1 design guidance, other factors might need to be considered in deriving the final design solution:

- *Fatigue* – where repeated loading is a design condition, e.g. for crane and bridge girders, fatigue resistance must be checked.
- *Brittle fracture* – notch ductile steel may have to be used for the tension flange, see BS 5400[3].
- *Large temperature range* – special bearings may have to be made at the supports to accommodate the expansion and contraction.
- *Deflection* – compliance with the appropriate deflection limit would mean either increasing girder stiffness or cambering the girder. Deflection limitations are more severe for crane girders.
- *Transportation and erection* – site splices are usually required for simply supported spans over 25–40 m. Special handling and lifting arrangements may be necessary as a result of the low torsional stiffness of plate girders. Wind loading might be a problem due to the large surface area of a girder.

Composite design of plate girders is not covered, but design guidance is given in reference [3]. Finally, always check that the required plate sizes are currently available.

STUDY REFERENCES

Topic	References
1. Plate information	British Steel Corporation
2. Plate girder design	BS 5950 *Structural Use of Steelwork in Building* Part 1: Code of practice for the design of simple and continuous construction: hot rolled sections (1985)
3. Plate girder design	BS 5400 *Steel, Concrete and Composite Bridges* Part 3: Code of practice for the design of steel bridges (1982)
4. Shear buckling	(1992) Plate girders, *Steel Designer's Manual*, pp. 464–466. Blackwell
5. Tension field action	**Porter D.M., Rockey K.C. & Evans H.R.** (1975) The collapse behaviour of plate girders loaded in shear, *Structural Engineer*, vol. 53 (Aug), pp. 313–25
6. Tension field action	(1992) Plate girders, *Steel Designer's Manual* pp. 449–454. Blackwell
7. Openings	**Lawson R.M. & Rackman, J.W.** (1989) *Design for Openings in Webs of Composite Beams.* Steel Construction Institute
8. Composite plate	**Owens G.W.** (1989) *Design of Fabricated Composite Beams in Buildings.* Steel Construction Institute

PART II

THE DESIGN OF STRUCTURAL STEEL FRAMEWORKS

The design of simple elements given in Part I is usually only part of the overall building concept. It is necessary to develop a spatial awareness of the structural framework in three dimensions. The action of the whole framework must be considered including its behaviour under lateral loading. In some cases frame action (continuous construction) may be preferred to connected element design (simple construction) from considerations of cost of appearance. Many of the design procedures used in Part II have been developed in Part I, and it is advisable for the student first to become familiar with element design.

12

DESIGN OF SINGLE-STOREY BUILDING – LATTICE GIRDER AND COLUMN CONSTRUCTION

12.1 INTRODUCTION

Over half of the total market share of the constructional steelwork fabricated in the United Kingdom is used in single-storey buildings. Therefore, it is almost certain that an engineer will, at some time, have to design or check such a building. Whereas previous chapters have introduced the design of various simple elements, the next three chapters extend the design concept of the overall design procedure of whole structures, i.e. three-dimensional structural arrangements. Essentially, most members in frameworks are positioned so as to transfer load in space to other members and eventually down to the ground, by the simplest, economic 'structural' route. The next two chapters will be devoted specifically to the various design aspects of single-storey structures. Though these structures represent the simplest form of three-dimensional frameworks, they will illustrate most of the structural design criteria which an engineer might encounter. This particular chapter outlines the different considerations used in designing all the structural members for a complete building, based on the main frame being of lattice girder and column construction (see Fig. 12.1). In the next chapter, the same building will be redesigned using portal frame construction. For brevity, only the main supporting frame will be redesigned as a portal frame, as the design of the remaining structural members is common to both forms of construction.

In developing the structural arrangement for a single-storey building or even a multi-storey building it should be borne in mind that the shorter the span of a structural member the more economic it becomes. However, the client/owner of a single-storey building frequently stipulates, as in this design exercise, that the floor area should be free of internal columns in order to obtain the greatest flexibility of space which can readily accommodate any future modifications to the usage of the floor area, without major structural

Fig. 12.1 Typical truss and column construction

alterations to the building. For economic reasons (such as having a flow-line production operation), most industrial buildings have a rectangular floor plan, and therefore always arrange, where possible, to span the main frames across the shorter distance, thereby minimizing member sizes (see Fig. 12.2).

Though the client usually gives the designer a free choice of structural arrangement, the best 'least cost' solution is nevertheless expected. In a paper[1] giving comparative costs of four different types of single-storey arrangements (roof truss, lattice girder, portal frame and space frame) the single-span portal frame seemingly did not produce the most economic answer for the **single-bay** frame, when assessed on initial cost. The study indicated that the lattice girder construction produces the most economic solution for the span being considered in the design example. However, when one takes into account the cost of maintaining the statutory minimum temperature within such buildings then the low roof construction of a portal frame could have a financial advantage over other forms of construction during the lifespan of a building (usually 50 years). This is probably why, together with its simple, clean lines, the most common form of single-storey building found on any modern industrial estate is that of portal frame construction, with over 90% of all single-storey buildings so constructed[1].

Fig. 12.2 Proposed main frame spacing

Despite the dominance of portal frame construction in the past, there is a growing demand from the hi-tech industries for higher quality and flexibility in the use of buildings. The structures of such buildings are frequently flat-roofed utilizing solid/castellated beams or lattice girders. The advantage of the lattice girder or castellated form of construction is that it allows services to be accommodated within the depth of the roof construction, at the expense of deeper roof construction when compared with portal frames.

In order to understand the overall design procedure for a building, the following design exercise will deal with the complete design of a single-storey building based on lattice girder and column construction. However, in Chapter 13 an alternative approach, using portal frames as the main supporting structure, will be considered; that is, the simple lattice girders with universal sections as column members will be replaced by a series of portal frames.

12.2 DESIGN BRIEF

A client requires a single-storey building, having a clear floor area, 90 m × 36.4 m, with a clear height to the underside of the roof steelwork of 4.8 m, with possible extension to the building in the future. The slope of the roof member is to be at least 5°. It has been specified that the building is to be insulated and clad with PMF metal sheeting profile Long Rib 1000R (0.70 mm thick, minimum necessary for roof to prevent damage during maintenance access). A substrata survey of the site, located in a new development area on the outskirts of Guisborough, North Yorkshire, has shown that the ground conditions are able to sustain a foundation bearing pressure of 150 kN/m² at 0.5 m below existing ground level.

12.3 PRELIMINARY DESIGN DECISIONS

The two common arrangements for open web (lattice) girders are illustrated by the diagrams in Fig. 12.3, i.e. the Warren or Pratt truss girders. The difference between the two types is basically that the Warren truss has pairs of diagonal members of approximately the same length, while the Pratt truss has short verticals and long diagonals. Under normal circumstances of gravity loading when there is no load reversal, the Pratt truss is structurally more efficient because the short vertical members would be in compression and the long diagonals in tension. However, when there is load reversal in the

Fig. 12.3 Two common forms of lattice girder

(a) Warren

(b) Pratt

diagonals then the Warren truss may prove to be the more economic arrangement. Also, the Warren truss may give larger access space for circular and square ducts and is considered to have a better appearance. Therefore, the Warren truss has been selected for this design exercise.

One of the cheapest forms of construction for the lattice girders would be if the angle sections, used for the diagonals (web members), were welded directly to the top and bottom chords, fabricated from T-sections, and this construction will be the basis of the main design. An alternative form of construction is to use tubular (hollow) sections, which are being increasingly incorporated into hi-tech steel buildings. Tubular sections have a good appearance and are efficient as compression members but are difficult to connect satisfactorily, particularly when subject to high loads, when stiffener plates are required to control bending of the section walls.

The slope of the top chord is chosen to reflect the minimum specified (5°) or thereabouts. Consequently, because the roof slope is less than 15° then the sheeting will need to be laid with special strip mastic lap sealers, in order to prevent capillary action and hence rain leaking into the building. Other practices for ensuring weathertightness are to increase the side and end laps and fastener frequency. Such details should be checked with the sheeting manufacturer's catalogue. The alternative is to use standing seam type sheeting with concealed fixings. As regards limiting deflection, there are no mandatory requirements, but commonly accepted criteria for a typical insulated building are $L/200$ for roofs and $L/150$ for vertical walls.

The required slope can be achieved by making the depth of the lattice girder at the eaves equal to 1/20 of the span and the depth at the centre of the span 1/10. Assuming the column depth to be 0.6 m then, with reference to Fig. 12.4:

Fig. 12.4 Structural arrangement of main frame members

Span between column centres $(L-36.4+0.6)$	$=37.0$ m
Spacing of lattice frames	$= 6.0$ m
Height to underside of girder	$= 4.8$ m
Depth of girder at eaves	$= 1.85$ m
Depth of girder at ridge (apex)	$= 3.70$ m
Actual slope of rafter $[\theta=\tan^{-1}(1.85/18.5)]$	$= 5.71°$

12.4 LOADING

Before any design calculations can be undertaken, the loads that can occur on or in a building have to be assessed as accurately as possible. The loads which normally govern the design of a single-storey building are dead loads[2], snow load[2] and wind loading[3,4]. In addition, the designer should give thought to the possibility of unusual loadings, such as drifting of snow[6], and overloading of a gutter if the downpipes become blocked or cannot cope with a large volume of rainwater during a deluge. Modern buildings may also be required to accommodate services, such as ducting or sprinkler systems. The weight of these items can be significant and it is advisable that advice from the suppliers is sought. (Note that the snow load information referred to in references (2) and (6) is to be incorporated in BS 6399: Part 3, *Code of Practice for Snow Loads*). Also, depending on the function of a building, dynamic loading from crane operations can be an extra design consideration.

12.4.1 Dead load

The dead loads affecting the design of the building result from the self weight of the sheeting (including insulation), the secondary members and main frames and will be included in the design calculations, as and when they occur. Estimating the selfweight of sheeting and secondary members is relatively easy as this information is contained in the manufacturers' catalogues. Assessing the selfweight of the main frame is more difficult, as this information is required before the design of the frame. Designers with experience can make rapid estimates. In this design example, a rough guide would be to make it about 15% of the total gravity load acting on the main frame. Clearly, the selfweight of frames spanning smaller distances, as a percentage of the total gravity loading, would be less and for larger spans the percentage is larger. The self-weight of the BSC profile Long Rib 1000R with insulation will be taken as $0.097 \, kN/m^2$.

12.4.2 Snow load

The relevant information regarding the snow loading is at present contained in BS 6399: Part 1[2]. For the proposed site (Guisborough) it is estimated that the relevant snow load is $0.75 \, kN/m^2$ (acting on plan) though it is anticipated that in Part 3 of BS 6399 there will be regional variations, as is already permitted in the farm building code BS 5502. The equivalent snow load acting along the inclined roof member is $0.75 \cos \theta = 0.75 \times 0.995 = 0.75 \, kN/m^2$. The use of an equivalent load makes due allowance for the purlin spacing being given as a slope distance. However, at this slope it is seen that the difference in load on slope and on plan is negligible.

12.4.3 Wind load

From CP3 Chapter V Part 2 (also to be incorporated into BS 6399), the wind loading on the building being designed can be established. Also, the reader is directed to reference (4) which deals more fully with wind loading on buildings and contains the background information on which reference (3) is based. As the site is located in Guisborough, then from reference (3), the basic wind speed is estimated as being 45 m/s (Fig. 1 of reference (3)); the factors S_1 and S_3 are both 1.0. From the information supplied, the ground roughness is assumed to be 3, and because the building is longer than 50 m, the 'building size' is designated as class C. From Table 3[3], knowing the building height is in the 5–10 m region, the factor S_2 is found to be 0.69, based on a height of 10 m. (A slightly lower value might be obtained by interpolation, as the actual height is 8.5 m). Hence the dynamic pressure (q) is:

$$q = 0.613 \ (1.0 \times 0.69 \times 1.0 \times 45)^2/1000$$
$$= 0.59 \text{ kN/m}^2$$

The **external pressure coefficients** for a building (with a roof slope of 5.7°, a building height ratio $h/w = 8.5/37.0 = 0.23$ and a building plan ratio $l/w = 90.0/37.0 = 2.43$) are obtained from Table 7 (walls) and Table 8 (pitch roofs)[3]. Though the dimensions used in this example are based on the centre lines of members, it is the usual practice to use the overall dimensions of a building. However, the latter would not materially affect the calculated values. The **internal pressure coefficients** are assessed from Appendix E[3], assuming that the two long faces of the building are equally permeable while the gable faces are not; that is, $+0.2$ when the wind is normal to a permeable face and -0.3 when normal to an impermeable face. However, it could be argued that one should allow for the occurrence of dominant openings, particularly if details of openings have not been finalized at the design stage; that is, the designer makes the appropriate decision, depending on the information available regarding openings in the buildings. Therefore, use the clause in reference (3) which allows the designer to take the more onerous of $+0.2$ and -0.3. This covers the possibility of a dominant opening, provided it is closed during a severe storm. The resulting wind loading conditions for the building (cf. Fig. 2.9 for frame only), are given in Fig. 12.5. The value of 0.95 is determined by interpolation. If there is no possibility of a dominant opening, then only wind cases B and C apply (see Fig. 12.5).

The diagrams show that the maximum pressure that can act on the sides and gables of the building is $1.0 \times 0.59 \text{ kN/m}^2$. However, the maximum **local** pressure, for which the sheeting has to be designed, is generally significantly higher. In assessing the local pressure on the **roof** sheeting, a revised value of S_2 has to be obtained from Table 3[3], noting that cladding is defined as class A, hence $S_2 = 0.78$ and the appropriate value of $q = 0.613 \ (1.0 \times 0.78 \times 1.0 \times 45)^2/1000 = 0.76 \text{ kN/m}^2$. Attention is drawn to the larger value of C_{pe} which occurs at the edge zones of roofs, as indicated in Table 8[3]. Therefore, the maximum design pressure (in this case, suction) acting on the roof sheeting is

$$(C_{pe} + C_{pi}) \ q = - \ (1.4 + 0.2)0.76 = \ -1.22 \text{ kN/m}^2.$$

Similarly for the **side** cladding where $S_2 = 0.70$ based on a height of 5 m (actual 6.65 m) and the local C_p value is 1.0[3], then the maximum pressure sustained by the side sheeting is

$$(1.0 + 0.2)0.61 = 0.73 \text{ kN/m}^2.$$

Fig. 12.5 External and internal wind coefficients for building

These wind loads, based on **pressure coefficients**, are used when determining the loads acting on a particular surface or part of the surface of a building, i.e. they are applicable in the design of the lattice girder or columns. However, in estimating the wind loads acting on the whole of the building, the **force coefficients** have to be used; that is, the total wind load on a building is calculated from:

$$F' = C_f q A_e$$

where C_f is obtained from Table 10[3] and A_e is the effective frontal area. This means that when the wind is blowing perpendicular to the longitudinal axis of the building, then $S_2 = 0.60$ and $q = 0.45 \text{ kN/m}^2$, hence:

$$F' = 1.0 \times 0.45 \times (6.0 \times 6.65) = 18.0 \text{ kN per main frame}$$

For the case of the wind blowing on the gables, then:

$$F' = 0.7 \times 0.59 \times [37.0 \, (6.65 + 1.85/2)] = 116 \text{ kN}$$

This force of 116 kN has to be distributed between the bracing systems (see Section 12.8).

In addition to direct wind pressure, there can be frictional drag forces. For rectangular clad buildings, these drag forces need to be taken into account only where the ratio of the dimension in the wind direction (d) compared with the dimension normal to the wind direction (b) is greater than 4. The drag force can be determined from[3]:

$$F' = C_{f'}[\text{(roof surface)}\ q_1 + \text{(wall surface)}\ q_2]$$

Consider the wind blowing parallel to the longitudinal axis of the building; then $d/b = 2.43$ and $d/h = 10.7$, i.e. drag force must be taken into account. Under this wind condition, the selected sheeting has ribs running across the wind and therefore $C_{f'} = 0.04$[3] and for the condition $h \leq b$, the code of practice[3] states that:

$$\begin{aligned} F' &= C_{f'}[b(d-4h)q_1 + 2h(d-4h)q_2] \\ &= 0.04[37.0(90.0 - 4 \times 8.5)0.59 + 2 \times 6.65(90.0 - 4 \times 6.65)0.45] \\ &= 49.0\ \text{(roof)} + 15.2\ \text{(walls)} = 64.2\ \text{kN} \end{aligned}$$

which has to be resisted by the braced bay(s) (see Section 12.8).

When the wind is deemed to blow in the lateral direction, then $d/b = 0.41$ and $d/h = 4.40$. As the ribs of the sheeting do not run across the wind direction, then $C_{f'} = 0.01$, and the drag force is determined from[3]:

$$\begin{aligned} F' &= C_{f'}[b(d-4h)q_1 + 2h(d-4h)q_2] \\ &= 0.01[90.0(37.0 - 4 \times 8.5)0.59 + 2 \times 7.58(37.0 - 4 \times 7.58)0.45] \\ &= 1.6 + 0.5 = 2.1\ \text{kN}\ (= 0.2\ \text{kN per main frame}) \end{aligned}$$

That is, when the wind is blowing in the longitudinal direction of a single-bay building, the drag force can be significant, while the drag force per frame in the lateral direction is comparatively small and is usually ignored.

12.5 DESIGN OF PURLINS AND SHEETING RAILS

One of the initial decisions that the engineer has to make is the spacing or centres of the main frames. Though the paper on costs[1] indicates that 7.5 m spacing would be more economic, it has been decided to use 6.0 m centres owing to the practical consideration of door openings. Also, if large brick panel walls are used in the side elevations instead of sheeting, it is advisable in any case to limit the frame centres to about 6.0 m or less to avoid having to use thicker than standard cavity wall construction.

As indicated in Section 1.5, the imposed loading acting on a single-storey structure is due to snow and wind, which is carried initially by the cladding and is transferred into the secondary members, purlins (roof) and side rails (vertical walls). These members, which are usually designed as double-span members, transfer the imposed loads plus their own self weight by flexural action on to the main frames as a series of point loads. Therefore, another decision to be made is the actual spacing of the purlins, which is dependent on the snow load and the profile and thickness of the metal sheeting selected for the cladding.

It has been shown[1] that the spacing of purlins has little effect on the total cost of purlins, though increased spacing would lead to an increase in cost of sheeting. Metal sheeting is generally used in long multi-span lengths to minimize the number of transverse joints. Such long lengths can be easily handled on the roof (as purlins give support), but not as side cladding because it is difficult to support vertically during erection, even with scaffolding. In this example, the client has specified the sheeting. Nevertheless, the maximum length over which the sheeting can span has to be established. The PMF sheeting profile Long Rib 1000R when used as roof cladding has to support a snow load of >0.75 kN/m^2 and/or a local wind suction of -1.22 kN/m^2. *Note that the high local wind pressures/suctions apply only to the design of the cladding.* From the PMF catalogue[6] and assuming that the length of sheet runs over at least two spans, it can be shown that the selected profile can sustain 1.65 kN/m^2 over a span of about 2.0 m, while complying with the deflection limitation for roof sheeting of $L/200$.

Today, the design of the secondary members is dominated by cold formed sections. Though there is a British Standard covering the design of cold formed members (BS 5950: Part 5[7]), the manufacturers tend to develop new profiles based on the results of extensive testing. There are a number of manufacturers of purlins and sheeting rails and therefore, in making a choice, one needs to consult the various manufacturers' catalogues.

The 'design' of cold formed members consists of looking up the relevant table for the chosen range of sections. The choice of a particular manufacturer's products is dependent on a client's or designer's experiences and preferences. Table 12.1 illustrates a typical **purlin** load table based on information from a manufacturer's catalogue (Ward Multibeam[8]) for the double-span condition. The loads shown in the table are based on lateral restraint being provided to the top flange of the purlin by the sheeting. Also, it should be noted that the loads quoted in Table 12.1 are for ultimate load condition, i.e. **factored**, and that the self weight of the purlin has already been deducted from the limiting values of load given in the table.

Assume the overall distance between the outer faces of the column members is 37.6 m, which if divided into 24 equal portions would give purlin centres about 1.570 m (on the slope). The gravity loading (dead plus snow) supported by the purlins is $1.6 \times 0.75 + 1.4 \times 0.097 = 1.34$ kN/m^2, while the maximum uplift on the purlins is $1.4 (0.097 - 1.15 \times 0.59) = -0.81$ kN/m^2. From Table 12.1(a), knowing the purlin length of 6.0 m, purlin spacing of 1.570 m and the gravity load to be supported by the purlin (1.34 kN/m^2), the P145155 section seems adequate. (Usually purlin spacing tends to be cost-effective in the range 1.8–2.0 m).

If the design load is limited to 1.34 kN/m^2 (factored), then the maximum spacing for this particular profile would be:

$$L_s = \frac{\text{u.d.l.}}{\text{span} \times \text{max. applied load}}$$

where
u.d.l. – see third column of Table 12.1(a) (13.07 kN)
span – purlin length, i.e. 6 m
applied load – 1.34 kN/m^2

$$L_s = \frac{13.07}{6.0 \times 1.34} = 1.626 \text{ m}$$

172 STRUCTURAL STEELWORK DESIGN TO BS 5950

Table 12.1 Double span factored loads (kN/m²) for selected Multibeam sections (based on information given in Ward Building Components manual, see Reference 8)

Span	Section	UDL kN	\multicolumn{9}{c}{Purlin Centres (mm)}
			1000	1200	1400	1600	1800	2000	2200	2400	2600
5.0	P145130	11.55	2.31	1.93	1.65	1.44	1.28	1.16	1.05	0.96	0.89
	P145145	13.95	2.79	2.33	1.99	1.74	1.55	1.40	1.27	1.16	1.07
	P145155	15.55	3.11	2.59	2.22	1.94	1.73	1.56	1.41	1.30	1.20
	P145170	17.83	3.57	2.97	2.55	2.23	1.98	1.78	1.62	1.49	1.37
	P175140	16.52	3.30	2.75	2.36	2.07	1.84	1.65	1.50	1.38	1.27
	P175150	18.55	3.71	3.09	2.65	2.32	2.06	1.86	1.69	1.55	1.43
	P175160	20.23	4.05	3.37	2.89	2.53	2.25	2.02	1.84	1.69	1.56
	P175170	22.47	4.49	3.75	3.21	2.81	2.50	2.25	2.04	1.87	1.73
6.0	P145130	9.74	1.62	1.35	1.16	1.01	0.90	0.81	0.74	0.68	0.52
	P145145	11.74	1.96	1.63	1.40	1.22	1.09	0.98	0.89	0.82	0.75
	P145155	13.07	2.18	1.82	1.56	1.36	1.21	1.09	0.99	0.91	0.84
	P145170	*14.19*	*2.37*	*1.97*	*1.69*	*1.48*	*1.31*	*1.18*	*1.08*	*0.99*	*0.91*
	P175140	13.98	2.33	1.94	1.66	1.46	1.29	1.16	1.06	0.97	0.90
	P175150	15.67	2.61	2.18	1.87	1.63	1.45	1.31	1.19	1.09	1.00
	P175160	17.07	2.85	2.37	2.03	1.78	1.58	1.42	1.29	1.19	1.09
	P175170	18.93	3.16	2.63	2.25	1.97	1.75	1.57	1.43	1.31	1.21
	P175200	23.56	3.93	3.27	2.80	2.45	2.18	1.96	1.78	1.64	1.51
	P205145	17.09	2.85	2.37	2.03	1.78	1.58	1.42	1.29	1.19	1.10
	P205155	19.03	3.17	2.64	2.27	1.98	1.76	1.59	1.44	1.32	1.22
	P205165	20.92	3.49	2.91	2.49	2.18	1.94	1.74	1.58	1.45	1.34
	P205180	23.69	3.95	3.29	2.82	2.46	2.19	1.97	1.79	1.65	1.52
	P205190	25.50	4.25	3.54	3.04	2.66	2.36	2.13	1.93	1.77	1.63
	P205200	27.28	4.55	3.79	3.25	2.84	2.53	2.27	2.07	1.89	1.75
7.0	P175160	14.71	2.10	1.75	1.50	1.31	1.17	1.05	0.96	0.88	0.81
	P175170	16.30	2.33	1.94	1.66	1.46	1.29	1.16	1.06	0.97	0.90
	P175200	*19.40*	*2.77*	*2.31*	*1.98*	*1.72*	*1.54*	*1.39*	*1.26*	*1.15*	*1.07*
	P205155	16.46	2.35	1.96	1.68	1.47	1.31	1.18	1.07	0.98	0.90
	P205165	18.06	2.58	2.15	1.84	1.61	1.43	1.29	1.17	1.08	0.99
	P205180	20.44	2.92	2.43	2.09	1.83	1.62	1.46	1.33	1.22	1.12
	P205190	21.99	3.14	2.62	2.24	1.96	1.75	1.57	1.43	1.31	1.21
	P205200	23.51	3.36	2.80	2.40	2.10	1.87	1.68	1.53	1.40	1.29
	P235170	22.76	3.25	2.71	2.32	2.03	1.81	1.63	1.48	1.35	1.25
	P235190	26.57	3.80	3.16	2.71	2.37	2.11	1.90	1.73	1.58	1.46
	P235200	28.42	4.06	3.38	2.90	2.54	2.26	2.03	1.85	1.69	1.56
	P235230	33.88	4.84	4.03	3.46	3.03	2.69	2.42	2.20	2.02	1.86

(a) Purlin load table (gravity loading)

Span	Section	UDL kN	\multicolumn{9}{c}{Rail Centres (mm)}
			1000	1200	1400	1600	1800	2000	2200	2400	2600
5.0	R130130	10.43	2.09	1.74	1.49	1.30	1.16	1.04	0.95	0.87	0.80
	R130140	11.63	2.33	1.94	1.66	1.45	1.29	1.16	1.06	0.97	0.89
	R130150	12.80	2.56	2.13	1.83	1.60	1.42	1.28	1.16	1.07	0.98
	R145130	11.94	2.39	1.99	1.71	1.49	1.33	1.19	1.09	1.00	0.92
	R145140	13.32	2.66	2.22	1.90	1.67	1.48	1.33	1.21	1.11	1.02
	R145150	14.67	2.93	2.45	2.10	1.83	1.63	1.47	1.33	1.22	1.13
	R145165	16.64	3.33	2.77	2.38	2.08	1.85	1.66	1.51	1.39	1.28
6.0	R145130	10.10	1.68	1.40	1.20	1.05	0.94	0.84	0.77	0.70	0.65
	R145140	11.26	1.88	1.56	1.34	1.17	1.04	0.94	0.85	0.78	0.72
	R145150	12.38	2.06	1.72	1.47	1.29	1.15	1.03	0.94	0.86	0.79
	R145165	*13.87*	*2.31*	*1.93*	*1.65*	*1.44*	*1.28*	*1.16*	*1.05*	*0.96*	*0.89*
	R175140	14.28	2.38	1.98	1.70	1.49	1.32	1.19	1.08	0.99	0.92
	R175155	16.44	2.74	2.28	1.96	1.71	1.52	1.37	1.25	1.14	1.05
	R175170	18.55	3.09	2.68	2.21	1.93	1.72	1.55	1.41	1.29	1.19
	R205145	17.41	2.90	2.42	2.07	1.81	1.61	1.45	1.32	1.21	1.12
	R205155	19.10	3.18	2.65	2.27	1.99	1.77	1.59	1.45	1.33	1.22
7.0	R175140	12.39	1.77	1.48	1.26	1.11	0.98	0.89	0.80	0.74	0.68
	R175155	14.24	2.03	1.70	1.45	1.27	1.13	1.02	0.92	0.85	0.78
	R175170	160.4	2.29	1.91	1.64	1.43	1.27	1.15	1.04	0.95	0.88
	R205145	15.12	2.16	1.80	1.54	1.35	1.20	1.08	0.98	0.90	0.83
	R205155	16.57	2.37	1.97	1.69	1.48	1.32	1.18	1.08	0.99	0.91
	R205170	18.69	2.67	2.23	1.91	1.67	1.48	1.34	1.21	1.11	1.03

(b) Cladding rail load table (pressure loading)

that is, the design spacing of 1.570 m is just within the capacity of the purlin section P145155. However, Table 12.1 shows that this section is near its deflection limit, i.e. not included for a 6.5 m span. Therefore, it might be advisable to select a deeper section, i.e. use P175140 profile. If a purlin 'safe load' table does not state the deflection limit, check with the manufacturer that the limit of $L/200$ is not exceeded.

For the majority of design cases, the design of the purlin section would now be complete. However, under some wind conditions the resultant uplift on the roof can produce a stress reversal in the purlin, thereby inducing compression in the outstand flange, e.g. in this exercise the uplift is 0.81 kN/m^2. As this flange is not laterally restrained by the cladding, then some form of restraint to the flange may be necessary; check with the manufacturer regarding any special restraint requirement for wind uplift. Indeed, under high wind loading, the wind uplift–no snow condition could result in a more severe loading for the purlin than that due to gravity loading. As the metal cladding is normally fixed by self-tapping screws (designer's choice) then the same load–span table (Table 12.1a) can be used for suction conditions, provided the anti-sag tie arrangements are adhered to (see next paragraph). Note that if the selected cladding had been metal or asbestos sheeting fixed with hookbolts, then a mid-span restraint would have been necessary[8], as the wind uplift condition exceeds this particular manufacturer's limit of 50% of the permissible gravity loading, i.e. $0.81/1.34 = 60\%$. Such limitations are dependent on individual manufacturer's recommendations.

Anti-sag ties at mid-span of the purlins spanning more than 6.1 m are recommended by the manufacturer. Such ties are required to prevent distortions and misalignment of purlins during the fixing of sheeting or where extreme axial loads exist. Under normal conditions, it would appear that sag bars are not required in this example, as the purlin span is less than 6.1 m. However, if any purlin forms part of the roof bracing system, then sag bars may become necessary.

Coming now to the design of the side rails for the vertical walls, as snow loading is not a design problem, the section is usually chosen independent of the purlin. The wind conditions for the sides/gables (Fig. 12.5) indicate that a pressure of $1.4 \times 1.0 \times 0.59$ kN/m^2 and a suction of $-1.4 \times 0.8 \times 0.59$ kN/m^2 are the appropriate design loading, which acts perpendicular to the sheeting (allowed implicitly by clause 4.12.4.4b; that is, it is assumed that the vertical panel of sheeting (connected to the side rails) behaves as a deep girder, thereby imposing negligible flexural action (due to self weight) on the side rails in the vertical plane. (Try bending a flat sheet of paper in the plane of the paper). However, care must be taken during erection to reduce any distortion that can occur in side rails before the cladding is attached. The reduction of such distortion is discussed in the next paragraph. To maximize the strength of the side rails, they are placed normal to the sheeting and column members. Wind load permitting, the side rails can be spaced further apart. In this example, one could use the same purlin size (P175140). However, the manufacturer of the Multibeam system produces special sections for cladding rails and reference to their catalogue[8] would indicate that the section size R145130 is suitable for the same reasons given in selecting the purlin profile.

There is also the possibility of having to restrain laterally the column member, which might cause the maximum spacing of some rails to be reduced.

The maximum factored wind suction $(-1.4 \times 0.8 \times 0.59 = -0.66 \, kN/m^2)$ acting on the sheeting would cause compression in the outstand flange, therefore mid-span restraints may become necessary. In using the cladding section (R145130), the manufacturer limits suction loading to 80% of the allowable wind pressure load. By coincidence, the suction coefficient is 0.8 and therefore the same section selected to withstand the wind pressure can be used.

It is essential during erection that any distortion, which can occur in side rails before the cladding is attached, is minimized. This can be achieved by employing the 'single strut' system (for use up to 6.1 m frame centres), as recommended by the manufacturers[8]; that is, any distortion and levelling is controlled by adjusting the diagonal ties before the placement of the sheeting (see Fig. 12.1). The inherent benefit of the single strut system is the mid-span restraint it provides.

In practice, the joints of the double-spanning purlins/rails are staggered across each frame, thereby ensuring that each intermediate main frame receives approximately the same total purlin loading; that is, the larger central reactions arising from the continuity are applied to alternative frames. The self weight of selected purlin section (P175140) is 0.035 kN/m.

The purlins/sheeting rails are attached to the primary structural members by means of cleats, bolted or welded to the main members. As an integral part of the Multibeam system, the manufacturer supplies special cleats. If a manufacturer does not supply cleats, then they have to be designed (see Chapter 4). Nevertheless, it is essential that any standard hole arrangements stipulated by the manufacturer are complied with. Otherwise, extra cost could be incurred for a non-standard arrangement.

12.5.1 Summary of secondary member design

Purlin size: — Ward Multibeam – **P175140**

    Actual spacing of purlins (on slope) = 1.570 m
    Actual spacing of purlins (on plan) = 1.562 m

Side rail size: Ward Multibeam – **R145130**

    Actual spacing of side rails: see Section 12.7

12.6 DESIGN OF LATTICE GIRDER

Taking the overall dimensions for the main frame as defined in Section 12.3, a good structural arrangement for the 'web' members is to make the inclination of these diagonals in the region of 45°–60° to the horizontal. By dividing the top chord member, over half the span, into five equal panel widths, then the diagonals in the end panel are at 45°, while those at the mid-

DESIGN OF SINGLE-STOREY BUILDING – LATTICE GIRDER AND COLUMN CONSTRUCTION

Fig. 12.6 Member numbering for lattice girder

span are at 65° (see Fig. 12.6). This means that the purlin positions along the top chord do not coincide with the member intersecting points (connections) of the frame panels. Therefore, in addition to the primary axial forces, the top chord member has to be designed to resist the bending action induced by the purlin loads.

In spite of the fact that the top and bottom chords will in practice be continuous members, a safe assumption is to analyse the lattice girder initially as a pin-jointed frame, thereby allowing the primary axial forces in the various members to be evaluated readily; that is, the flexural action in the top chord caused by the purlin loads can be ignored for the purpose of calculating the primary axial forces. Indeed, clause 4.10 permits such a procedure.

Consequently, any purlin load needs to be redistributed so that it is applied only at the panel points. This is simply done by dividing the total load acting on the girder by the number of panels in the top chord, i.e. as there are 10 panels, then the panel load is the total load/10. Note that the two outermost panel nodes carry just over half the load, as they support only half a panel width of roof, plus any sheeting overlap. It can be shown that this apparent redistribution of load does not materially affect the magnitudes of the axial forces in the members. There is an implied assumption that the self weight of the girder is uniformly distributed throughout the frame. This approximation would have a negligible effect on the outcome of the design of a girder of this size.

Having decided the geometry of the girder and the different patterns of loading required, the next stage is to calculate the unfactored loads acting on the girder; see Table 12.2. The noted wind loads (w_w) are based on a wind coefficient of −1.0; wind loads for other wind coefficients are obtained by multiplying the noted values by the appropriate coefficient. Total self weight of the girder is estimated, based on the previously suggested figure of 15% of the total dead load; that is, the dead load, excluding self weight, is

20.1 + 7.0 + 166.5 = 193.6 kN,

then the estimated self weight is

193.6 × 15/(100 − 15) = 34.2 kN, i.e. 0.92 kN/m.

Consideration of the various load combinations (Section 2.7f)

Table 12.2 Unfactored loads (kN) on lattice girder

			Total load	Panel load	Purlin load
Dead load (w_d)					
sheeting and insulation	$0.09 \times 6.0 \times 37.18$	=	21.6	2.16	0.912 (on plan)
purlins	$26 \times 0.035 \times 6.0$	=	5.5	0.55	0.232 (on plan)
self weight	0.92 (est) $\times 37.0$	=	34.0	3.40	1.435 (on plan)
Snow load (w_i)	$0.75 \times 6.0 \times 37.0$	=	166.5	16.65	6.966 (on plan)
Wind load (w_w)	$0.59 \times 6.0 \times 37.18$	=	131.6	13.16	5.557 (on slope)

indicates that there are only two load conditions for which the girder needs to be designed, i.e.

- maximum gravity loading: $1.4\,w_d + 1.6\,w_i$
- maximum uplift loading: $1.0\,w_d + 1.4\,w_w$

As wind loading on the girder always produces an uplift condition for this example, then the load combination $1.2w_d + 1.2w_i + 1.2w_w$ will produce conditions between the combinations A and B (see Fig. 12.5) and therefore need not be considered for the design of the girder.

12.6.1 Forces in members

Knowing the 'panel' loads, an elastic analysis of the girder can now easily be undertaken manually, by resolving forces at a joint or by other well-known methods[9]. Note that **gravity loading** (dead + snow) acts in the **vertical** direction, while the **wind loading** acts **perpendicular** to the inclined roof members.

Alternatively, the primary axial forces in the members can be determined by a computer analysis. However, the kind of analysis program to which the designer may have access can vary, and the following observations might prove helpful:

- Always minimize the maximum difference between **adjacent** node numbers, e.g. in Fig. 12.7 the maximum difference is 2. By this simple rule, computing costs are kept to a minimum.
- If the analysis program is capable of handling **pin-jointed** structures, then the axial forces can be evaluated by making the assumption that all members have equal areas. Though this assumption is not correct (as will be demonstrated clearly by the final member sizes for the girder) it has no effect on the magnitudes of the primary axial forces. The assumption only affects the lattice girder deflections.
- However, if the only program available is a **rigid frame** analysis package, then check the program specification to see whether or not it has a facility for handling pin-ended members:

DESIGN OF SINGLE-STOREY BUILDING – LATTICE GIRDER AND COLUMN CONSTRUCTION

Fig. 12.7 Node numbering for lattice girder

- If it has this facility, then make all the diagonals pinned at both ends; make top and bottom chords continuous, except for the ends (adjacent to the verticals) which should be pinned (see Fig. 12.7). Do not also make the end verticals pin-ended, as this produces numerical instability. Make the chords relatively stiff by putting the second moment of area (inertia) for the chords equal to, say, 100 cm^4. This will ensure that any moments generated in these members are nominal. Such nominal moments are induced owing to the small relative movements of the panel nodes. (A similar effect would occur with slight settlement of the supports for a continuous beam). Again make all members have equal area.
- If the rigid frame analysis program does not have this facility, then make the second moment of area of the web members very small, say 0.01 cm^4, and make all members have equal area. An analysis will result in the same numerical values obtained from other analyses. This simple 'device' of using virtually zero inertia, in fact, prevents the web members from attracting moment, thereby producing effectively a pin-ended condition for the members so designated; that is, although the top chord is continuous, the loads are being applied only at the nodes (panel points) and because the connected web members are made to act as pin-ended, then only nominal moments can be induced into the top chord. The chord behaves essentially as a series of pin-ended members between panel points.

Owing to the symmetry of the components of vertical loading (dead and snow) on the girder, the girder need only be analysed for panel loads equivalent to the condition 1.0 w_d (= 2.16 + 0.55 + 3.40 = 6.11 kN). The resulting member forces can then be proportioned to give the appropriate forces due to 1.4 w_d (8.554 kN) and 1.6 w_i (26.64 kN). However, separate analyses are required for wind loading owing to the non-symmetrical nature of this type of loading.

The results from the different elastic analyses are shown in Figs. 12.8(a) (dead load), 12.8(b) (wind on side) and 12.8(c) (wind on end). The axial forces in each member, duly factored, are summarized in Table 12.3. Though the analysis for wind on the sides is executed for wind blowing from left to right, it should be borne in mind that the wind can blow in the reverse direction, i.e. right to left. Therefore, if wind affects the design (as in this

Fig. 12.8 Lattice girder, – unfactored member loads (kN)

(a) Dead load

(b) Wind on side walls

(c) Wind on end walls

case), members must be designed for the worst condition, irrespective of wind case, i.e. only the **worse load** from either Fig. 12.8b or Fig. 12.8c is recorded in Table 12.3.

The various members making up the Warren truss have been grouped, so that a common member size for any individual group of members can readily be determined.

Having established the individual factored forces, the design loads for each member (resulting from the two load combinations being considered) can be obtained; see columns (A) and (B) in Table 12.3.

An assessment must now be made of the flexural action in the top chord caused by the purlin loading being applied between panel joints. By taking account of the fact that the top chord in the half-span will be fabricated continuous, then the bending moments in the top chord can be assessed either manually by the moment distribution method or by a continuous beam computer program. This is achieved by assuming that the continuous member is 'supported' at panel points. Alternatively, the top chord or even the whole truss can be reanalysed with a rigid frame computer program, with additional nodes being introduced at the loaded purlin positions, if a facility for

DESIGN OF SINGLE-STOREY BUILDING – LATTICE GIRDER AND COLUMN CONSTRUCTION

Table 12.3 Member forces for gravity and wind loads (kN m)

	Members	$1.0w_d$ (1)	$1.4w_d$ (2)	$1.6w_i$ (3)	$1.4w_i$ (4)	(2)+(3) (A)	(1)+(4) (B)
Top chord	1,10	+25.1	+ 35.2	+109.5	− 77.3	+144.7	− 52.2
	2,9	+59.0	+ 82.7	+257.4	−181.3	+340.1	−122.3
	3,8	+75.7	+106.0	+330.2	−232.5	+436.2	−156.8
	4,7	+81.3	+113.8	+354.3	−249.6	+468.1	−168.3
	5,6	+79.2	+110.9	+345.2	−245.5	+456.1	−166.3
Bottom chord	11,21	0.0	0.0	0.0	+5.1	0.0	+5.1
	12,20	−45.8	− 64.2	−199.8	+144.4	−264.0	+98.6
	13,19	−69.8	− 97.8	−304.4	+210.2	−402.2	+140.4
	14,18	−80.2	−112.3	−349.6	+241.1	−461.9	+160.9
	15,17	−81.5	−114.1	−355.2	+242.7	−469.3	+161.2
	16	−76.4	−106.9	−333.0	+229.4	−439.9	+153.0
Web compression	22,43	+30.6	+ 42.8	+133.2	− 92.6	+176.5	− 62.0
	24,41	+32.5	+ 45.6	+141.9	− 99.2	+187.5	− 66.5
	26,39	+19.1	+ 26.7	+ 83.1	− 56.1	+109.8	− 37.0
	28,37	+ 9.1	+ 12.8	+ 39.8	− 27.0	+ 52.6	− 17.9
	30,35	+ 1.2	+ 1.7	+ 5.4	+ 6.9	+ 7.1	+ 8.1
	31,34	+ 5.5	+ 7.7	+ 24.1	− 23.4	+ 31.8	− 17.9
Web tension	23,42	−35.3	− 49.5	−154.1	+107.6	−203.6	+ 72.3
	25,40	−20.2	− 28.3	− 88.0	+ 59.4	−116.3	+ 39.2
	27,38	− 9.5	− 13.3	− 41.5	+ 28.5	− 54.8	+ 19.0
	29,36	− 1.3	− 1.8	− 5.5	− 9.7	− 7.3	− 11.0
	32,33	− 5.4	− 7.6	− 23.5	+ 23.9	− 31.3	+ 18.5

accepting point loads within a member length is not available. See Fig. 12.9 for positions of purlins relative to panel joints along the top half-chord member.

The self-weight of the top chord can be ignored in the determination of the bending moments along the chord as it will have minimal effect on the moments for this size of frame. Also, for the analysis it is assumed that the ends of the half-chord are pin-ended, which again is a safe assumption, as it could be argued that though the 'apex' end is continuous with the other half-chord, there remains the possibility of a site connection at the apex. Therefore the end might not achieve full continuity, depending on fabrication details.

Alternatively, the two separate computer operations (for the primary axial forces and for the moments in the top chord) could be combined to run as one loading condition, i.e. purlin loads being applied at correct positions, with the top and bottom chords made continuous and all web members pin-jointed.

Fig. 12.9 Positioning of purlins along top chord member

Fig. 12.10 Bending moments and axial load in top chord member

(a) Net bending moments (kN m)

Values shown: 2.45, 0.28, 5.68, 9.57, 2.03, 10.69, 0.92, 13.94, 14.17, 10.91, 5.77, 7.75, 4.65, 2.23, 11.14

(b) Axial loads (kN): 144.7, 340.1, 436.2, 468.1, 456.1

(c) Axial loads (kN): 148.0, 146.8, 145.5, 346.0, 344.5, 343.1, 441.7, 440.1, 438.7, 437.2, 472.4, 470.3, 469.5, 460.6, 458.1, 457.6

Where the exact positions of the purlin loads relative to panel points are not known, clause 4.10 allows the local bending moment to be taken as equal to $WL/6$. Fig. 12.10(a) shows the bending moment diagram for the half-chord as a result of the analysis for gravity loading, together with a diagram indicating the variation of primary axial load (based on panel load analysis) in each member (panel) length of the top chord (Fig. 12.10b). On the other hand, when a computer analysis is undertaken for the case where **vertical** loads are applied **at the purlin** positions, then there will be minor variations along a member length, coincident with purlin positions – see Fig. 12.10c. This is due to the small component of each vertical load parallel to the roof slope. Clearly, with wind loading, which acts normal to the slope, there is no variation of axial force within a member length, when applied at purlin positions.

12.6.2 Member sizes

One of the decisions which will affect the design of some members is the actual construction of the girder. It has already been noted (Section 12.3) that an economic form for lattice girders is to weld the diagonal members (angles) to the chord members (T-sections). This eliminates gusset plates, apart possibly for site bolted joints, which are needed to assist transportation of what might have been a long girder. Assuming that the girder will be delivered in two sections, then site connections near joints 11, 12 and 13 (see Fig. 12.7) will be necessary.

Having analysed the girder and determined the member forces, the next step in the design process is to select suitable member sizes for the various

DESIGN OF SINGLE-STOREY BUILDING – LATTICE GIRDER AND COLUMN CONSTRUCTION

members. The properties of the section sizes chosen are obtained from the relevant tables in the SCI guide[10] and the steel to be used is grade 43 steel, with $p_y = 275$ N/mm². Basically, the design of the lattice girder reduces to the individual design of member elements, and follows the principles outlined in Part 1.

12.6.2.1 TOP CHORD MEMBER

With reference to Table 12.3 and Fig. 12.6, Table 12.4 summarizes the forces acting on the various parts of the top chord (based on a manual analysis), owing to the design conditions (A) and (B). Later, the slight variations in axial forces arising from a computer analysis will be discussed in the light of the design objectives. Only the worst load condition from design case (B) has been tabulated, i.e. for members 1 and 10, the force -52.2 kN will govern the design, not -31.1 kN (see Table 12.3), and similarly for the moments.

Table 12.4 Forces in top chord for both design conditions

Members	Axial load (kN) (A)	(B)	Moment (kNm) (A)	(B)
1,10	+144.7	−52.2	−14.17	+6.04
2,9	+340.1	−122.3	−14.17	+6.14
3,8	+436.2	−156.8	−10.69	+4.63
4,7	+468.1	−168.3	−13.94	+6.04
5,6	+456.1	−166.3	−13.94	+6.04

As the top chord is to be fabricated in one length, i.e. continuous, then the most critical portion of that length is member 4 or 7, which has an axial compression of 468.1 kN and a bending moment of 13.94 kNm for the design case (A). The member is primarily a **strut** and the design will be based on that fact, i.e. design for load case (A) and check for case (B). The design procedure is essentially a trial and error method, i.e. a section is selected and then its adequacy checked against various design criteria. If the section size is inadequate or too large, then a different size is chosen.

Try **191 × 229 × 41 Tee** (cut from 457 × 191 × 82 UB).

(a) Classification This check has to be done from first principles as T-sections are not classified in the SCI guide[10]. The worst design conditions for the top chord occur at a panel point, i.e. the stem of the T-section is in compression for the design case (A). This means that the special limitations, noted in BS table 7, for stems of T-sections, apply.

$$\varepsilon = \sqrt{275/275} = 1.0$$

BS table 7
$$\frac{b}{t} = \frac{191.3}{2} \times \frac{1}{16.0} = 6.0 \leq 8.0\varepsilon \quad plastic$$

BS table 7
$$\frac{d}{t} = \frac{230.1}{9.9} = 23.2 > 19\varepsilon \quad slender$$

i.e. the T-section is **slender** and is governed by clause 3.6. Referring to clause 3.6.4 in particular, then the design strength (p_y) has to be modified by the strength reduction factor given in BS table 8, for stems of T-sections:

$$Factor = \frac{14}{\frac{d}{t\varepsilon} - 5} = \frac{14}{23.2 - 5} = 0.77$$

Hence, reduced design strength for the section is $0.77 \times 275 = 212$ N/mm². This reduced value of p_y (p'_y) has to be used for the T-section whenever the **stem** is in compression. However, the code deems that such a reduction is not necessary when designing connections associated with the stem.

(b) Local capacity check The design criterion to be satisfied is:

clause 4.8.3.2a
$$\frac{F_c}{A_y} + \frac{M_x}{M_{cx}} \leq 1.0$$

$$A_g = 52.3 \text{ cm}^2$$

clause 4.2.5
$$M_{cx} = p'_y Z = 0.212 \times 142 = 30.10 \text{ kNm}$$

$$\frac{468.1 \times 10}{52.3 \times 212} + \frac{13.94}{30.10} = 0.422 + 0.463 = 0.885 < 1.0$$

Section OK

(c) Member buckling check

clause 4.8.3.3.1
$$\frac{F_c}{A_g p_c} + \frac{mM_x}{M_b} \leq 1.0$$

This is the simplified approach; one really has no choice as a more exact solution cannot be considered owing to the reduced plastic moment of a T-section not being readily available.

In order to derive p_c and M_b, the slenderness of the chord member about the major and minor axes has to be evaluated. In assessing the effective lengths in these directions, i.e. in-plane (x–x axis) and out-of-plane (y–y axis), it should be understood that the top chord will attempt to buckle in plane between the panel connections and out of plane between purlin positions, as shown in Fig. 12.11. This point is reinforced by clause 4.10. This clause further states that for the purpose of calculating the effective lengths of members, the fixity of connections and the rigidity of adjacent members may be taken into account.

For in-plane buckling it can be seen from Fig. 12.1a that the web members effectively hold each panel length of the top chord in position at the connections and supply substantial end restraint to these lengths. Therefore the effective length in the x–x direction can conservatively be assumed to be $0.85 \times$ panel length (on slope) (BS table 24). Out-of-plane, the buckling mode (Fig. 12.11b) is such that the member behaves as though it is pin-ended between purlin positions which hold the chord effectively at those points, i.e. effective length is $1.0 \times$ distance between purlins (on slope).

DESIGN OF SINGLE-STOREY BUILDING – LATTICE GIRDER AND COLUMN CONSTRUCTION 183

(a) In-plane buckling of chord

(b) Out-of-plane buckling of chord

Fig. 12.11 Buckling modes of top chord member

$$\frac{L_{Ex}}{r_x} = \frac{0.85 \times 3718}{68.9} = 46$$

$$\frac{L_{Ey}}{r_y} = \frac{1.0 \times 1570}{42.3} = 37$$

These values of slenderness comply with clause 4.7.3.2a, which states that for members resisting loads other than wind loads, their slenderness should not exceed 180.

According to clause 4.7.6, the compressive strength p_c depends on the larger of the two slenderness values, i.e. 46, the reduced design strength of 212 N/mm² and the relevant strut table. In the strut selection table (BS table 25) it states that for T-sections, table 27(c) has to be used, irrespective of the axis about which buckling occurs. Referring to table 27(c) reveals that the lowest design strength tabulated is 225 N/mm². One can either extrapolate down to a value of 212 N/mm² in order to obtain p_c or alternatively p_c may be calculated from the formulae given in appendix C:

by extrapolation $p_c = 180$ N/mm²
by appendix C $p_c = 179.8$ N/mm²

Note that the value of E used in the formulae has the units N/mm², not kN/mm² as defined in section 3.1.2 of the code. Next p_b has to be evaluated, in order to define M_b.

clause 4.3.7.5 Now $\lambda_{LT} = nuv\lambda$.

In evaluating n, it must be recognized that the member is loaded within its unrestrained length between adjacent restraints (purlins) as indicated in Fig. 12.12. Also, the load does not have a destabilizing effect on the member. BS table 13 states that under these conditions the value of n can be derived from BS table 15 or 16. It can be shown by reference to Fig. 12.12 that the load does not lie within the middle fifth of the member and therefore table 15 does not apply. The problem with table 16 is that the diagram associated with it indicates a moment distribution, representative of a uniformly distributed load, whereas the case being considered is 'peaky'. Guidance might be obtained from BS table 20, though there is no direct reference to it in clause

Fig. 12.12 Local bending moment distribution near node 10

4.3.7.5. The design case lies approximately half-way between that for a load at mid-span and that at the quarter point. Therefore, taking a mean of 0.86 and 0.94 (see table 20) as the value of n, then:

BS table 13

$n = 0.90$
$m = 1.00$

From the SCI guide (10) the following values have been obtained:

$u = 0.584$
$x = 15.5$
$\lambda/x = 37/15.5 = 2.39$

Note that λ is defined in clause 4.3.7.5 as the **minor axis** slenderness, i.e. 37. With reference to BS table 14, it can be seen that, because the stem of the T-section is in compression for the critical part of the length of member being considered, then $N = 0.0$; hence the interpolation:

$v = 2.76$

Note that in this region of table 14 it is recommended that the value of v should be evaluated from appendix B2.5. In this example, the difference between the two values is not significant.

$\lambda_{LT} = 0.90 \times 0.584 \times 2.76 \times 37.18 = 54$

DESIGN OF SINGLE-STOREY BUILDING – LATTICE GIRDER AND COLUMN CONSTRUCTION

From BS table 11, by extrapolation down to $p_y = 212\,\text{N/mm}^2$ (or by using BS appendix B):

$$p_b = 185\,\text{N/mm}^2$$
$$M_b = p_b S_x$$
$$= 0.185 \times 251 = 46.44\,\text{kNm}$$

$$\frac{468.1 \times 10}{52.3 \times 180} + \frac{1.0 \times 13.94}{46.44} = 0.497 + 0.300 = 0.797 < 1.0$$

Section is OK for lateral torsional buckling.

Now, if the more accurate value of axial load (472.4 kN), obtained from a computer analysis (see Fig. 12.10c), is used then the axial load ratio would marginally increase from 0.497 to 0.502; that is, the small differences in axial forces between different analyses are not significant and can be ignored.

(d) Reverse load condition As a result of the wind suction on the girder, the design case (B) causes the top chord to act as a tie, i.e. 168.3 kN tension and a moment of −6.04 kNm which produces compression in the table of T-section.

clause 4.8.3.2a
$$\frac{F_t}{A_e p_y} + \frac{M_x}{M_{cx}}$$

As the connections are welded at the panel point at which worst design forces occur, then $A_e = A_g$. Also, because the stem of the T-section is not in compression, then the design strength is $275\,\text{kN/m}^2$, hence:

$$M_{cx} = p_y Z = 0.275 \times 142 = 39.05\,\text{kN m}$$
$$\frac{168.3 \times 10}{52.3 \times 275} + \frac{6.04}{39.05} = 0.117 + 0.155 = 0.272 < 1.0$$

This means that the member is more than adequate to cope with the tension arising from reversal of load conditions.

Use **191×229×41 Tee**

Clearly, when the primary mode of a member is as a strut, then those loading conditions that produced the compression govern the design.

12.6.2.2 BOTTOM CHORD MEMBER

Coming to the design of the bottom chord member, it can be seen that its primary function is a **tie**, with reversal of stress producing compression. One finds that in spite of the primary tension being significantly larger than the compression, the latter will probably govern. Even more so with a member like the bottom chord, where seemingly there are no secondary members to hold the chord at intervals along its length (37 m), which could result in an extremely large slenderness. However, clause 4.7.3.2c states that any member normally acting as a tie, but subject to reversal of stress resulting from the action of wind, is allowed to have a slenderness up to 350. Nevertheless, in order to comply with this requirement, the bottom chord would need a substantial member size and therefore would prove uneconomic.

This apparent problem can be overcome by the use of longitudinal ties, which run the full length of the building to a braced point and hold the bottom chord at selected positions. In this exercise, assume longitudinal ties occur at connections 9 and 15, i.e. the length is divided roughly into thirds (see Fig. 12.7). This means that the unrestrained lengths in the y–y direction are 12.95 m for the outer lengths (nodes 1–9 and 15–23) and 11.10 m for the middle length (nodes 9–15). In the x–x direction, the member will buckle in-plane between panel points (3.70 m), similar to the top chord. The longitudinal ties are designed in Section 12.8.3.4.

The bottom chord sustains only axial load under either design case (A) or (B). From Table 12.3, the maximum design forces in the outer lengths are 461.9 kN (tension) and 160.9 kN (compression) and for the middle length 469.3 kN and 161.2 kN, respectively. Though the outer lengths have slightly smaller loads than that of the middle length, their length is longer, giving a larger slenderness and hence a lower compressive strength. Therefore, design for compression in the outer lengths and if satisfactory, check the design size against the conditions that prevail for the middle length.

Outer length of bottom chord In choosing the section for the bottom chord at this stage, it should be borne in mind that the end lattice girders form part of the braced bays (Section 12.8.3.2) and therefore attract additional loading.

Try **254×127×37 Tee** (cut from 254 × 254 × 73 UB).

(a) Classification

$$\frac{b}{T} = \frac{254.0}{2} \times \frac{1}{14.2} = 9.0 \leq 9.5\varepsilon \quad compact$$

$$\frac{d}{t} = \frac{127.0}{8.6} = 14.8 \leq 19.5\varepsilon \quad semi\text{-}compact$$

The T-section is **semi-compact** and therefore the design strength $p_y = 275$ N/mm².

(b) Check compression resistance Assume the connections to the columns at nodes 1 and 23 give some restraint in the y–y direction to one of the ends of each outer length, i.e. make $L_{Ey} = 0.95L$, hence:

$$\frac{L_{Ey}}{r_y} = \frac{0.95 \times 12950}{64.6} = 190 < 350$$

$$\frac{L_{Ex}}{r_x} = \frac{0.85 \times 3700}{30.0} = 105$$

As the section selected is a T-section, use BS table 27(c), from which:

$$p_c = 46 \text{ N/mm}^2$$
$$A_g = 46.4 \text{ cm}^2$$
$$P_c = A_g p_c$$
$$= 46.4 \times 46/10 = 213.4 \text{ kN} > F_c \ (=160.9 \text{ kN})$$

(c) Check tension capacity Now check the selected member section for the primary function as a tie carrying 461.9 kN. As there are no site splices in this

length, then:
$$A_e = A_g$$
$$P_t = A_g P_y$$
$$= 46.4 \times 275/10 = 1276 \text{ kN} > F_t \text{ (461.9 kN)}$$

Section OK

Clearly, the smaller compression force dominates the design of the bottom chord, the primary function of which is as a tie.

Middle length of bottom chord Use the same section for the middle length of bottom chord as for the outer lengths, i.e. $254 \times 127 \times 37$ Tee. The design loads have been noted as 469.3 kN (tension) and 161.2 kN (compression).

(a) Check compression resistance From the previous check on the compression resistance, it is clear that only the slenderness about y–y axis needs to be examined, as the value of the x–x slenderness will be lower.

$$\frac{L_{Ey}}{r_y} = \frac{1.0 \times 11\,100}{64.4} = 172$$

BS table 27c
$$p_c = 54 \text{ N/mm}^2$$
$$P_c = 46.6 \times 54/10 = 250.6 \text{ kN} > F_c (161.2 \text{ kN})$$

(b) Check tension capacity As site splices occur in this portion of the bottom chord, it is anticipated that the T-section is connected through both its table (flange) and stem (see Fig. 12.26b). BS clause 4.6.3.3 indicates that the derivation of the **net** area of a T-section connected in this manner is governed by clause 3.3.2, i.e.:

clause 3.3.2
$$A_{net} = 46.4 - (2 \times 24 \times 8.6 + 2 \times 24 \times 14.2)10^{-2} = 35.5 \text{ cm}^2$$

However, clause 3.3.3 states that the **effective** area A_e of each element at a connection where fastener holes occur, may be taken as K_e times its **net** area, but no more than its gross area. K_e for grade 43 steel is 1.2, therefore:

$$A_e = 1.2 A_{net}$$
$$= 1.2 \times 35.5 < A_g \text{ (46.4 cm}^2\text{)}$$
$$= 42.6 \text{ cm}^2$$
$$P_t = 42.6 \times 275/10 = 1172 \text{ kN} > F_t \text{ (469.3 kN)}$$

Section OK

Use **254×127×37 Tee**

If the T-section had been bolted only through its flange, then clause 4.6.3 would apply. As that particular clause defines the effective area of a T-section, then the modification allowed by clause 3.3.3 cannot be used. An alternative approach for the bottom chord is to use a lighter section, but this would require additional longitudinal ties, which would produce a less pleasing appearance and probably cost more overall.

12.6.2.3 WEB MEMBERS

The diagonal members are to be fabricated from angle sections. From practical considerations the **minimum size** of angle used should be **50×50×6**. These diagonals will be welded direct to the stem of the T-sections forming the top and bottom chords (see Fig. 12.21). Though the nominal lengths of members range from 2.89 m to 4.14 m, there is a greater variation in the axial forces (Table 12.5) and therefore the diagonals will be designed according to the magnitude of the member load. Table 12.5 summarizes the various axial forces in the diagonal members. The members carrying the heaviest compressive force will be designed first, followed by the other members in descending order of loading, until the minimum practical size is reached.

DIAGONAL MEMBERS 24, 41

Table 12.5 Axial forces in the diagonal members (kN)

Members	Length (m)	Load case (A)	Load case (B)	Members	Length (m)	Load case (A)	Load case (B)
23,42	2.62	−203.6	+72.3	24,41	2.89	+187.5	−66.5
25,40	2.89	−116.3	+39.2	26,39	3.18	+109.8	−37.0
27,38	3.18	− 54.8	+19.0	28,37	3.49	+ 52.6	−17.9
29,36	3.49	− 7.3	−11.0	30,35	3.81	+ 7.1	+ 8.1
32,33	4.14	− 31.1	+18.5	31,34	3.81	+ 31.8	−17.9

These particular members have to sustain an axial compression of 187.5 kN or a tension of 66.5 kN. Because the members are discontinuous, then an equal angle is preferred as it is structurally more efficient (weight for weight) than an unequal angle and therefore more economic. Using the SCI guide[10], refer to p.101, which gives a table (for members welded at ends) listing the compression resistance of equal angle struts for different nominal lengths. Given the nominal length is 2.89 m, the table indicates that a 100 × 100 × 12 angle can support 192 kN over a nominal length of 3.0 m. This section size is selected, because it provides sufficient compression resistance for least weight, i.e. compare areas of other sections. Although a check on the adequacy of the 100 × 100 × 12 angle is not necessary, one is now undertaken to illustrate the validity of the values given in the SCI guide.

(a) Check compression resistance Since the strut is connected directly to another member by welding, then for single angle members (clause 4.7.10.2) the slenderness λ should not be less than:

$$0.85L/r_{vv} \quad \text{or} \quad 0.7L/r_{aa} + 30$$

where r_{vv} and r_{aa} are defined in BS table 28 for discontinuous angle struts (see also Fig. 6.5).

$$\lambda = \frac{0.85 \times 2890}{19.4} \quad \text{or} \quad \frac{0.70 \times 2890}{30.2} + 30 = 127 \quad \text{or} \quad 97$$

As with the T-section, BS table 27c must be used for rolled angle sections, irrespective of the buckling axis. Therefore, knowing that $\lambda = 127$ and $p_y = 275\,\text{N/mm}^2$, p_c is evaluated as $89\,\text{N/mm}^2$, hence:

$$P_c = 22.7 \times 89/10 = 202.0\,\text{kN} > F_c\ (187.5\,\text{kN})$$

However, the strut tables for equal angles, given in the SCI guide[10], can be used directly and by interpolation of the tabulated values a very good estimate of compression resistance can be obtained. Thus, referring to p.101[10]:

$$P_c = (est) = 262 - \frac{(2.89 - 2.50)(252 - 192)}{(3.00 - 2.50)} = 205.2\,\text{kN}$$

Such interpolation gives a slight overestimate of the value of P_c. Therefore, if the interpolated value of P_c is within 2–3% of F_c, one might have to resort to first principles, using BS 5950.

(b) Check tension capacity Now check the tension capacity of the selected member. With reference to clause 4.6.3.1, the effective area of a single angle is given by:

$$A_e = a_1 + \frac{3a_1 a_2}{3a_1 - a_2}$$

where a_1 is the net area of connected leg
 a_2 is the gross area of unconnected leg

As one leg of the angle section is connected by welds to the chords, then there are no holes to be deducted in the connected leg, i.e.:

$$a_1 = \left(b - \frac{t}{2}\right)t = (100 - 6)12 = 11.28\,\text{cm}^2$$
$$a_2 = a_1 = 11.28\,\text{cm}^2$$

which means for this special case for equal angles then:

$$A_e = 1.75 a_1 = 1.75 \times 11.28 = 19.74\,\text{cm}^2$$
$$P_t = 19.74 \times 275/10 = 543\,\text{kN} > F_t\ (66.5\,\text{kN})$$

Section OK

This is the identical value for P_t for a $100 \times 100 \times 12$ angle noted in the SCI guide; see table for tension capacity of equal angle ties connected with one leg given on p. 111[10].

Use $100 \times 100 \times 12$ angle

The **rest of the diagonals** have been proportioned direct from the guide[10] and the relevant design details and member sizes are summarized in Table 12.6. Note that the **end vertical members** are in fact the upper portions of the column members, as can be seen from Fig. 12.13.

Table 12.6 Design details of diagonal members

Members	Length (m)	Design loads Compression (kN)	Tension (kN)	Max. λ	Angle size	Allowable Compression (kN)	Tension (kN)	Actual λ
24,41	2.89	187.5	66.5	180	110×100×12	205.2	543	127
26,39	3.18	109.8	37.0	180	100×100×8	118.3	370	138
23,42	2.62	72.3	203.6	350	90× 90×6	85.0	251	125
28,37	3.49	52.6	17.9	180	90× 90×6	55.3	251	167
25,40	2.89	39.2	116.3	350	90× 90×6[2]	73.6	251	138
31,34	3.81	31.8	17.9	180	90× 90×6	53.0	251	182[1]
32,33	4.14	18.5	31.1	350	90× 90×6[3]	41.8	251	198
27,38	3.18	19.0	54.8	350	90× 90×6[3]	64.0	251	152
30,35	3.81	7.1[4]	—	180	90× 90×6	53.0	251	182[1]
29,36	3.49	—	11.0	—	90× 90×6	—	251	—

[1] The slenderness of 182 just exceeds the limitation of 180. It can be argued, owing to practical considerations, that the nominal length (distance between intersections) would be reduced by at least 10 mm at each end of the member, which would result in $\lambda < 180$.

[2] A 70 × 70 × 8 angle section would suffice for this member, but a 90 × 90 × 6 section has the same area. The latter has been selected in an attempt to standardize on section sizes wherever possible as this leads to economy. The benefit is that the larger angle has improved properties.

[3] Though an 80 × 80 × 6 could have been used, a 90 × 90 × 6 has been preferred in order to standardize.

[4] The 7.1 kN load as noted is for the design case (A) and is less than the 8.1 kN load for case (B); see Table 12.3. However, the slenderness limitation of 180 for case (A) as opposed to 350 for (B) will control the design of the member.

12.6.3 Additional design checks

When a lattice girder forms part of a braced bay (see Section 12.8), then it has to sustain additional bracing forces. After the relevant forces have been determined (Section 12.8.2), the top and bottom chords have to be checked to see whether or not the chosen member sizes require modification (see Sections 12.8.3.1 and 12.8.3.2). Also, dependent on the gable framing chosen (see comments in Section 12.8.1.1), one or two lines of sheeting rails may have to be supported by some of the diagonal members in the end girders. The transference of loads from the rail(s) would induce additional forces into these diagonals, in which case the appropriate members need to be checked (see Section 12.10).

12.6.4 Self weight of girder

Having designed the various members of the lattice girder, the designer

DESIGN OF SINGLE-STOREY BUILDING – LATTICE GIRDER AND COLUMN CONSTRUCTION   **191**

is now in a position to check the estimated total self weight of the girder, i.e. 34.0 kN, used in the determination of the dead load forces. All that is needed is a rapid assessment, i.e:

chords:	$(41+37) \times 37.0$	$= 2886$
diagonals:	$2 \times 9.6 \times 2.6$	$= 50$
	$2 \times (17.8+8.3) \times 2.9$	$= 151$
	$2 \times (12.8+8.3) \times 3.2$	$= 135$
	$2 \times (8.3+8.3) \times 3.5$	$= 116$
	$2 \times (8.3+8.3) \times 3.8$	$= 126$
	$2 \times 8.3 \times 4.1$	$= 68$
	total	$= 3532 \text{ kg} = 34.6 \text{ kN}$

There has been a small underestimate of the self weight, i.e. 0.6 kN, which represents an error of 0.25% in a total dead load of 227.6 kN. A quick scan of the girder design indicates that member sizes would not be affected. Though there is no need to revise the calculations, 0.3 kN must be added to both vertical reactions.

12.7 DESIGN OF COLUMN MEMBERS

The columns, in supporting the lattice girders in space, transfer the load from the ends of the lattice girder down to the ground and into the foundation. In addition to the roof loads and the weight of the wall cladding, the column has to be designed to resist the side wind loading (see Fig. 12.5). Assuming that the same sheeting rail (R145130) is used for both side and end walls and with the probability of the gable post spacing being in the 6.0–6.5 m range (see Section 12.8.1 for exact positioning), the maximum rail spacing is limited to about 2.0 m[8]. As a sheeting rail is required at the bottom chord level to provide restraint to the girder, then Fig. 12.13 shows a possible arrangement of the rails along the side elevations, while Fig. 12.20 indicates the rail spacing for the gable walls.

Fig. 12.13 Arrangement of sheeting rails

12.7.1 Forces in column members

The axial load acting on either column is the cumulative total of:

- the end reactions from lattice girder, depending upon which design case is being considered (see next paragraph)
- weight of vertical cladding: $0.13 \times 6.0 \times 6.8 = 5.3$ kN (including insulation + liner panel)
- weight of side rails: $5 \times 0.045 \times 6.0 = 1.4$ kN
- self weight of column: $50 \times 6.8 \times 9.81/1000 = 3.3$ kN
- weight of gutter[11,12]: $0.15 \times 6.0 = 0.9$ kN

The end reactions from the lattice girder and the wind induced moment in the column are interrelated, being dependent on the design case, i.e.

(A) $1.4\,w_d + 1.6\,w_i$
(B) $1.0\,w_d + 1.4\,w_w$
(C) $1.2\,w_d + 1.2\,w_i + 1.2\,w_w$

and the wind pressure distribution (see Fig. 12.5). In determining the side wind load acting on an individual column, the combined wind load acting on both columns is shared in proportion to their stiffnesses (I/L), i.e. in this case equally. The end reactions of the lattice girder can be evaluated readily by assuming the girder is a simply supported beam (see Fig. 12.14 for different roof loading cases). Note that the loads shown for the wind case represent the vertical components of the applied wind loads.

Fig. 12.14 Unfactored vertical loads acting on roof (kN)

(a) 61.7 kN — Dead load
(b) 166.5 kN — Imposed load
(c) −75.3 kN | 1.15 | 0.60 | −39.3 kN — Wind on side wall
(d) −42.6 kN | 0.65 | 0.10 | −6.6 kN — Wind on side wall
(e) −131.0 kN | 1.00 — Wind on end wall

12.7.1.1 DESIGN CASE (A)

The vertical load is based on the total load figures noted in Table 12.2, with 0.6 kN added to the dead load value to compensate for the underestimation of the girder's own weights (see Fig. 12.14a). Therefore, the cumulative axial load in the column member is:

$$F_c = 1.4\,(10.9 + 61.7 \times 0.5) + 1.6\,(166.5 \times 0.5) = 191.7\,\text{kN}$$

As there is no wind loading in the design case, there is no associated moment, i.e:

$$M_x = 0.0$$

12.7.1.2 DESIGN CASE (B)

For this design case, select the wind condition that maximizes the uplift force on the roof, while maximising the wind moment in the column member, i.e. $1.0\ w_d + 1.4\ w_w$. Loads due to wind on side walls are shown in Fig. 12.14(c) and the loads due to wind on end walls in Fig. 12.14(e).

(i) For wind blowing on side walls With reference to Fig. 12.14(a) and (c):

$$F_t = 1.0(10.9 + 61.7 \times 0.5) - 1.47(75.3 \times 0.75 + 39.3 \times 0.25)$$

$$= -51.1\text{ kN (i.e. uplift)}$$

The **combined** wind pressure acting on the main building elevations is:

$(0.5 + 0.45)q$ or $(1.00 - 0.05)q$ (Fig. 12.5), therefore

$$F = 1.4(0.95 \times 0.59 \times 6.0 \times 6.8) = 32.0\text{ kN}$$

giving a base shear of $32.0/2 = 16.0$ kN per column.

Though it is assumed that the base is 'pinned', the main frame will be subject to 'portal action'. Advantage can therefore be taken of clause 5.1.2.4b which allows a nominal base moment of 10% of maximum column moment. Hence the maximum wind moment, acting at bottom chord level (see Fig. 12.15) is:

$$M_x = 90\% \text{ (base shear} \times \text{moment arm)}$$
$$= 0.90(16.0 \times 4.8) = 69.1\text{ kNm}$$

Fig. 12.15 Bending moment in column member (leeward side)

(ii) Wind blowing on end walls (gables) With reference to Fig. 12.14(a) and (e):

$$F_t = 1.0(10.9 + 61.7 \times 0.5) - 1.4(131.0 \times 0.5) = -50.0\text{ kN}$$

Any lateral wind load acting on the end column is taken by vertical bracing in the side walls (see Section 12.8.1.3). However, wind on the gables causes a suction of $0.2q$ or $0.7q$ acting simultaneously on both side walls. This leads to a zero combined wind load at bottom chord level. Nevertheless, the columns are subject to bending due to the suction. Assume that the column acts as a simply supported member between the base and bottom chord level, with the maximum moment occurring approximately mid-span. The effect of the 10% base moment is to reduce the mid-height moment by 5%, i.e.

$$M_x = 0.95[1.4(0.7 \times 0.59 \times 6.0 \times 4.8^2/8)] = 9.5\text{ kNm}$$

12.7.1.3 DESIGN CASE (C)

For this design case, choose the wind condition that minimizes the uplift forces on the roof, while maximizing, where possible, the wind moment on the column, i.e. $1.2(w_d + w_i + w_w)$. See Fig. 12.14(d) and (e) for the appropriate roof loads.

(i) Wind blowing on side walls With reference to Fig. 12.14(a), (b) and (d):

maximum axial compression occurs in the leeward column, as the effect of the wind suction is smaller on the leeward side of the roof (see Fig. 12.14d):

$$F_c = 1.2[10.9+(61.7+166.5)0.5-(42.6 \times 0.25+6.6 \times 0.75)]$$
$$= 131.3 \text{ kN}$$
$$F = 1.2(0.05 \times 0.59 \times 6.0 \times 6.8) = 27.4 \text{ kN}$$
$$M_x = 0.90(27.4/2) \times 4.8 = 59.2 \text{ kNm}$$

(ii) Wind blowing on end walls With reference to Fig. 12.14(a), (b) and (e):

$$F_c = 1.2[10.9+(61.7+166.5-131.0)0.5] = 71.4 \text{ kN}$$
$$M_x = 0.95[1.2(0.7 \times 0.59 \times 6.0 \times 4.8^2/8)] = 8.2 \text{ kNm}$$

This case is generally not critical.

(iii) Wind blowing on end walls With reference to Fig. 12.14(a)(b) and (e), but considering the wind condition of $-0.2q$ that occurs on the leeward part of the roof, i.e. the load given in Fig. 12.14(e) is multiplied by 0.2, then:

$$F_c = 1.2[10.9+(61.7+166.5-0.2 \times 131.0)0.5] = 134.3 \text{ kN}$$

However, the corresponding side suction is $-0.2q$, hence:

$$M_x = -0.95[1.2(0.2 \times 0.59 \times 6.0 \times 4.8^2/8)] = 2.3 \text{ kNm}$$

12.7.2 Column member section

Table 12.7 summarizes the different design loadings for the column member and the design checks for **25 × 146 × 37 UB**, which is a **plastic** section. The design checks are based on the following:

LOCAL CAPACITY CHECK

$$\frac{F_c}{P_z} + \frac{M_x}{M_{cx}} \leq 1.0 \quad \text{or} \quad \frac{F_t}{P_t} + \frac{M_x}{M_{ex}} \leq 1.0$$

OVERALL BUCKLING CHECK

(a) For the combination of **axial compression** and **moment**, the following condition must be satisfied:

$$\frac{F_c}{P_c} + \frac{mM_x}{M_b} \leq 1.0$$

(b) For the combination of **axial tension** and **moment** clause 4.8.1 implies that the member buckling check is based solely on the following condition:

$$\frac{mM_x}{M_b} \leq 1.0$$

The noted values of P_t (=P_z as there are no deductions for holes) and M_{cx} are listed in the left-hand column under the appropriate section size, on p.171 of the SCI guide[10]. Also, from the tables on p.85 and p.134, the respective values of P_c and M_b are extrapolated, knowing the effective length about the y–y axis, i.e.

$$L_{Ex} = 1.5 \times 4800 = 7.2 \text{ m}$$
$$L_{Ey} = 1.0 \times 4800 = 4.8 \text{ m}$$

It is assumed that the side rails do not restrain the inside flange of the column, except at bottom chord level (see Section 12.8.3.5). (If the designer needs to account for the restraint offered by the rails to the tension side of the column, then conditions outlined in BS appendix G govern). The value of 0.85 indicated in BS appendix D1, for calculating the effective length about the y–y axis, is not used as the base is pinned, i.e. conditions at the column base do not comply with D1.1c. However, this condition is covered by D1.2d.

Thus, L_{Ey} is 4.8 m and by interpolation of the appropriate values listed in the table on p.134 of the SCI guide[10], M_b is evaluated to be 63 kN m. The value of m is defined by BS table 18, based on the ratio of the end moments (β) being -0.10, i.e. 0.54.

The values in Table 12.7 depicted in bold indicate those allowable values used in the member buckling check. Clearly, the wind on the side walls governs the design. The **254 × 146 × 37 UB** section appears to be more than adequate and is adopted. (Note that a 254 × 146 × 31 UB fails to meet the buckling check for the worst design condition, i.e. 1.085). However, when the wind blows on the end walls, additional forces are produced and induced into the bracing system (see next section) which have to be transmitted down to the foundations. As the columns in the end bays form an integral part of the bracing system, the selected section has to be checked for the additional forces when they are calculated.

Table 12.7 Details of column member design

	Axial load (kN)	Wind moment (kNm)	Allowable P_z (kN)	P_c (kN)	M_{cx} (kNm)	M_b (kNm)	Local capacity check	Member bucling check
Design case (A)	191.7	0.0	1310	**408**	—	—	0.146	0.470
Design case (B):								
wind on side	−51.1	69.1	**1310**	—	133	63	0.559	0.592[1,2]
wind on end	−50.0	9.5	**1310**	—	133	63	0.116	0.081[1,2]
Design case (C):								
wind on side	131.3	59.2	1310	**408**	133	63	0.545	0.829[2]
wind on end	71.4	8.2	1310	**408**	133	63	0.116	0.245[2]
wind on end	134.3	2.3	1310	**408**	133	63	0.120	0.349[2]

[1] For this design condition of axial tension plus moment, clause 4.8.1 implies that the member buckling check is based solely on the moment component, i.e. mM_c/M_b. The combined effect of tension plus moment is covered by the local capacity check.

[2] Note that in the member buckling check the moments are multiplied by m (=0.54).

12.8 OVERALL STABILITY OF BUILDING

The designer must always ensure the structural stability of the building. At this stage, the main frames have been designed to cater for in-plane stability, particularly with respect to side wind loading. However, in order to give stability to the building in the longitudinal direction, all frames need to be connected back to a braced bay. Generally, the **end bay(s)** of the building are braced, so that the wind loads acting on the gables can be transferred to the foundations as soon as possible and thereby the rest of the structure is not affected. Another function of the braced bay is that it ensures the squareness and verticality of the structural framework, both during and after erection.

The bracing system for a single-storey building usually takes the form of rafter bracing in the plane of the roof (Fig. 12.16a). Occasionally another wind girder becomes necessary at bottom chord level, when the roof steelwork is deep, as with this design (Fig. 12.16b). Vertical bracing (located in the side walls) conveys the rafter bracing/wind girder reactions from eaves/wind girder level down to the foundations (Fig. 12.16c). It is assumed that each end of the building is braced.

Fig. 12.16 Details of bracing systems

This means that the top chords of the appropriate lattice girders form part of the rafter bracing, and the bottom chords are part of the wind girder. Hence, the only members that have to be designed are the diagonals. However, the top and bottom chords of the lattice girders must be checked against any additional effect from the bracing forces. Also, the main column members associated with the vertical bracing system must be checked for the effect of bracing forces.

In this example, the client has indicated that the building might be extended (design brief – Section 12.2), hence normal main frames are to be used at the gables. Consequently, there is no need for vertical bracing in the plane of the gable walls (cf. 'gable framing' of portal frame building, Section 13.10). The inherent in-plane stiffness of the main frame supplies the necessary stability to the walls, apart from any stressed skin action that might exist.

12.8.1 Arrangement of gable posts and bracing systems

12.8.1.1 GABLE POSTS

The loading on, and the structural arrangement of, the different bracing systems are dependent on the positioning of the gable posts and the relationship between the posts and the lattice girders. The gable posts, in addition to supporting the side rails and cladding, resist the wind loads acting on the gables. There are two basic choices when dealing with gable posts:

- either the posts run past the lattice girder and are connected sideways to the top and bottom chord members of the girder; or the posts run up and connect to the underside of the bottom chord member. This means that those lines of sheeting rails which occur within the depth of the girder have to be supported (connected) by some of the diagonals of the girder, thereby inducing flexural action in those members owing to both the eccentricity of the vertical loading, relative to the centroidal axes of the diagonals, and the horizontal wind loading. However, it has to be remembered that the end lattice girders carry only half the load, compared with any intermediate girder, and therefore it might be possible to accommodate these additional moments without changing any of the member sizes of the end girders.

Both schemes allow the gable cladding, sheeting rails and posts to be removed in the future and relocated, with minimum disturbance to the 'end' main frames. For this design example, it has been decided that the shorter gable posts are to be used, i.e. connected to underside of girder. Next, the spacing of the gable posts has to be decided. If there had been any dominant openings in the gable(s), then such openings could influence the positioning of the posts. As it has been assumed that there are no openings in the gables, arrange the posts to be positioned at approximately equal distances, i.e. the posts placed at 5.25 m, except for the end posts which are 6.0 m from the main frame columns (see Figs. 12.17(c) and 12.20). Thus, the span of the side rails on the gable is 6.25 m and with maximum rail centres of 1.85 m, the rail section is adequate [8].

12.8.1.2 RAFTER BRACING

Though it is not essential, it is good practice to make the joints of the bracing system coincident with purlin lines. By doing so, severe minor axis bending (arising from restraint forces in the purlins) is not introduced into the top chords of the lattice girders. (If the gable posts run past the outside of the end girders then it is desirable that posts coincide with the rafter bracing intersections.) Thus, within the constraint(s) just outlined, a rafter bracing system can be readily defined; see the configuration in Fig. 12.16(a).

12.8.1.3 WIND GIRDER

Coming to the wind girder bracing, located at bottom chord level, it is desirable that the girder connections coincide with the gable posts, thereby giving the posts positional restraint, and, at the same time, preventing bending action on the bottom chord. On the internal side of the girder, it is equally important that the longitudinal ties are connected into the bracing system, as clause 4.10d states that such ties must be 'properly connected to an adequate restraint system'. This can result in the wind girder configuration like that shown in Fig. 12.16(b).

12.8.1.4 VERTICAL BRACING

There are a number of ways of providing vertical bracing in the side elevations, as discussed in Chapter 10. Current practice for single-storey buildings is to use single member struts, instead of cross bracing. Figure 12.16(c) shows how the diagonal members have been arranged, so that wind pressure on a gable causes the members to go into tension. This means that under the smaller wind suction condition the members have to resist proportionally smaller compression forces; that is, bracing is arranged to minimize the compression that diagonal members have to sustain, leading to economic sizes.

If the client designates that openings are required in the end bay, then the vertical bracing arrangement may have to be modified if any bracing member intrudes into the space reserved for an opening. The vertical bracing can be located in another bay, remembering that an eaves strut would be required to transfer the gable forces to the vertical bracing system. In an extreme case, the designer may need to provide 'portal' framing in the lateral direction (see Chapter 10).

DESIGN OF SINGLE-STOREY BUILDING – LATTICE GIRDER AND COLUMN CONSTRUCTION

Fig. 12.17 Rafter bracing and wind girder – unfactored member loads (kN)

(a) Rafter bracing – wind on gable

(b) Rafter bracing – wind drag

(c) Wind girder and longitudinal ties – wind on gable

12.8.2 Forces in the bracing members

Having configured the rafter bracing, wind girder and vertical bracing, the next step is to determine the **unfactored** loads acting in the various 'bracing' members. This is easily done, by assessing the proportional share of the wind load acting on the gable taken by:

- the rafter bracing at roof level;
- the wind girder at bottom chord level; and
- the foundation at base level.

Taking the wind coefficient as being 1.0 (assuming no dominant openings subject to storm conditions) and referring to Fig. 12.16(a), then the unfactored load applied at wind girder level, for, say, the post adjacent to the central post:

$$F = (\text{height of post} \times \text{post spacing} \times \text{wind pressure})/2$$
$$= [4.8 + 1.85(2 - 6.25/18.5)]6.25 \times 0.59/2 = 14.52 \text{ kN}$$

and the unfactored load applied at rafter level for the same post:

$$F = (\text{girder depth} \times \text{post spacing} \times \text{wind pressure})/2$$
$$= 1.85(2 - 6.25/18.5)6.25 \times 0.59/2 = 5.67 \text{ kN}$$

from which the unfactored load applied at the foot of the column can be deduced, i.e.

$$F = 14.52 - 5.67 = 8.85 \text{ kN}$$

Thus, all wind loads acting on the rafter bracing and wind girder can be calculated in a similar manner and applied to the appropriate joints. The bracing systems are then analysed, assuming the diagonals are pin-jointed, to determine the unfactored member forces; see Figs. 12.17(a) (rafter bracing), 12.17(c) (wind girder) and 12.18(a) (vertical bracing). Member forces for different wind coefficients (other than unity) can be obtained by proportion.

In addition, both rafter bracing and vertical bracing have to resist **wind drag forces**, arising from wind friction across the surfaces of the building. As there is a braced bay at each end of the building, the unfactored drag force of 49 kN for the **roof** (see Section 12.4) is divided equally between the two braced bays. The rafter bracing is analysed for the wind drag force, assuming that the force is uniformly distributed across the rafter bracing system (see Fig. 12.17b). Next, analyse the vertical bracing for both the roof drag forces from the rafter bracing and the forces due to frictional drag on the side walls

Fig. 12.18 Vertical bracing – unfactored member loads (kN)

(a) Wind on gable

(b) Wind drag

(=15.2/4=3.8 kN per bracing; see Section 12.4.3). The unfactored member forces (due to wind drag) for the vertical bracing are given in Fig. 12.18(b).

12.8.3 Design of bracing members and ties

Structural hollow sections usually prove to be the economic solution for bracing members, owing to their structural efficiency in both the x–x and y–y directions. An alternative profile is an angle section, which is much less efficient but easier to connect. For members resisting very small loads in tension only (no stress reversal), then solid round bars could be used.

The slenderness for members resisting self weight and wind loads only has to be less than 250 (clause 4.7.3.2b). This is particularly relevant when designing the diagonal members in the bracing systems.

12.8.3.1 RAFTER BRACING

First, the diagonal members are considered separately from the main booms, which are in fact the top chords of the end two lattice girders and therefore subject to multiple loading. Where possible, standardize on a common size for groups of members, so that an economic solution is produced. Discussion with a fabricator may establish preferred sizes.

The maximum forces in the diagonals occur when the wind loading on the gables is a maximum, i.e. when a wind coefficient for a gable is +1.0, together with the effect of wind drag. Therefore, design the diagonals based on the member with the largest combined compression force. From the unfactored loads indicated in Fig. 12.17(a) and (b), the design load is calculated as:

$$F_c = 1.4(14.66 + 12.54) = 38.1 \text{ kN}$$

which is sustained by a member with a 'discontinuous' length of 6.66 m.

Looking at the table for circular hollow sections (CHS) on p.89, SCI guide[10], it can be seen that a **88.9×4.0 CHS** satisfies both design criteria — strength and stiffness. However, the loads in the diagonals decrease towards the centre of the bracing. Therefore, apart from the four outermost diagonals, the section size can be reduced to **88.9×3.2 CHS**.

Check top chord of lattice girder Next, check to see whether or not the 'bracing' forces induced in top chord members of the penultimate lattice girder modify the member size. The penultimate girder is chosen, as it is more heavily loaded than the end girder, which carries only half the roof loading. In examining the bracing loads that can be induced into this particular top chord, it should be borne in mind that the design of the chord was based on the maximum compression occurring in that member; see Table

12.4 and Section 12.6.2.1. This compression arose from the gravity load only, design case (A). Now, only suction on the gable ($-0.8q$) appears to induce further compression into the top chord of the penultimate girder. However, this suction exists only as a result of the wind uplift on the roof, which in this example represents a maximum uplift condition. The resulting tension in the chord (from the uplift) more than nullifies the compression from the bracing system; that is, in this particular design example, the bracing forces do not create a more severe design situation than those already considered.

The other condition which could be examined is the cumulative tension in the top chord which arises from design case (B) (Table 12.4), plus tension bracing forces. The latter occur in the top chord of the penultimate girder when there is wind pressure on the gable. However, the wind loads involved are lower than the 'compression' condition just examined. Therefore, there is no need for further calculations.

12.8.3.2 WIND GIRDER

Again, the diagonals are treated separately from the boom members of the wind girder, which are the bottom chords of the lattice girders. From Fig. 12.17(c), it can be seen that the maximum compression load for a diagonal is 49.74 kN (unfactored), hence:

$$F_c = 1.4 \times 49.74 = 69.9 \text{ kN}$$

for a member having a discontinuous length of 8.49 m. Again, with reference to p.89[10], the selection is a **114.3×6.3 CHS** for the two outermost diagonals. A **114.3×5.0 CHS** is suitable for the other diagonals, even for the reversed load conditions of $-0.8q$. For example, the member carrying 33.75 kN (unfactored) tension for wind pressure on a gable (see Fig. 12.17), has to sustain 27.0 kN (unfactored) compression for wind suction.

Check bottom chord of lattice girder This time, it is the maximum uplift conditions for the roof which produce compression in the bottom chord of the lattice girder. Consequently, the effect of the additional compression in the bottom chord of the penultimate lattice girder (resulting from suction on the gable) must be checked. The forces noted on Fig. 12.17(c) need to be multiplied by 1.4(−0.8) to obtain the required bracing forces in the bottom chord; see column (4) in Table 12.8.

On checking the design calculations for the bottom chord in Section 12.6.2.2, it can be seen that the compression resistance of both the outer and middle lengths is just adequate, i.e. 200.3/213.4 = 0.939 and 237.1/250.6 = 0.946, respectively. Therefore, adopt the **254×127×37 Tee** for **all** bottom chords of lattice girders.

Table 12.8 Bracing forces in bottom chord

Members	Axial forces (kN)				
	Roof loading		Design load	Gable wind	Design load
	1.0_d (1)	$1.4w_w$ (2)	(1)+(2) (3)	$1.4w_w$ (4)	(3)+(4) (5)
11,21	0.0	+ 5.1	+ 5.1	+39.4	+ 44.5
12,20	−45.8	+144.4	+ 98.6	+39.4	+138.0
13,19	−69.8	+210.2	+140.4	+39.4	+179.8
14,18	−80.2	+241.1	+160.9	+39.4	+200.3
15,17	−81.5	+242.7	+161.2	+75.9	+237.1
16	−76.4	+229.4	+153.0	+75.9	+228.9

12.8.3.3 VERTICAL BRACING

Figure 12.18(a) shows the unfactored forces in the members due solely to the gable wind with a coefficient of +1.0, while Fig. 12.18(b) indicates the unfactored member forces arising from wind drag forces on both the roof and side walls, acting in the same direction as the wind. Figure 12.19 gives the factored forces (with $\gamma_f = 1.4$) in the members for those wind conditions which might affect the design of the members carrying bracing forces. The load factor of 1.4 is used as the loads in the non-vertical members are entirely due to wind. Also, these members must have a limiting slenderness of 250. The following information summarizes the design of these non-vertical members, based on the maximum loads that can arise from wind on a side, with insignificant drag forces (Fig. 12.19(a)) or the combined effect of wind on a gable plus drag (Figs. 12.19(b), (c) and (d)).

Horizontals   6.0 m long; 47.7 kN compression; 26.8 kN tension
Use 88.9×4.0 CHS[10]

Upper diagonal   6.3 m long; 25.4 kN compression; 40.9 kN tension
Use 88.9×3.2 CHS[10]

Lower diagonal   7.7 m long; 80.9 kN compression 127.5 kN tension
Use 114.3×5.0 CHS[10]

Check column member Now the penultimate column member has to be checked for the maximum axial compression load from the bracing forces, $45.14 + 15.68 = 60.8$ kN (unfactored). This load arises from a wind condition which causes a pressure of $+1.0q$ on the gable and $-0.2q$ uplift on the roof. In fact, this is the design case (C) (iii) noted in Section 12.7.1.3 and Table 12.7. Therefore, total axial compression is:

$$F_c = 1.2[10.9 + 61.7 + 166.5 - 0.2 \times 131.0)0.5 + 60.8] = 207.2 \text{ kN}$$
$$M_x = 2.3 \text{ kN (Section 12.7.1.3(iii))}$$

Fig. 12.19 Vertical bracing – factored member loads (kN)

(a) Wind suction due to pressure on side
(b) Wind suction on gable and drag
(c) Wind pressure on gable + drag
(d) Wind pressure on gable + drag

giving

$$\frac{207.2}{408} + \frac{0.54 \times 2.3}{63} = 0.508 + 0.020 = 0.528 < 1.0$$

Though wind suction on the gable and the associated wind drag can cause additional tension for design case (B), noted in Section 12.7.1.3, the effect is marginal in terms of the local check. As the buckling check does not apply for the 'tension' case, the adopted section remains satisfactory.

12.8.3.4 LONGITUDINAL TIES

As stated, the purpose of the longitudinal ties is to restrain the bottom chord member of the lattice girders (see Fig. 12.16b). Assume that the 'ties' need to provide a restraining force equal to at least 2% of the maximum load in the chord, i.e. $0.02 \times 469.3 = 9.4$ kN, which may cause compression in the 'ties'. Over a discontinuous length of 6 m and requiring a slenderness not exceeding 250, a **76.1×3.2 CHS** is suitable[10].

12.8.3.5 EAVES TIES

These members give positional restraint to the column heads and also provide restraint to the ends of the top chords of the lattice girder. Use the same size as that for the longitudinal ties, i.e. **76.1 × 3.2 CHS**. Restraint at bottom chord level is to be provided by diagonal braces from the nearby sheeting rail to the inner flange of the column.

12.9 DESIGN OF GABLE POSTS

The design of the gable posts is straightforward. They are self-supporting and are only connected to the main frame at top and bottom chord level (see Fig. 12.16(a) or (b)). Having established the spacing of the posts, i.e. 6.25 m, the posts are assumed (for simplicity) to be simply supported between the column base and the connection at wind girder level, i.e. 4.8 m span. Apart from the gravity load of the sheeting, plus insulation, plus liner, plus sheeting rails and the column's own weight (which produces a nominal axial load in the post), the main loading is a bending action, due to wind blowing on the gable.

Design the central post; see Fig. 12.20. The axial load includes that for cladding plus self weight of the side rails and post, i.e.

$$F_c = 1.4[(\text{cladding} + \text{insulation}) + (\text{post} + \text{side rails})]$$
$$= 1.4[0.13 \times 5.72 \times 6.25 + 0.5(\text{est.}) \times 4.8] = 9.9 \text{ kN}$$

Wind load on a 4.8 m length of post is:

$$F = 1.4 \times 1.0 \times 0.59 \times 4.8 \times 6.25 = 24.8 \text{ kN}$$
$$M_x = 24.8 \times 4.8/8 = 14.9 \text{ kNm}$$

Select a **203 × 133 × 25 UB**. From the SCI guide[10], $P_z = 779$ kN and $M_b = 29$ kN m, from which a buckling check can be undertaken, i.e.

$$\frac{9.9}{779} + \frac{14.9}{29} = 0.013 + 0.514 = 0.527 < 1.0$$

This neglects (for simplicity) any lateral restraint from sheeting rails.

Use 203 × 133 × 25 UB.

Fig. 12.20 Gable framing

12.10 DESIGN OF CONNECTIONS

Having designed the members that make up the lattice girder there remains the design of the connections to be completed. Basically, the function of a connection is to transfer loads efficiently from one member to another without undue distress, while minimizing the cost of fabrication and ensuring ease of erection.

When designing connections, it is always useful to draw connection details to scale. For bolted joints, such details can demonstrate readily whether or not the bolts can be inserted easily at the appropriate locations. In the case of welded joints, it is essential to check that the welder is able to deposit the weld metal without impedance.

Though the girder was analysed assuming that the diagonals were pin-ended, connecting the diagonals to the chords results in a degree of fixity. Independent of whether or not the connection is designed bolted (with at least two bolts) or welded, clause 4.10 implies that secondary stresses could be ignored, provided that the slenderness of the chord members in the plane of the girder is greater than 50 and that of the web members is greater than 100. In this example, the slenderness of the chord is just below 50 (46), while the web members have slendernesses well above 100. Therefore, it could be deemed that such stresses are insignificant.

By arranging alternate diagonal members to be connected on the opposite side of the steam of the T-section, the number of ends cut on a skew is reduced and hence cost of fabrication. The other benefit is that each half girder can be manufactured in an identical manner (Fig. 12.21) and when spliced together, the alternate pattern for the diagonals continues across from one half to the other half; that is, all diagonals sloping one way are connected to one side of the T-section, while the remaining diagonals sloping in the opposite direction are attached to the other side of the chord. This arrangement allows any sheeting rail(s) in the gable, located within the depth of the end girders, to be supported by the outstand legs of every other diagonal member. However, the disadvantage is that the additional loading

Fig. 12.21 General details of lattice girder

imposed on these members might cause their sizes to be modified (see Section 12.10.5). The design of typical connections for the lattice girder is now examined in detail.

12.10.1 Typical diagonal to chord connections

The typical internal connection chosen to be designed is node 3, where member 23 (90 × 90 × 6 Angle) and member 24 (100 × 100 × 12 Angle) intersect with the bottom chord (254 × 127 × 37 T-section); see Fig. 12.22(b). When designing welds, it is immaterial whether the axial load is tension or compression (apart from fatigue or brittle fracture considerations, which do not apply to this example). The welds have to be designed to transfer the largest possible load occurring in the member under any loading condition, i.e. member 23 carries a force of 204 kN and member 24 a force of 188 kN (see Table 12.5). Almost invariably with this form of construction (using angle sections), the centre of the weld group is eccentric to the centroidal axis of angle section, i.e. the weld group is not balanced. Therefore, in addition to

Fig. 12.22 Details of typical internal connections

Fig. 12.23

the axial load, the weld group has to be designed to resist the in-plane torsional moment generated by this eccentricity, as will be illustrated by the ensuing calculations.

For example, take the connection between member 23 and the bottom chord: the maximum possible lengths along the toe and heel of the angle are 135 mm and 55 mm, respectively (see Fig. 12.22b). Also, the weld across the end of the angle is 90 mm. As the stem of the T-section is relatively short it is probably advisable to weld the angle to the edge of the stem, i.e. on the reverse side to the other welds producing the weld group shown in Fig. 12.23.

First, determine the distance of the longitudinal axis of the weld group from the heel by taking moments of the weld lengths about the heel. The inclined weld length (120 mm) is treated in the same manner as the other welds.

$$\bar{x} = (90^2/2 + 135 \times 90 + 120 \times 45)/(55 + 135 + 120 + 90) = 54.0 \text{ mm}$$

The lateral axis of the weld group, relative to the end, is determined in a similar manner, i.e.

$$\bar{y} = (55^2/2 + 135^2/2 + 120 \times 95)/(55 + 135 + 120 + 90) = 55.1 \text{ mm}$$

From the SCI guide[10] the centroidal axis is 24.1 mm from the heel of the angle. Hence, the eccentricity of the load is $54.0 - 24.1 = 29.9$ mm, resulting in a moment of $204 \times 29.9/1000 = 6.10$ kNm. Thus, the weld group for member 23 has to be designed for an axial load of 204 kN and an in-plane moment of 6.10 kNm. Therefore the equivalent polar second moment of area (inertia) of the weld group needs to be evaluated.

The general practice is to ignore the local inertia about their own axes for welds parallel to the global axis being considered – a conservative assumption. Note that the inertia about its local axis for the component of an inclined weld (of length d), perpendicular to the global axis, is significant and must be taken into account, i.e.

$$I'_{cg} = \frac{dh^2}{12}$$

Fig. 12.24

where h is the projected length of the inclined weld, normal to the global axis about which the equivalent inertia is being determined (see Fig. 12.24). Therefore, the total inertia for an inclined weld about a global axis, say the x–x axis, is:

$$I'_x = \frac{dh^2}{12} + dy^2$$

where y is the distance from the centroid of the inclined weld to the global axis, normal to the axis being considered (see Fig. 12.24). The total inertia for an inclined weld about the other global axis can be obtained in a similar manner, as demonstrated in the following calculations for the weld group shown in Fig. 12.23.

$$I'_p = I'_x + I'_y$$

$$\begin{aligned}I'_x &= 90 \times 55.1^2 + 2 \times 55.1^3/3 + (55.0 - 55.1)^3/3 + (135 - 55.1)^3/3 \\ &\quad + 120 \times 80^2/12 + 120(95 - 55.1)^2 \quad = 810 \times 10^3 \text{ mm}^4\end{aligned}$$

$$\begin{aligned}I'_y &= 54.0^3/3 + (90 - 54.0)^3/3 + 55 \times 54.0^2 + 135(90 - 54.0)^2 \\ &\quad + 120 \times 90^2 + 120(54.0 - 90/2)^2 \quad = 494 \times 10^3 \text{ mm}^4\end{aligned}$$

$$I'_p = (810 + 494) \times 10^3 \qquad = 1304 \times 10^3 \text{ mm}^4$$

The shear per unit length due to axial load (F_S) is:

$$F_S = 204/(55 + 135 + 120 + 90) = 0.510 \text{ kN/mm}$$

and the shear per unit length due to the eccentric moment (F_T) is obtained by considering that part of the weld furthest away from the centre of weld group, i.e. point A in Fig. 12.23:

$$F_T = MR_{max}/I'_p$$

where $\quad R_{max} = \sqrt{[(90-54.0)^2 + (135-55.1)^2]} = 87.6 \text{ mm}$

and $\quad \cos\theta = (90-54.0)/87.6 = 0.411$

$$F_T = 6.10 \times 87.6 / 1304 = 0.410 \text{ kN/mm}$$

Now

$$\begin{aligned}F_R &= \sqrt{F_S^2 + F_T^2 + 2F_S F_T \cos\theta} \\ &= \sqrt{[0.510^2 + 0.410^2 + 2 \times 0.510 \times 0.410 \times 0.411]} \\ &= 0.816 \text{ kN/mm} < 0.903 \text{ kN/mm} \quad \text{(p. 205 reference 10)}\end{aligned}$$

Use 6 mm FW.

Now examine the connection at node 4, where member 24 ($100 \times 100 \times 12$ Angle) and member 25 ($90 \times 90 \times 6$ Angle) intersect with the top chord ($191 \times 229 \times 41$ Tee); see Fig. 12.22(a). Designing the connection between member 24 and the chord, it is noted that the stem of the top chord is deeper than that for the bottom chord, with the result that the longitudinal welds can have lengths up to 240 mm (heel) and 130 mm (toe), while the end weld is 100 mm long (Fig. 12.25). In this case, the weld group will be designed without the inclined weld length as it might not be necessary.

The determination of the weld size is identical to previous calculations, except that there is no inclined weld. For brevity, only the results are given:

$$\bar{x} = 38.6 \text{ mm}$$
$$\bar{y} = 71.4 \text{ mm}$$
$$I'_x = 1885 \times 10^3 \text{ mm}^4$$
$$I'_y = 877 \times 10^3 \text{ mm}^4$$
$$I'_p = 2762 \times 10^3 \text{ mm}^4$$

Fig. 12.25

Relative to the centre of the weld group, the eccentricity of the axial load is $38.6 - 29.0 = 9.6$ mm. Thus the weld group for member 24 has to be designed for an axial load of 188 kN (Table 12.5) and an in-plane moment of $188 \times 9.6/1000 = 1.80$ kNm.

The shear per unit length due to axial load is:

$$F_S = 188/(120 + 100 + 220) = 0.427 \text{ kN/mm}$$

$$R_{max} = 153.5 \text{ mm}$$

$$\cos \theta = 0.251$$

Therefore, the shear due to the torsional moment is:

$$F_T = 1.80 \times 153.5/2762 = 0.100 \text{ kN/mm}$$
$$F_R = \sqrt{[0.427^2 + 0.100^2 + 2 \times 0.427 \times 0.100 \times 0.251]}$$
$$= 0.462 \text{ kN/mm} < 0.602 \text{ kN/mm (p.205 reference 10)}$$

In fact, a 6 mm fillet weld (FW) is the minimum size for welds.

Use 6mm FW.

Other internal connections can be designed in a similar way.

12.10.2 Lattice girder to column connections

The design of the welded connection between member 1 and member 23 at node 2 (see Fig. 12.26a) follows the same method already outlined in

Fig. 12.26 Lattice girder to column connections

Section 12.10.1. There remains the transfer of the vertical shear 176 kN from the end-plate connection into the column flange. In fact, this is a site connection, which is generally bolted (more cost-effective) and, anticipating other site connection details, 22 mm diameter bolts (grade 4.6) are used. As the single shear strength of this bolt size is 48.5 kN[10], then the number of bolts required is 176/48.5 = 3.6, i.e. four bolts (see Fig. 12.26a). Bearing strength requirements indicate a minimum plate thickness of 6 mm[10]. This is more than satisfied by using an end-plate thickness of 12 mm, which would restrict any excessive out-of-plane deformation of the plate. Also, the shear per unit length for the weld, joining the end-plate to the top chord member, is

$$176/(2 \times 190 + 1 \times 210) = 0.220 \text{ kN/mm}$$

Therefore use the minimum size, i.e. 6 mm FW.

A nominal site connection is required at node 1 between the bottom chord member and the main column flange, as there is little load being transferred between the two elements, apart from the axial load due to 'portal action' (Fig. 12.15), i.e.

$$69.1/1.85 = 37.5 \text{ kN (Section 12.7.1.2(i)).}$$

Again, use a 12 mm end-plate and 6 mm FW; see Fig. 12.26(b) for details.

12.10.3 Site splice connections

Owing to the overall length of the lattice girder, it was decided to arrange the girder to be delivered in two halves. The girder is assembled on the ground from two halves, prior to being lifted aloft during the erection process. In the design of the lattice girder (Section 12.6.2) it had been contemplated that site splices were to be positioned at or near nodes 11, 12 and 13. However, it could be argued that instead of making splices near both nodes 11 and 13, a single splice at mid-length of the bottom chord could prove more economic. The minor advantage of having the two splices in member 16 is that it allows a greater flexibility in adjustment should a camber be required (see Section 12.12).

The site splices illustrated in Figs. 12.27(a) (apex) and 12.27(b) (bottom chord) show one half of each splice as being site bolted. Depending on the relative costs and flexibility required during assembly of the lattice girders, the other half of each splice can be either shop welded or also site bolted. The benefit of the splices being totally bolted is that each half girder would then become identical, ideal from the fabrication point of view; that is, if a designer can build in repetition during the design stage, then fabrication costs will tend to be lowered as a direct result.

Figure 12.27(a) gives the details of the connection at the apex (node 12). In order to have square ends the centroidal axes of the diagonals 32 and 33 are offset from the apex intersection by 50 mm. As the load in these members is relatively small, the slight eccentricity should not have a significant effect. Also, though one bolt would suffice to transmit the load of 31 kN from the

Fig. 12.27 Details of internal splices

(a) Apex connection splice

(b) Bottom chord splice

diagonal to the chord, it is good practice to use a minimum of two bolts. Note that in the case of the lower bolt in the diagonal, its load is transferred via the gusset plates and back into the stem of the chord.

The size and shape of the gusset plates has been arranged so that it can be square cut from a standard plate section. It makes good sense to avoid too many cuts in a gusset plate, otherwise costs will increase. Too many cuts indicates a lack of appreciation of fabrication processes and costs. The splice has to transmit 456 kN between the two halves of the chord. Using grade 4.6 bolts, the number required is 456/48.5 = 9.4, say 10. This is accomplished by locating (in one half of the splice) 6 bolts in the table of the T-section and the other 4 bolts in the stem, roughly in proportion to the areas of the table and stem. Figure 12.27(a) shows a possible arrangement for the 10 bolts.

Coming to the bottom chord splices, the design of them is not too dissimilar from that of the apex splice, i.e. 10 bolts are also required on each side of the splice in order to transfer 439 kN (see Fig. 12.27(b) for details of splice).

12.10.4 Typical bracing connections

The details of a connection between the diagonals of the wind girder and the bottom chord of the penultimate girders (given in Figs. 12.28 and 12.29) illustrate the design basis for other bracing connections, including those for the rafter bracing at top chord level. By careful detailing, the tee-pieces welded to the ends of the diagonal members of the wind girder and the longitudinal ties can be standardized. These tee-pieces not only seal the ends of these circular hollow section, but also provide a simple means for bolting these members to a gusset plate (see Fig. 12.28). Again, a minimum of two bolts per joint is used.

These tee-pieces can be either made by welding two plate elements together or manufactured from a cutting from a suitable I-section. The plate element (welded directly to the tubular members) must be stiff enough, otherwise dishing can occur, causing high stress concentrations at the junction between the two plate elements comprising the tee-piece. Such high stress can lead to material rupture in the form of tearing, particularly if the elements are welded together. It is recommended that, for the size of the hollow section selected, at least a 12 mm thick plate should be used for the table of the tee-pieces.

Figure 12.28 shows the gusset plate (12 mm thick) bolted to the table of the bottom chord, for ease of removal if and when the building is extended. Though no holes were deducted in the design of the outer lengths of the bottom chord (Section 12.6.2.2), an examination of those particular calculations indicate the reduced area is more than adequate to resist the design loads in tension. Alternatively, this plate could be welded into position, though this might cause some problems should removal be required.

The connection between the diagonals of the wind girder and the bottom chord of the end girders is similar to the connection just outlined, except that there is an additional factor. A previous design decision was to run the gable posts up to the underside of the end girders, coinciding with the nodal points of the wind girder (see Fig. 12.17c). Each gable post is to be connected into the wind girder system via a plate welded underneath and to the gusset plate (see Fig. 12.29). However, the end lattice girder (even though carrying only half the load) deflects vertically under imposed and/or wind loading and could cause problems in the gable posts. Therefore the plate to which the web of the post is bolted has slotted holes which allow the girder to deflect the calculated movement due to imposed loading and/or wind uplift.

Finally, the connection detailed in Fig. 12.30 is typical of those connections in which two bracing systems acting in different planes intersect at a common point. In the example shown, the intersection of the wind girder and vertical bracing with the column member is considered. To achieve a correct transfer of forces, the various bracing members should intersect at a common point, otherwise moments are induced. However, practical considerations dictate that the horizontal diagonal be offset by 25 mm.

Fig. 12.28 Wind girder, longitudinal ties and bottom chord connection

Fig. 12.29 Wind girder, gable post and bottom chord connection

Fig. 12.30 Wind girder, vertical bracing and column connection

12.10.5 Check diagonals of the end girders

As the gable posts run up to the underside of the end girders, this means that the sheeting rails located within the depth of an end lattice girder have to be supported via the alternate diagonal members of the girder. Owing to the eccentricity of the rails relative to the vertical plane of the girder (Fig. 12.31), additional loading in the diagonals is induced. Therefore the sizes of these diagonals need to be checked for their adequacy to withstand this extra loading. It is noted that the end girder carries only half the load for which the intermediate girders were designed (Table 12.6).

Assuming that the rails are supported by every other diagonal, it can be seen from Fig. 12.32 that, if the vertical spacing of sheeting rails is continued at 1.55 m, then the span of these rails is about 3.7 m. (Doubling the span would result in making the section R145130 unsuitable.) Examination of the central diagonal (member 33), shows that the additional loading arises from two sheeting rails. This load is mainly due to the wind suction on the gable $(-0.8q)$ in the design case (B); that is, referring to Fig. 12.32:

Horizontal load at A due to wind

$$= 1.4 \times 0.8 \times 0.59 \times 1.55 \times 3.7 = 3.79 \text{ kN}$$

Vertical load at A due to cladding, insulation, liner and rails

$$= 1.4(0.13 \times 1.55 + 0.043)3.7 = 1.28 \text{ kN}$$

The loads at B can be derived by proportion, i.e:

$$= [1.55 + 0.50(\text{av.})]/(2 \times 1.55) = 0.66$$

hence the horizontal load at B

$$= 0.66 \times 3.79 = 2.50 \text{ kN}$$

and the vertical load at B

$$= 0.66 \times 1.28 = 0.84 \text{ kN}$$

The resulting moment in the diagonal due to the horizontal load is noted in Table 12.9, based on the dimensions indicated. The moment due to the vertical load is not significant (2%) and is ignored. The appropriate loads can be determined for all the other diagonal members supporting sheeting rails (see Table 12.9).

In checking the adequacy of member 33, it is assumed that the combined out-of-plane stiffness of the cladding and sheeting rails constrains the diagonal member (via the rail attachments) to bend about its x–x axis, rather than its y–y axis.

The axial load is $9.3 + 1.28 + 0.84 = 11.4 \text{ kN}$ and the maximum bending moment in the member is 4.54 kNm. The SCI guide[10] indicates that the member size ($90 \times 90 \times 6$ Angle) is **slender**, therefore the strength reduction factor (BS table 8) is the lesser of:

$$\frac{11}{\dfrac{d}{T\varepsilon} - 4} \quad \text{or} \quad \frac{19}{\dfrac{b+d}{T\varepsilon} - 4}$$

$$= 1.00 \quad \text{or} \quad 0.73$$

Fig. 12.31

Fig. 12.32

Table 12.9 Additional loading on diagonal members

Member	Size	Length (m)	Distance from bottom chord along diagonal member (A)	(B)	Max. moment (kNm)	Axial load (kN)
23	90 × 90 × 6	2.62	2.20	—	0.83	+36.2
25	90 × 90 × 6	2.89	2.02	—	1.65	+19.6
27	90 × 90 × 6	3.18	1.90	—	2.42	+ 9.5
29	90 × 90 × 6	3.49	1.83	—	3.26	− 5.5
31	90 × 90 × 6	3.84	1.77	3.54	3.84	+ 9.0
33	90 × 90 × 6	4.14	1.73	3.46	4.54	+ 9.3
35	90 × 90 × 6	3.84	1.77	3.54	3.84	+ 4.1
37	90 × 90 × 6	1.83	—	3.26	—	− 9.0
39	100 × 100 × 8	3.18	1.90	—	2.42	−18.5
41	100 × 100 × 12	2.89	2.02	—	1.65	−33.3

This gives a reduced design strength of $0.73 \times 275 = 201$ N/mm^2, hence the local capacity check for the 90 × 90 × 6 Angle section is:

$$\frac{F_c}{P_z} + \frac{M_x}{M_c} \leq 1.0$$

$$\frac{11.4 \times 10}{10.6 \times 201} + \frac{4.54}{0.201 \times 12.2} = 0.054 + 1.851 > 1.0$$

Therefore, the angle is not adequate – the section needs to be increased substantially or its properties enhanced by compounding it with another angle. The latter solution is the one adopted, as it has the advantage of providing support to the sheeting rails at the required distance from the vertical plane of the girder, i.e. 135 mm. If an additional angle is bolted to the diagonal (see Fig. 12.33), then it can easily be removed in the future, at which time the original end girder in fact becomes an intermediate frame.

A 125 × 75 × 8 Angle is selected, resulting in the compound section shown in Fig. 12.33. The properties of the compound section now have to be calculated, i.e:

$$\bar{x} = \frac{15.5 \times 4.14 + 10.6(13.5 - 2.41)}{15.5 + 10.6} = 6.96 \text{ cm}$$

$$I_x = [247 + 15.5(6.96 - 4.14)^2 + [80.3 + 10.6(13.5 - 2.41 - 6.96)^2]$$
$$= 631 \text{ cm}^4$$

$$Z_x = 631/6.54 = 96.5 \text{ cm}^3$$

$$r_x = \sqrt{\frac{I_x}{A}} = \sqrt{\frac{631}{26.1}} = 4.92 \text{ cm}$$

$$\bar{y} = \frac{15.5(1.68 + 0.6) + 10.6(2.41)}{26.1} = 2.33 \text{ cm}$$

$$I_y = [67.6 + 15.5(2.33 - 1.68 - 0.60)^2 + [80.3 + 10.6(2.41 - 2.33)^2]$$
$$= 148 \text{ cm}^4$$

$$r_y = \sqrt{\frac{I_y}{A}} = \sqrt{\frac{148}{26.1}} = 2.38 \text{ cm}$$

Fig. 12.33 Details of compound section for diagonal member

DESIGN OF SINGLE-STOREY BUILDING – LATTICE GIRDER AND COLUMN CONSTRUCTION 217

Local capacity check

$$\frac{11.4 \times 10}{26.1 \times 201} + \frac{4.54}{0.201 \times 69.5} = 0.022 + 0.234 < 1.0$$

Member buckling resistance

$$\lambda_x = L_{Ex} = \frac{0.85 \times 4140}{49.2} = 72$$

$$\lambda_y = L_{Ey} = \frac{1.0 \times 1730}{23.8} = 73$$

As p'_y is 201 N/mm², then by extrapolation from Table 27(c), BS 5950, $p_c = 139$ N/mm². Owing to the nature of the compound section and in order to keep the calculations simple, several safe assumptions are made, so that p_b can be quickly evaluated:

BS table 14

BS table 11

$$n = m = 1.0$$
$$u = 1.0$$
$$x = D/T = 135/6 = 23$$
$$\lambda/x = 73/23 = 3.2 \text{ and assuming that } N = 0.7$$
$$v = 0.81$$
$$\lambda_{LT} = 1.0 \times 1.0 \times 0.81 \times 73 = 59$$
$$p_b = 168 \text{ N/mm}^2$$

$$\frac{11.4 \times 10}{26.1 \times 139} + \frac{4.54}{0.168 \times 69.5} = 0.031 + 0.389 < 1.0$$

Though wind pressure on the gable $(1.0q)$ would generate a larger moment, it can be seen that the compound section has adequate reserves of strength and therefore no further check is necessary. It can be shown that the other diagonals when compounded with a $125 \times 75 \times 8$ Angle are more than adequate.

The only other check that has to be made is to ensure that the net sectional area of each alternate diagonal (which would result in the future with the removal of the $125 \times 75 \times 8$ Angles) can support the loads noted in Table 12.6. Such a check indicates that the members are satisfactory.

An alternative to this solution would be to increase the size of the diagonals in the end girders and then to use short lengths of angles to connect the sheeting rails to the diagonals. Also, the reader is reminded that the gable posts could have been run up past the end girders, but this has the disadvantage of constraining the deflections of the end lattice girders relative to the other girders.

12.10.6 Corner columns

Unlike the other main column members, there is lateral loading on the corner columns from the gable sheeting rails. However, the end frames carry only half the load compared with that carried by the intermediate frames. A check would show that the total loading regime acting on the corner columns represents a less severe condition than that for which the columns were originally designed.

12.10.7 Design of column bases

The column base has to be designed for the factored loads, arising from the three design cases:

(A) $F_h = 0.0$ kN; $F_v = 191.7$ kN; $M = 0.0$ kNm
(B) $F_h = 16.0$ kN; $F_v = -51.1$ kN; $M = 7.7$ kNm
(C) $F_h = 13.7$ kN; $F_v = 131.3$ kN; $M = 6.6$ kNm

The values of moment in cases (B) and (C) represent the nominal 10% base moment allowed by the code, see clause 5.1.2.4b, BS 5950.

12.10.7.1 DESIGN OF COLUMN BASE PLATE

The column member size is a 254 × 146 × 37 UB. If the base has to sustain a substantial moment, then the base plate size should have a minimum protection of 100 mm beyond the column section's overall dimensions of 256 mm × 146 mm, to allow bolts to be positioned outside the flanges. However, the base carries only a nominal moment and the usual detail in these circumstances is either to place two holding down bolts along the neutral axis of the column section, at right angles to the column web, or to position four bolts just inside the section profile (see Fig. 12.34). The latter detail is to be used as it affords a certain amount of moment resistance which would prove useful in case of fire, as well as helping the erectors in positioning the columns. Therefore, make the base plate wide enough for the plate to be welded to the column member, i.e. 275 mm × 160 mm.

Consider initially the loading from design case (A). Using a concrete mix for the foundation which has a cube strength of $f_{cu} = 30$ N/mm², the bearing pressure should not exceed $0.4 \times 30 = 12$ N/mm² (clause 4.13.1, BS 5950):

$$\text{Bearing pressure} = \frac{191.7 \times 10^3}{275 \times 160} = 4.4 \, \text{N/mm}^2$$

Fig. 12.34 Column base detail

The plate thickness, t, may be determined from the formula given in clause 4.13.2.2, i.e.

$$t = \sqrt{\left[\frac{2.5w}{p_{yp}}(a^2 - 0.3b^2)\right]}$$

As the projections of the base plate beyond the profile of the column section are minimal, the theoretical value of t would be small. In these circumstances, it is recommended that the base plate thickness > column flange thickness, say 15 mm thick plate (grade 43 steel). The welds connecting the column member to the base-plate need to transfer 191.7 kN. Assuming that the column is fillet welded all the way round its profile then the weld length would be about 1 m, hence the required design strength of weld is 192/1000 = 0.19 kN/mm, i.e. nominal size required – use 6 mm FW.

Use 160 mm × 15 mm plate × 275 mm long
6 mm FW

12.10.8 SIZING OF HOLDING DOWN BOLTS

Where axial load is transmitted by the base plate (without moment) then nominal holding down bolts are required for location purposes (Section 8.2) and reference (13)). In this example, the bolts have to resist an uplift of 51.1 kN, coupled with a moment of 7.7 kNm and a nominal horizontal shear of 16.0 kN. Assume **four** 24 mm diameter bolts, grade 4.6 steel; a smaller diameter is more prone to damage.

Use four 24 mm diameter bolts (grade 4.6 steel)

12.11 DESIGN OF FOUNDATION BLOCK

Generally, the design of the foundations for any structure is dependent on the ground conditions that exist on site, the maximum load conditions that can arise from any combination of loads and the strength of concrete used. Therefore it is important that the engineer has data regarding soil conditions or some reasonable basis for estimating the soil capacity. In this example, the soil bearing pressure has been stated as 150 kN/m^2 (Section 12.2). This is a **permissible** pressure and current practice for foundation design is based on serviceability conditions, i.e. working load level. Therefore, for the two load cases (A) and (C) (see Section 12.7.1), the partial load factors are made equal to unity, i.e. the loads for these cases become:

(A) $1.0\ w_d + 1.0\ w_i$ – maximum bearing pressure under vertical load

(C) $1.0\ w_d + 1.0\ w_i + 1.0\ w_w$ – maximum bearing pressure under combined vertical and horizontal loads

However, the third load case (B) must be examined, because it produces the maximum uplift condition. The partial load factor for the wind load must be the largest possible, so that a foundation block of sufficient weight can be selected to counter balance any uplift force, i.e.

(B) $1.0\ w_d + 1.4\ w_w$ – maximum uplift condition

Clause 2.4.2.4, BS 5950 states that the design of foundations should be in accordance with CP 2004 and be able to accommodate all forces imposed on them. Usually, foundation design is governed by gravity loading, but in this example, the uplift force in the main columns is significant. Clause 2.4.2.2, BS 5950 indicates that factored loads should not cause the structure (including the foundations) to overturn or lift off its seating; that is, the weight of the foundation block must be sufficient to counter balance any uplift force. Therefore, it might be prudent to proportion a **mass concrete** foundation block, based on the worst uplift condition, i.e. design case (B) (i) (wind blowing on the side walls) (see Section 12.7.1.2) and then check the resulting foundation block size against the other critical design cases. The loads for design case (B) are:

$$F_v = 1.0\ (10.9 + 61.7 \times 0.5) - 1.4(75.3 \times 0.75 + 39.3 \times 0.25)$$
$$= -51.1\text{ kN (uplift)}$$
$$F_h = 16.0\text{ kN}$$

Taking the specific weight of concrete as 23.7 kN/m^3, the minimum volume of mass concrete necessary to prevent uplift of the column member is $51.1/23.7 = 2.16\text{ m}^3$. A base of $1.7\text{ m} \times 1.7\text{ m} \times 0.9\text{ m}$ thick, which has a volume of 2.60 m^3 and weighs 61.6 kN, provides sufficient mass. Assuming a spread of the vertical load of $45°$ through the concrete from the edges of the base plate to the substrata, then the block provides $1.7 \times 1.7 = 2.89\text{ m}^2$ of bearing area. The $45°$ spread line should cut the vertical sides of the block, otherwise the depth of the block has to be increased. In this example the block appears to be adequate, i.e. $0.85 - 0.14 = 0.61\text{ m} < 0.90\text{ m}$ (see Fig. 12.35a).

Fig. 12.35 Foundation block detail

(a) Initial concrete base size

(b) Design case (A)

(c) Design case (B)

(d) Design case (C)

However, the bearing pressure has to be checked for the complete design case (B), i.e. uplift, together with the moment generated by the horizontal shear (16 kN) at the bottom of the block. If the resultant soil bearing force lies within the middle third of the block, i.e. not more than $L/6$ (=0.283 m) from the centre line of the base, then tension will not occur at the concrete/soil interface. In order to achieve this condition, the column member needs to be offset a horizontal distance y from the centre line of the base, i.e.

$$\frac{16.0 \times 0.9 - 5.1 \times y}{61.6 - 51.1} \leq 0.283 \text{ m}$$

$$y \geq 0.224 \text{ m say } 0.250 \text{ m}$$

This means that the 45° spread does not cut one of the vertical sides of block, i.e. $(0.85 + 0.25 - 0.14) = 0.96$ m > 0.9 m. Therefore, the depth of the block is increased to 1.0 m, which results in a block weight of 68.5 kN.

By taking moments about the centre line of the base, the point at which the resultant force acts at the concrete/soil interface can be determined, i.e.

$$(16.0 \times 1.0 - 51.1 \times 0.25)/(68.5 - 51.1) = 0.185 \text{ m}$$

from the base centre line (see Fig. 12.35c) and therefore the base is adequate for this loading case. Figures 12.35(b) and 12.35(d) show the loading for the other two cases (A) and (C).

Check that the proposed size is satisfactory with respect to these other design cases. First, examine the design case (A), the loads for which are:

$$F_v = 1.0(10.9 + 61.7 \times 0.5) + 1.0(166.5 \times 0.5) + 68.5$$
$$= 125.0 + 68.5 = 193.5 \text{ kN}$$
$$F_h = 0.0$$

As this represents the maximum vertical load condition, the minimum soil bearing area required is $186.6/150 = 1.24$ m². The block provides $1.7 \times 1.7 = 2.89$ m² of bearing area, which is more than adequate. The resultant force on the base acts at a point $125.0 \times 0.25/193.5 = 0.161$ m (see Fig. 12.35b).

Finally, the design case (C) (i) (wind blowing on the side walls), which produces the maximum horizontal shear condition, has to be checked:

$$F_v = 1.0[10.9 + (61.7 + 166.5)0.5] - 1.0(42.6 \times 0.25 + 6.6 \times 0.75) + 68.5$$
$$= 109.4 + 68.5 = 177.9 \text{ kN}$$
$$F_h = 1.0(0.95 \times 0.59 \times 6.0 \times 6.8)/2 = 11.5 \text{ kN per column}$$
$$M = 6.6/1.4 = 4.7 \text{ kN m}$$

For this loading case, the resultant force acts through a point $(109.4 \times 0.25 - 4.7 + 11.5 \times 1.0)/177.9 = 0.192$ m from the centre line (see Fig. 12.35d) and therefore the foundation block is satisfactory.

It should be noted that the assessment of the forces in the vertical bracing system (Section 12.8.3.3) indicates an additional factored uplift force in the penultimate columns (see Fig. 12.19a), i.e. -50.6 kN. The foundation blocks for these particular columns have to be made larger than that just calculated for the normal conditions, but again the uplift condition controls the design, i.e. $-(51.1 + 50.6) = -101.7$ kN. Use a 2.2 m × 2.0 m × 1.0 m base, which weighs 104.3 kN. However, in this case the column need only be offset 0.15 m from the base centre line (see Fig. 12.36). Also, this non-standard

Fig. 12.36 Foundation block – penultimate frame

block has to be checked for the normal conditions (excluding bracing force) in anticipation of the building being extended. Further calculations would show that all the design conditions are satisfied. The foundation pads for the gable posts can be designed in a similar manner.

12.12 OTHER CONSIDERATIONS

Both examples of single-storey construction reviewed in detail deal essentially with the basic design of the structural members for standard arrangements, i.e. lattice girder and column construction in this chapter and portal frame construction in Chapter 13. In practice, it can be said that generally there are no 'standard' buildings, as each client has different requirements. Though they will not be discussed in detail, the reader should be aware of other considerations which might affect the basic design.

For example, many single-storey buildings serve to house goods/materials or enclose large stores requiring the movement of large volumes of goods, being both delivered and dispatched. The design examples do not specifically deal with framing of doorways, large access openings, or the effect of loading bays (with or without canopies). It is essential that clients define their requirements as early as possible, preferably no later than the design stage, otherwise late alterations can radically affect the structural arrangements and therefore costs.

12.12.1 Loading

Common types of loads which can be imposed on to a single-storey building, and which could significantly influence the design, include ventilation extractors, crane loading and drifted snow. Extractor units are usually located on the roof, near the ridge. Generally, these extractor units require a special framing (trimmer) detail in order to transfer their own local load to the purlins and hence to the main frame. The main frame itself is rarely affected.

Occasionally monorails are required to be hung from the roof members, thereby imposing local concentrated loads on particular members. Light to medium crane girders can be supported from the main column by means of a bracket (see Section 8.6). However, for medium to heavy girders it may be more economic to support the girders directly; that is, another column is positioned immediately underneath the crane girder(s) and battened back to the main column leg, thereby eliminating a large moment from acting on the main column.

A trend in recent years has been the use of a façade or parapet at eaves level, in order to hide the sloping roof and gutter detail. The disadvantage is that it provides an effective barrier whereby drifting snow can cause a local accumulation of snow in that region. Multi-bay structures also create conditions that lead to snow drifting into the valleys between bays; see reference (5).

When designing the gutters, the engineer must ensure that the gutter size is adequate and that the outlets can cope with the expected volume of water. Badly designed outlets have been known to cause overflow problems; see references (11) and (12) for guidance.

12.12.2 Deflections

BS 5950 recommends limitations on deflections for most buildings, but specifically does not give guidance for portal frames. Though deflections for the normal pitched portal frame used in practice do not usually cause any problems, the designer should be wary of non-standard frames, especially if they are relatively tall. The reason for deflection limits is to prevent damage to the cladding or other secondary members and to avoid psychological unease of the client or employees. 'Excessive' deflection does not necessarily indicate that the building is unsafe in a structural sense, i.e. deflections not acceptable to the client might be acceptable to the structural engineer. The designer can always offset a predicted downward deflection due to the effect of dead + imposed loads, by deliberately introducing an upward camber at the fabrication stage. The deflection itself is, of course, unchanged.

Another problem that possibly could arise is **ponding**. Ponding is associated with shallow flexible roofs, where even 'dead load' deflections are sufficient to create a hollow on the roof surface, increasing in size as rainwater is collected. This can lead to leakage of the water into the building, which might prove costly. Also, if a building is subjected to constant buffeting by wind forces, then the cladding might suffer from low cycle failure resulting from cracking of the cladding along its crests or troughs. Asbestos or asbestos substitute corrugated sheets in existing buildings are more prone to this form of damage, as the sheets become brittle with age.

12.12.3 Fire

Normally fire protection is not required by Building Regulations for single-storey buildings, unless there is a potential fire risk arising from a particular use of the building or the building is located within the fire boundary of another person's property. If it becomes necessary to control the spread of fire to structural members, then the SCI guidance regarding fire boundary and fire protection should be read[14], being particularly relevant to portal frames. The function of any fire protection is to allow the occupants to escape within a specified period of time, e.g. one hour. There are various methods of producing the required fire rating for a building, which is dependent on the use of the building. Most of these methods are passive, i.e. they delay the effect of fire on structural members. However, one method which might have long term financial benefits is the water sprinkler system. As this method is an active system (attempts to control the fire), lower insurance premiums might be negotiated. The disadvantage is that the building has to be designed to accommodate the additional loading from the pipe network supplying the water to the individual sprinklers in the roof zone. Also, if the pipes have to be supported eccentric to the plane of the supporting members, then the torsional moments acting on the supporting member must be taken into account.

12.12.4 Corrosion

Generally, for most single-storey structures, corrosion is not a problem as the steelwork is contained within the cladding envelope. The minimum statutory temperature requirements are such that the ambient conditions inside a modern building tend to be dry and warm, which are not conducive to the propagation of corrosion and therefore any internal steelwork only needs a nominal paint specification, unless the function of the building is such as to generate a corrosive atmosphere[15].

12.12.5 General

The designer should always bear in mind the erection process and eliminate at the design stage any difficulties which could arise[16]. Also, he/she should be aware of the limitations on length, width and height of structural steelwork which has to be transported by sea, rail or road[17].

Finally, the reader is advised to consult copies of references [8] and [18], in which numerous construction details for different forms of single-storey buildings are graphically illustrated.

STUDY REFERENCES

Topic	Reference
1. Comparative costs	**Horridge J.F. & Morris L.J.** (1986) Comparative cost of single-storey steel framed structures, *Structural Engineer*, vol. 64A (no.7), pp.177–81
2. Loading	BS 6399 *Loading for Buildings* Part 1: *Dead and Imposed Loads* (1984) Part 2: *Wind Loads* (1995)
3. Wind loading	British Standards Institute CP3 Chapter V Part 2
4. Wind loading	**Newberry & Eaton K.** (1974) *Wind Loading on Buildings*. Building Research Establishment
5. Snow drifting	Building Research Establishment (1984) *Loads on Roofs from Snow Drifting against Vertical Obstructions and in Valleys*, Digest 332
6. Cladding	(1993) *Roof and wall cladding/decking*, Precision Metal Forming Ltd, Cheltenham
7. Cold formed	BS5950 *Structural Use of Steelwork in Buildings* Part 5: *Design of Cold-formed Sections* (1987)
8. Multibeam purlins	Ward Building Components (1986) *Multibeam Structural Products Handbook*. Sherburn, North Yorkshire

9. Truss analysis	**Coates R.C., Coutie M.G. & Kong F.K.** (1980) Analysis of plane trusses, *Structural Analysis*, pp.30–40. Van Nostrand Reinhold
10. Section properties	(1985) *Steelwork Design* vol. 1, Section properties, member properties. Steel Construction Institute
11. Roof drainage	Building Research Establishment (1976) *Roof Drainage – Part 1*, Digest 186
12. Roof drainage	Building Research Establishment (1976) *Roof Drainage – Part 2*, Digest 187
13. Holding down bolts	(1980) *Holding Down Systems for Steel Stanchions.* Steel Construction Institute
14. Fire boundary	(1980) *The Behaviour of Steel Portal Frames in Boundary Conditions.* Steel Construction Institute
15. Painting	**Haigh I.P.** (1982) *Painting Steelwork*, CIRIA report no. 93
16. Lack of fit	**Mann A.P. & Morris L.J.** (1981) *Lack of fit in Steelwork*, CIRIA report no. 89
17. Transportation	(1980) *Construction Guide.* British Steel Corporation
18. Design Details	(1984) *Design Manual – Single-Storey Steel Framed Buildings.* British Steel Corporation/Conder, Steel House, Redcar

13

DESIGN OF SINGLE-STOREY BUILDING – PORTAL FRAME CONSTRUCTION

13.1 INTRODUCTION

An alternative, economic solution to the design of a single-storey building is to use portal frame construction[1] (see Fig. 13.1). Therefore, the building designed in the previous chapter will be redesigned, replacing the lattice girder and column framework by a portal frame. However, the selection of cladding, purlins and sheeting rails, together with the design of bracing, will not be undertaken in detail as these member sizes will be similar to the corresponding members for the previous solution based on lattice girder construction. The only fundamental change to the structural arrangement is the direct substitution of the lattice girder and column by a portal frame. This chapter deals basically only with the design of the portal frame together with the gable framing, which is different from that used in Chapter 12. Therefore, detailed design of other structural members in the building can be obtained by referring to the appropriate section in the previous chapter.

Though the steel portal frame is one of the simplest structural arrangements for covering a given area, the designer probably has to satisfy at least as many different structural criteria as for more complex structures. In essence, the portal frame is a rigid plane frame with assumed full continuity at the intersections of the column and rafter (roof) members. In the following example, it will be assumed that the columns are pinned at the bases. This is the normal practice as the cost of the concrete foundation for fixed bases (owing to the effect of large fixing moments) more than offsets the savings in material costs that result from designing the frame with fixed feet[1]. Indeed, the pin-based portal frame represents the economic solution in virtually all practical design cases. However, fixed bases may become necessary with tall portal frames as a means of limiting deflections.

Fig. 13.1 Typical portal frame construction

13.2 DESIGN BRIEF

A client requires a similar single-storey building to that described in the previous chapter (Section 12.2), i.e. a clear floor area, 90 m × 36.4 m (see Fig. 12.2), with a clear height to underside of the roof steelwork of 4.8 m. However, in this case the client has decided that there will be no extension of the building in the future. The building is also to be insulated and clad with PMF profiled metal sheeting, Long Rib 1000R, and the slope of the rafter member is to be at least 6°. The site survey showed that the ground conditions can support a bearing pressure of 150 kN/mm² at 0.8 m below the existing ground level.

13.3 DESIGN INFORMATION

Though the design brief specified at least 6° it has been decided to make the slope at least 10° so that the sheeting can be laid without special strip mastic lap sealers, which are necessary for shallower slopes in order to prevent capillary action and rain leakages into the building. If it is assumed at this stage that the depth of the column is 0.6 m and the distance from the underside of haunch to the column/rafter intersection is 0.7 m, then with reference to Fig. 13.2:

Span of portal frame	$(L = 36.4 + 0.6)$	$= 37.0$ m
Centres of portal frames		$= 6.0$ m
Height to eaves intersection	$(h_1 = 4.8 + 0.7)$	$= 5.5$ m
Height to eaves of apex	$[h_2 = (37.0 \tan 10°)/2]$	$= 3.4$ m (say)
Actual slope of rafter	$[\theta = \tan^{-1}(3.4/18.5)]$	$= 10.41°$

The imposed load (snow) is 0.75 kN/m² on plan, which gives an equivalent load of $0.75 \times \cos\theta = 0.75 \times 0.984 = 0.74$ kN/m² on the slope along the rafter.

Fig. 13.2 Design information for proposed portal frame

The use of an equivalent load makes due allowance for purlin spacing which is usually calculated as a slope distance. As this building is to be located on the same site as the building designed in Chapter 12, then the design wind pressure is identical, i.e. $0.59\,kN/m^2$. The self weight of the PMF profile Long Rib 1000R, plus insulation, is taken as $0.097\,kN/m^2$.

13.4 DESIGN OF PURLINS AND SHEETING RAILS

The design of the purlins and sheeting rails is virtually identical to the corresponding members in the lattice girder and column construction, apart from the equivalent snow load along the rafter (roof member) being $0.74\,kN/mm^2$, giving a factored combined load (dead + snow) of $1.32\,kN/m^2$. Therefore, the same cold formed section, Ward Multibeam P175140 has been selected for the purlins and the R145130 for the sheeting rails (Table 12.1), for the reasons noted in Section 12.5. However, a slightly reduced purlin spacing of 1.5 m (on slope) from that used in the previous example is assumed in anticipation of an effective restraint being required at haunch/rafter intersection (see Fig. 13.12).

Again, the joints of the double spanning purlins/rails are assumed to be staggered across each frame, thereby ensuring that each intermediate portal frame receives approximately the same total loading via the purlins. The self weight of the selected purlin section is found to be 0.035 kN/m and hence the 'average' load (unfactored) being transferred by each purlin is:

Snow load	$0.74 \times 6.0 \times 1.5 = 6.66\,kN$
Sheeting and insulation	$0.097 \times 6.0 \times 1.5 = 0.87\,kN$
Self weight	$0.035 \times 6.0\ \ \ \ \ \ \ \ \ = 0.21\,kN$
Total load supported by purlin	$= 7.74\,kN$

Therefore, the 'average' end reaction per purlin is 3.86 kN.

13.5 SPACING OF SECONDARY MEMBERS

Spacing of purlins (on slope)   1.500 m
Spacing of purlins (on plan)   1.475 m
Spacing of sheeting rails – see Section 13.8.1.1

In this example it has been assumed that the client did not require a parapet round the perimeter of the building. Frequently nowadays, a parapet is provided to mask the visual effect of a long sloping roof, thereby making the building appear to be box-shaped, i.e. flat topped. This usually done by extending the side cladding above the gutter level, say 1.0 m. Vertical members attached to the top of the external portal legs support this cladding and would be designed to resist the wind load on the cladding. The disadvantage of the parapet is that it can cause snow drifting local to the eaves/haunch area of the roof. Snow drifting can also occur in the valleys of multi-bay portal frames or when one frame is higher than its adjacent frame. The additional snow load due to drifting can be evaluated by using the guidance outlined in reference (7). This snow load has to be transferred to the main frames in the eaves/haunch area via the roof sheeting and purlins. The usual solution is to decrease the purlin spacing, so that the same sheeting and purlin size is maintained over the entire roof. The effect of this additional snow load on the portal frame is minimal, i.e. it causes virtually no change in the free moment diagram.

13.6 DESIGN OF PORTAL FRAME

Since the mid 1950s, portal frame construction in the United Kingdom has been widely based on the principles of plastic design developed by Baker and his team at Cambridge[2]. By taking advantage of the ductility of steel, plastic design produces lighter and more slender structural proportions than similar rigid frames designed by elastic theory [3,4].

13.6.1 Gravity load condition

Most portal frame designs are governed by gravity (dead + snow) loading, and by applying plastic design principles, general expressions for the plastic moment capacity (M_p) of a uniform frame can be derived[4,5] for both pinned and fixed base conditions. In the design of the modern haunched portal frame with pinned bases, the initial step is to estimate the plastic moment capacity of a **uniform** frame (no haunching) with **pinned bases**.

With reference to Section 13.3 and Fig. 13.3, the appropriate expression for a pinned base frame[5] is:

$$M_p = \frac{W^*L}{8} \frac{1}{[(1+k/2)+\sqrt{(1+k)}]}$$

where $k = h_2/h_1 = 3.4/5.5 = 0.618$
 w_d = self weight (i.e. rafter) + cladding + purlins (on slope)
  $= [0.88(\text{est}) + 0.097 \times 6.0 + 0.035 \times 6.0/1.5] = 1.60\,\text{kN/m}$
 w_i = imposed load (i.e. snow) **(on slope)**
  $= 0.74 \times 6.0 = 4.44\,\text{kN/m}$
 $W^* = (1.4w_d + 1.6w_i)L/\cos 10.41°$ **(on plan)**
  $= (1.4 \times 1.60 + 1.6 \times 4.44)\,37.0/0.984 = 351\,\text{kN}$

$$M_p = \frac{351 \times 37.0}{8} \frac{1}{[(1+0.309)+\sqrt{(1+0.618)}]} = 629\,\text{kNm}$$

Fig. 13.3 Typical moment distribution for haunched portal frame

(a) Geometry of a typical frame

(b) Net bending moment diagram

It is important to note that the value of M_p has been determined for a rigid-jointed plane frame without requiring any prior knowledge of section properties, unlike elastic design. The design of the structural steelwork is to be based on grade 43 steel being used throughout the project. (The use of higher grades of steel would produce more slender frames, which could lead to difficulties with respect to stability and deflections). Therefore, at this stage of calculation assume that the design strength (p_y) is 275 N/mm^2; then the plastic modulus required is:

$$S_x \text{ required} = M_p/p_y$$
$$= 629/0.275 = 2287 \text{ cm}^3$$

This would indicate that a 533 × 210 × 92 UB section ($S_x = 2370$ cm^3) could be chosen, assuming that the frame has a constant section throughout its length. However, the section must be checked to see whether or not it can sustain **plastic action**. Referring to the SCI guide, p.130[6], the section is designated as a 'plastic' section, i.e. the section is capable of adequate rotation at a plastic hinge position without local flange buckling developing.

However, the most cost-effective arrangement for portal frames with spans greater than 15–20 m is to haunch the rafter member in the eaves region[1], thereby allowing a lighter (in weight) section to be used for the rafter than the column; that is, the rafter size, for the major part of the rafter, can be reduced below the section based on the frame having a constant cross-section, while the column section would probably need to be increased in size to compensate for the increased moment at the eaves. The resulting design becomes more structurally efficient. Also, as the rafter member is generally much longer than the combined length of the column members, the frame is more economic.

There are a number of different methods by which portal frames can be analysed using plastic theory[3,4]. One method commonly adopted by the construction industry for frames with pinned bases[3] is used for this example. The design procedure is to select a suitable member size for the rafter. Bearing in mind the section previously obtained for a uniform frame (533 × 210 × 92 UB), try the section 457 × 191 × 67 UB ($S_x = 1470 \text{ cm}^3$), noting that this is also a 'plastic' section[6]. (Note that when the only hinge in a member is the last hinge to form, then the member need only be compact, i.e. it requires the capability to form a hinge, but rotation capacity is not essential). At this stage, the effect of axial load on the moment capacity of a rafter member for a normal pitched roof portal frame can be neglected. (For other forms of frame, such as a tied portal[3], it could become significant.) The plastic moment capacity of the rafter member is $1470 \times 0.275 = 404.3 \text{ kNm}$.

It should be noted that the expression for M_p, given previously, was derived by assessing the true position of the 'apex' hinges (see Fig. 13.3), based on the assumption that the total factored vertical loading is uniformly distributed[5]. (For a frame with at least 16 purlin points, this assumption is, for all intents and purposes, accurate.) In practice, however, the actual position of the 'apex' plastic hinges in the rafter is controlled by the purlin spacing, since these hinges will develop at particular purlin supports, i.e. coincident with the application of the point loads on the rafter member. If the haunch is to remain elastic (by doing so, it could help with member stability in that region), assume that the hinge positions occur at the column/haunch intersection (in the column) and near the apex.

For the gravity loading condition (dead + snow), experience has found that the 'apex' hinge usually occurs at the first or second purlin support down from the apex purlin (cf. wind condition, Section 13.6.2).

Assuming that the hinge will form at the second purlin support from the apex (point X), then take moments about, and to the left of the left-hand X. As the plastic moment at X is known (404.3 kNm), then with reference to Fig. 13.3, the value of R (the horizontal thrust at the pinned base) can be calculated from the resulting equilibrium equation:

$$-M_{pr} = -\frac{w^*}{2}[(L/2)^2 - x^2] + [h_1 + (1 - 2x/L)h_2]R$$

$L = 37.0 \text{ m}$
$h_1 = 5.5 \text{ m}$
$h_2 = 3.4 \text{ m}$
$x = 2.95 \text{ m}$
$w^* = 351/37.0 = 9.50 \text{ kN/m}$

$$-404.3 = -\frac{9.50}{2}[18.5^2 - 2.95^2] + [5.5 + (1 - 5.9/37.0)3.4]R$$
$$= -1584.4 + 8.358R$$
$$R = (1584.4 - 404.3)/8.358 = 141.2 \text{ kN}$$

Making the assumption that a plastic hinge will occur at position B_s, i.e. level with the underside of the haunch, see Fig. 13.3, then the required plastic modulus for the column member can be readily determined by taking moments about and below B_s, i.e.

$$M_{ps} = (h_1 - a)R$$

where a is the distance from B to $B_s \approx 1.4 D_b$ (estimate) $= 0.64$ m
$M_{ps} = (5.5 - 0.64)141.2 = 686.2$ kNm
$S_{xs} = 686.2/0.275 = 2495$ cm^3

Though there seems to be a choice between a $533 \times 210 \times 101$ UB or a $610 \times 229 \times 101$ UB (both are 'plastic' sections [6]), the flange thickness of the former section (17.4 mm) is such that the design strength would have to be reduced to 265 N/mm^2 (see BS table 6). In any case, the second choice gives a better plastic modulus and increased stiffness for an identical weight, which could prove useful in controlling eaves deflections. Now check the adequacy of the frame by using a **457 × 191 × 67 UB** for the rafter and a **610 × 229 × 101 UB** for the column member.

First, the full plastic modulus of the column section needs to be reduced to take account of the axial load in the column member, i.e.

$$S_{xs} = 2280 - 3950\, n^2 \quad \text{(reference 6)}$$
$n = \alpha f_c / p_y$
$\alpha =$ adequacy factor (see below) – estimate 1.1

As the weight of side cladding and rails generally affects the lower portions of the column, it is not included in F.

$f_c = F/A = 351 \times 10/(2 \times 129) = 13.6$ N/mm^2
$n = 1.1 \times 13.6/275 = 0.05$
$S_{xs} = 2880 - 3950(0.05)^2 = 2870$ cm^3
$M_{ps} = 2870 \times 0.275 = 789.3$ kNm

The adequacy factor (α) indicates the strength of the frame **as designed** compared with the minimum design strength necessary to produce a safe design, and hence must not be less than unity. Therefore it is essential to check the value of the adequacy factor. This is done by setting up the equilibrium equations by the method outlined in references (2) and (3), i.e. the frame is cut at the apex to give three internal releases M, R and S (see Fig. 13.4). Each half frame acts as a cantilever with its free end at the apex. It can be shown that due to the symmetry of both the loading on, and the geometry of, the frame, the internal vertical force S is zero[3].

This leaves three unknowns, M, R and α to be determined; therefore three independent equilibrium equations are necessary. As the net moments at the positions A, B_s and X are known, i.e. zero (pinned base), M_{ps} and M_{pr}, respectively, equate them to the combined effect of the internal and external moments acting at those points. Also, knowing both the column and rafter

Fig. 13.4 Internal forces acting in a portal frame

sizes and anticipating that the haunch depth (D_h) will be 900 mm (see Fig. 13.18), then the correct value of a can be determined by the following expression, i.e.

$$a = D_h - \left[\frac{D_b + D_c \sin\theta}{2\cos\theta}\right] = 900 - \left[\frac{453.6 + 602.2 \sin 10.41°}{2\cos 10.41°}\right] = 0.61 \text{ m}$$

Hence the exact position of B_s can be used in the appropriate equilibrium equation. Remeber that X is the second purlin from the apex purlin.

A: $\quad 0.0 = \dfrac{351 \times 37.0}{8}\alpha - M - 8.90R$

B_s: $\quad 789.3 = \dfrac{351 \times 37.0}{8}\alpha - M - 4.01R$

X: $\quad -404.3 = \dfrac{351 \times 2.95^2}{37.0 \times 2}\alpha - M - 0.54R$

Solving these equations gives:

$\alpha = 1.1083$
$M = 362.56 \text{ kNm}$
$R = 161.41 \text{ kN}$

Note that the values of M and R include the adequacy factor.

This particular design produces a 10.8% overstrength in the frame; this overstrength arises owing to the range of discrete sizes of sections available to the designer. This enhanced capacity of the frame will be used throughout the remaining calculations, i.e. the design will include the factor 1.1083 so that the frame will effectively be capable of carrying $1.1083 \times 351 = 389$ kN. This allows the client to take advantage of this overstrength in the future without any additional cost. However, it can be argued (usually by designer/ fabricators) that in spite of the overstrength of the frame, all subsequent design checks (member stability, connection design) should be undertaken at the design load level, i.e. $\alpha = 1.0$. This approach is acceptable, provided the correct moment distribution for that load factor is used in the design calculations. It is incorrect to use the moments determined by taking the moment distribution obtained at failure ($\alpha = 1.1083$) and dividing by 1.1083.

The general expression for calculating the net moment at any position along the rafter for the half frame, shown in Fig. 13.4, is:

$$M_x = \frac{1.1083 \times 351}{37.0 \times 2} x^2 - 362.56 - \frac{3.4 \times 161.41}{18.5} x$$

$$= 5.2569 x^2 - 29.665 x - 362.56$$

Now the assumed position of the 'apex' hinge has to be checked by evaluating the net moments at the purlin supports immediately adjacent to, and on either side of, the second purlin (hinge position), using the expression for M_x, i.e.

$$M_2 = 5.2569 \times 1.475^2 - 29.665 \times 1.475 - 362.56 = -394.9 \text{ kNm}$$
$$M_4 = 5.2569 \times 4.425^2 - 29.665 \times 4.425 - 362.56 = -390.9 \text{ kNm}$$

As the moments at the purlin supports on either side of the second purlin are less than the moment (404.3 kN m) at the assumed plastic hinge position near the apex, then the assumption is correct.

Next, the length of the haunch needs to be determined so that the haunch/rafter intersection remains just elastic, i.e. the moment at the intersection is to be less than or equal to the **yield moment** M_y ($= p_y Z_x$):

$$0.275 \times 1300.0 = 5.2569 x^2 - 29.665 x - 362.56$$

Hence $\qquad x = 14.86$ m.

Therefore the minimum length of haunch from the column centre-line is $18.50 - 14.86 = 3.64$ m. By increasing the required haunch length by 0.11 m, it allows the haunch/rafter intersection to coincide with a purlin support position (some 14.75 m on plan from apex) which might prove useful when checking the adequacy of the haunched rafter against member instability. This means the actual haunch length, as measured from the column flange/end plate interface is $3.75 - 0.30 = 3.45$ m.

Now draw the bending moment diagram to check that nowhere does any moment exceed the local moment capacity of the frame. Also, the resulting information will be required to enable checks on member stability to be undertaken. The resulting net bending moment diagram for the gravity loading condition (dead + snow) is plotted on the tension side of the frame in Fig. 13.5.

Fig. 13.5 Moment distribution for dead plus imposed load condition

13.6.2 Wind load condition

In elastic analysis, the analyses from different loading cases can be added together. However, because plastic analysis deals with the final collapse state of a structure, each loading pattern generates its own unique failure mode. Therefore moments and forces from individual analyses cannot be added together. As a result, the dead + wind loading condition has to be analysed separately.

The basic wind pressure (q) of $0.59\,\text{kN/m}^2$ has already been established (Section 12.4.3). Knowing the slope of the rafter is $10.41°$ and that $h/w = 8.9/37.0 = 0.24$ and $l/w = 90.0/37.0 = 0.43$, then the external coefficients for the roof are -1.2 (windward) and -0.4 (leeward)[8]. These coefficients, when combined with the more onerous of the internal pressure coefficients (Section 12.4.3) result in roof wind suctions of $1.4q$ and $0.6q$ (Fig. 13.6).

Fig. 13.6 Net wind pressure coefficients

The design case to be considered in order to maximize the wind uplift is $1.0w_d + 1.4w_w$ (BS table 2). It is now necessary to calculate the various purlin and sheeting rail loads, in order to establish the total loading pattern acting on the portal frame. For example, the vertical and horizontal components of load for typical purlins on the roof are:

Dead load vert. load $= 1.0 \times 1.60 \times 1.475/\cos 10.41° = 2.40\,\text{kN}$

Wind load (windward):
  vert. component $= 1.4 \times 1.4 \times 0.59 \times 1.475 \times 6.0 = 10.23\,\text{kN}$
  horiz. component $= 10.23 \times 3.4/18.5$ $= 1.88\,\text{kN}$

Wind load (leeward):
  vert. component $= 1.4 \times 0.6 \times 0.59 \times 1.475 \times 6.0 = 4.39\,\text{kN}$
  horiz. component $= 4.39 \times 3.4/18.5$ $= 0.81\,\text{kN}$

The wind loads for other purlins and sheeting rails can be determined in a similar manner, using the net pressure coefficients given in Fig. 13.6. The factored loads (dead + wind) for the complete frame are shown in Fig. 13.7.

Assume the collapse mechanism shown in Fig. 13.8; that is, hinges are deemed to occur at the fourth purlin (X) down from the apex purlin on the left-hand side, and at the rafter/haunch intersection (D_r) on the right-hand side of the frame.

Unlike the previous analysis, four independent equilibrium equations are required, as the loading is non-symmetrical and therefore S will not be zero:

A: $0.0 = -1039.1\alpha + M + 8.900R - 18.50S$
X: $-404.3 = -96.5\alpha + M + 1.084R - 5.90S$
D_r: $404.3 = -157.8\alpha + M + 2.710R + 14.75S$
E: $0.0 = -339.8\alpha + M + 8.900R + 18.50S$

Fig. 13.7 Loading on frame for dead + wind load condition

Fig. 13.8 Collapse mechanism for dead + wind load condition

Solving these equations gives:

$\alpha = 2.375$
$M = -192.68 \text{ kNm}$
$R = -162.31 \text{ kN}$
$S = 44.89 \text{ kN}$

Clearly, the adequacy factor for this load condition is higher than that of the dead + snow load condition. This means that, in order to be consistent with the load level achieved with dead + snow loading, checking the portal frame design for the wind case need only be executed at a load level of 1.108. Figure 13.9 indicates the load levels at which hinges develop for the mechanisms associated with the two loading cases being discussed. It is clear from the diagram that at a load level of 1.108, no hinges will have formed in the frame for the dead + wind case, i.e. the frame remains elastic at that load level.

Hence the moments and forces which are required for the purpose of checking member stability and connection design can be obtained from an elastic analysis, using a design loading equal to 1.108 × factored loads. The bending moment diagram from such an analysis is represented by the solid curve in Fig. 13.10, as well as the bending moment diagram at collapse ($\alpha = 2.375$) – see broken line.

Fig. 13.9 Load–apex deflection for both load cases

Fig. 13.10 Moment distribution for dead + wind

Column moments — Moment values for α = 1.108 — Column moments

— Elastic state   α = 1.108
---- Collapse mode   α = 2.382

The bending moment diagram illustrates that the frame is subjected to reverse conditions under the dead + wind load case, compared with those for gravity load condition (cf. Fig. 13.5). Therefore, member stability and connection design will also need to be checked for the wind loading condition.

13.6.3 Computer analysis

Nowadays, current practice is to analyse and design portal frames by means of dedicated computer software. A plastic analysis of a portal frame can readily be executed by a number of commercial systems on the market, as well as numerous versions developed by individuals for their own use. Those particular programs that analyse the final collapse state directly, i.e. linear programming (optimization) methods or predetermined pattern of hinges solved by equilibrium equations, will give a correct solution for the conditions analysed. However, there is a class of plastic analysis programs based on the modified stiffness matrix method[9], which unless *correctly programmed* can produce false solutions.

With the modified stiffness method, the factored loads are increased proportionally from zero until somewhere in the frame the elastic moment at a potential hinge position equals the local M_p value. The stiffness matrix of the member in which the 'hinge' forms is then modified by replacing the appropriate 'fixed end' condition by a real pin, sustaining an internal moment equal to M_p[9]. This simple device allows the 'hinge' to rotate while sustaining a moment of M_p. A shape factor of unity is implied, i.e. the hinge position behaves elastically up to the moment M_p, when it becomes instantly completely plastic, as is assumed in simple plastic theory[3,4]. It should be noted that this internal moment remains 'locked' at this local value of M_p for the rest of the analysis.

The analysis continues with increasing load, modifying the stiffness of appropriate members every time a new hinge forms, until such time as there are sufficient hinges to produce a mechanism. The advantage of this method is that the analysis provides the sequence of hinge formation, while indicating the load level at which the individual hinges occur. The load level at which a mechanism is produced is in fact equal to the adequacy factor, α.

However, unless the modified stiffness method has been programmed to allow the moment at any defined hinge to unload, i.e. as the loading continues to increase, the moment at the 'hinge' position *might* reduce below the 'locked-in' value of M_p, then **false** solutions can result. The dead + imposed loading condition for the current design will be used to demonstrate this potential problem.

The proportions of the frame, particularly the haunch length, together with the selected member sizes are such that the first hinges to 'form' during a plastic analysis (using the modified stiffness method) occur at the haunch/rafter intersections. The two hinges form simultaneously at B_r and D_r owing to the symmetry of both the frame and loading, as indicated by Fig. 13.11(a).

Fig. 13.11 Transient plastic hinge condition

(a) Mechanism – false solution

(b) Mechanism – true solution

Fig. 13.12 Transient hinge condition – moment distribution

------ hinges at B_r, D_r
—·—· hinges at B_S, B_r, D_r, D_S – false solution
——— hinges at B_S, X, X, D_S – true solution

At this stage of the analysis, the relevant moment distribution (chain-dot line, Fig. 13.12) is correct, as these hinges, being transient hinges, would actually develop in the frame before unloading at a higher load level. Loading continues to increase until further hinges develop in the column members at the column/haunch intersections and the analysis stops, indicating a mechanism. However, inspection of the signs of the moments at the defined hinge positions show that all hinges are rotating in the same sense (see appropriate moment distribution (broken line) in Fig. 13.12). This is a physical impossibility as a mechanism cannot form under these conditions, i.e. it is a false solution.

This situation arises because the moments at the first hinge positions (in the rafter) are 'locked' and remain constant in spite of the effect of the additional load. As the loading increases, the frame tries to fail as a member mechanism between B_r and D_r and the moment in the central region of the rafter responds by taking more moment. But as the moments at the already defined 'hinge' positions remain at a fixed value, then the 'moment line' is forced to pivot about the points B_r and D_r, causing the moments at the eaves to increase disproportionately, and hence the corresponding moments in the column members (Fig. 13.12). The moments at the column/haunch intersections reach their local values of M_p before the hinges can form in the central portion of the rafter, thereby preventing the development of a member mechanism spanning between B_r and D_r.

However, both the ratio of the haunch length to rafter length and the ratio of M_{pr} and M_{ps} are such that the correct solution is a member mechanism forming over the length between B_s and D_s (Fig. 13.11b). From an examination of the relevant moment distribution (solid-line, Fig. 13.12), it can be seen that the moments at B_r and D_r have become lower than their local M_p values. What would actually happen is that hinges would initially develop at the haunch/rafter intersections, but as the load continued to increase these hinges would start to 'unload', and the corresponding moments would actually reduce. This unloading would be accompanied by a reversal in the small rotation that had occurred at these intersections. This 'hinge' unloading allows the designated hinges to develop at the column/haunch positions, followed by the 'apex hinges', thereby producing a true mechanism.

This example has highlighted the fact that transient hinges can develop at an intermediate load level and subsequently disappear from the final mechanism. A plastic analysis by direct methods, though giving the correct solution, would not indicate the presence of any potential transient hinges and alert the designer to possible member instability problems at a load level lower than the design load level, as illustrated by the portal frame design to date. Referring to Fig. 13.12, it can be seen that the moment conditions in the haunched rafter for the intermediate stage when the transient hinges have just formed (chain-dot line) are more severe than those at failure (solid-line). It would be advisable under these circumstances to restrain the transient hinge positions.

The analytical difficulty experienced with the standard modified stiffness method can be overcome. This can be achieved by the simple device of increasing the local value of M_{pr} at the intersections sufficiently to force the first hinges to form in the column members and not in the rafters. This device allows the correct mechanism to develop, but the solution is acceptable only if the final moments at the haunch/rafter intersections are below the appropriate M_p values for the rafter. Also, since the publication of the first edition, commercial analysis/design programs have become available that allow the unloading of transient hinges to take place, so that the true mechanism can be predicted with its accompanying moment distribution.

It should be noted that portal frame designs based on false solutions are inherently safe in terms of strength, as a false solution would indicate 'failure' at a load factor lower than the correct value. Clearly, this example, based on the dead + imposed loading condition, could be used as a test case for checking plastic analysis programs.

Note that, as with the lattice girder analysis executed by computer program, the axial load in the rafter changes at the purlin positions as each component of purlin loading parallel to the rafter member takes effect. In this case, the variation of axial load along the rafter in the vicinity of the hinge position is of the order 1–2%, which would not affect significantly the local rafter capacity.

13.6.4 Elimination of transient hinges

It has already been noted that the occurrence of a transient hinge during the formation of a collapse mechanism does not affect the load level at which a frame 'fails'. Nevertheless, as the example demonstrates, a more severe stress condition might exist at a lower load level (at which transient hinges develop) than that which occurs at failure, resulting possibly in premature member instability. Although a transient hinge would not be expected to develop significant rotation capacity prior to the unloading process starting, nevertheless, the haunch/rafter intersection should be restrained. A member stability check of the haunched region (based on the method outlined in Section 12.8.1.2) would indicate that the haunch is unstable, even when effectively torsionally restrained at the ends.

Therefore, in order to reduce any potential stability problems, the frame is redesigned to eliminate the transient hinges. This can be achieved by increasing the rafter size and/or reducing the haunch length, which would inherently result in lower stress levels in the haunch. The rafter size is to be increased to a **457 × 191 × 74 UB** section. The revised portal needs to be reanalysed using the same procedures as outlined for the original frame configuration.

13.6.5 Analysis of revised frame

Member sizes:

column section	**610 × 229 × 101 UB**
rafter section	**457 × 191 × 74 UB**
haunch	457 × 191 × 74 UB

$$M_{pr} = 1660 \times 0.275 = 456.5 \text{ kNm}$$
$$n = 1.14 \times 13.6/275 = 0.056$$
$$S_{xs} = 2880 - 3950(0.056)^2 = 2868 \text{ cm}^3$$
$$M_{ps} = 2868 \times 0.275 = 788.7 \text{ kNm}$$

By increasing the rafter size, the haunch will become shorter in length, therefore the original purlin spacing is adjusted from 1.500 m to 1.525 m on slope (1.500 m compared to 1.475 m on plan). Assuming the same hinge positions, as for the original frame, and with

$$a = 0.61 \text{ m}$$

the equilibrium equations become:

$$A: \quad 0.0 = \frac{351 \times 37.0}{8}\alpha - M - 8.90R$$

$$B_s: \quad 788.7 = \frac{351 \times 37.0}{8}\alpha - M - 4.01R$$

$$X: \quad -456.5 = \frac{351 \times 3.000^2}{37.0 \times 2}\alpha - M - 0.551R$$

which on solving gives:

$\alpha = 1.141$
$M = 416.33 \text{ kNm}$
$R = 161.29 \text{ kN}$

The general expression for moment at any position along the rafter becomes,

$M_x = 5.4106 x^2 - 29.642 c - 416.33$

Check the moments at the purlin positions adjacent to the 'apex' hinge, i.e.

$M_2 = 5.4106 \times 1.500^2 - 29.642 \times 1.500 - 416.33 = -448.6 \text{ kNm}$
$M_4 = 5.4106 \times 4.500^2 - 29.642 \times 4.500 - 416.33 = -440.2 \text{ kNm}$

As these moments are less than the hinge moment (456.5 kNm) then 'apex' hinge position is correct. Next, determine the minimum length of haunch so that the moment at the haunch/rafter intersection remains in the elastic range:

$0.275 \times 1460.0 = 5.4106 \, x^2 - 29.642 \, x - 416.33$

hence $x = 15.335$ m.

Therefore the minimum length of haunch from the column centre-line is $18.50 - 15.335 = 3.115$ m. Increasing the required haunch length by 0.385 m, allows the haunch/rafter intersection to coincide with a purlin support position (some 15.000 m on plan from apex). This also ensures that a transient hinge does not develop. This means the actual haunch length, as measured from the column face is $3.500 - 0.30 = 3.200$ m.

A plastic analysis of the revised would indicate that a transient hinge would not occur within the haunched length. For the dead+wind uplift condition, at load level $\alpha = 1.141$, the frame is shown to be elastic, see Section 13.6.2, and therefore the moment distribution for the wind case can be determined from an elastic analysis. The revised moment distributions for both the dead+snow load condition and the dead+wind condition are given in Figs. 13.13(a) and 13.13(b), respectively.

Fig. 13.13 Revised moment distributions for both load cases

All design calculations will now be based on the conditions appertaining to the revised frame.

13.7 FRAME STABILITY

Before checking the frame for member stability and designing the connections, it is prudent to check the selected member sizes against the possibility of frame instability (clause 5.5.3), as new sections need to be chosen if these criteria are not satisfied.

13.7.1 Sway stability of frame

First, the member sizes have to be checked against frame sway instability (clause 5.5.3.2). This condition, modified to take the effect of the haunches into account, has to be satisfied in the absence of a rigorous analysis of frame stability[3], i.e.

$$\frac{L-b}{D} \leq \frac{44\,L}{\Omega\,h} \frac{\rho}{(4+\rho L_r/L)} \frac{275}{p_{yr}}$$

$$\rho = \frac{2I_c}{I_r} \frac{L}{h} \quad \text{for single-bay frame}$$

$$\Omega = \alpha W^* L / (16 M_{pr})$$

$L = 37.0$ m
$L_r = 37.0/\cos 10.41° = 37.62$ m
$D = 0.457$ m
$h = h_1 = 5.50$ m
$b =$ average eaves haunch length $= 3.500$ m
$I_c = 75\ 700$ cm^4
$I_r = 33\ 400$ cm^4

$\rho = 2 \times 75\ 700 \times 37.0/(33\ 400 \times 5.5) = 30.5$
$\Omega = 1.141 \times 351 \times 37.0/(16 \times 456.5) = 2.03$

$$\frac{37.0 - 3.50}{0.457} \leq \frac{44}{2.03} \frac{37.0}{5.5} \frac{30.5}{(4 + 30.5 \times 37.62/37.0)} \frac{275}{275}$$

$$73 < 127$$

Frame is stable.

13.7.2 Snap-through buckling of rafter

The condition of snap-through buckling of the rafters (clause 5.5.3.3) is unlikely to apply to single-bay frames unless roof slope is shallow. Such a possibility increases with multi-bay frames owing to the accumulative spread of the eaves which may be sufficient to allow the rafters of the internal bays to snap-through, see reference (10).

Research by Davies[11] indicates that the BS 5950 formulae for frame instability can produce unsafe solutions. The in-plane frame stiffness of multi-span portal frames, particular when slender internal columns of valley beams are used, requires careful consideration[12].

13.8 MEMBER STABILITY – LATERAL TORSIONAL BUCKLING

BS 5950 caters specifically for different moment gradients and geometry of member for portal frames and the appropriate methods will be adequately illustrated when checking the lateral torsional resistances of both column and haunched rafter members. Guidance is given in those situations which are not specifically covered by the code or where it is felt that the code recommendations are not appropriate. First, the various parts of the frame are checked for stability under gravity loading condition and then rechecked for the dead + wind loading condition.

13.8.1 Stability checks for dead + snow loading condition

As noted in Section 13.6.2, member stability is to be checked at a load level $1.141 \times$ factored loads; see Fig. 13.13(a) for the relevant bending moments.

13.8.1.1 CHECK COLUMN MEMBER BUCKLING

The design of the frame assumes a plastic hinge forms at the top of the column member ($610 \times 229 \times 101$ UB), immediately below the haunch level. This plastic hinge position must be **torsionally restrained** in position by diagonal stays; see later comment regarding size of stays (Section 13.8.3). With the hinge position restrained, check the plastic stability of the length of column from the hinge position, between the two sheeting rails, S_4 and S_3 (see Fig. 13.14). This is done by using the expression for uniform members, with no tension flange restraint between effective end torsional restraints, given in clause 5.3.5(a), which is based on the stability curves given in reference (13).

$$L_m = \frac{38 r_y}{\sqrt{\left[\frac{f_c}{130} + \left(\frac{p_y}{275}\right)^2 \left(\frac{x}{36}\right)^2\right]}}$$

$f_c = 1.141 \times 351 \times 10/(2 \times 129) = 15.5\,\text{N/mm}^2$
$x = 43.0$ (reference 6)

$$L_m = \frac{38 \times 47.5 \times 10^{-3}}{\sqrt{\left[\frac{15.5}{130} + \left(\frac{275}{275}\right)^2 \left(\frac{43.0}{36}\right)^2\right]}} = 1.452\,\text{m}$$

This indicates that the sheeting rail S_3, immediately below the rail at the hinge position S_4, is required to restrain the column member on both flanges,

Fig. 13.14 Member stability – column member

i.e. by means of diagonal stays which must be placed not more than 1.452 m from the stay at the hinge position. As a consequence, the side rails are positioned from ground level at 0.15 m (S_1), 1.80 m (S_2), 3.45 m (S_3) and 4.89 m, say 4.90 m (S_4); see Fig. 13.14.

Now check the elastic stability of the column length between sheeting rail position S_3 and the pinned base A. Under the action of gravity loading (dead + snow) the inner flanges of the column members are in compression. Now:

$$\frac{F}{P_c} + \frac{\bar{M}}{M_b} < 1$$

where
$F = 1.141 \times 351/2 = 200.2$ kN
$M_{max} = 556.5$ kNm (max. moment occurs at S_3)

As the moment at base A is zero, then $\beta = 0$, hence $m = 0.57$ (BS table 18) and:

$$\bar{M} = mM_{max} = 0.57 \times 556.5 = 317.2 \text{ kNm}$$

It is assumed that the effective length for the column member buckling about the x–x axis is the distance S_4–A, i.e. 4.90 m, hence:

$$L/r_x = 4900/242 = 20$$

BS table 27(a) Hence $p_c = 273$ N/mm².

At this stage, it is assumed that there are no further restraints to the inner (compression) flange below S_3 and therefore the effective length of the compression flange about the y–y axis is S_3–A, i.e. 3.45 m.

BS table 27(b)
$L/r_y = 3450/47.5 = 73$
$p_c = 196$ N/mm²
$P_c = 196 \times 129/10 = 2530$ kN

BS table 13 $n = 1.0$
$u = 0.863$ (reference 6)
$x = 43.0$ (reference 6)
$\lambda = 73$
$\lambda/x = 73/43.0 = 1.70$
BS table 14 $v = 0.974$

As the universal section has equal flanges, then $N = 0.5$ and with $\lambda/x = 1.70$ hence:
$$\lambda_{LT} = nuv\lambda$$
$$= 1.0 \times 0.863 \times 0.974 \times 73 = 61$$

BS table 11 $p_b = 212\,\text{N/mm}^2$
$M_b = 0.212 \times 2830 = 611\,\text{kNm}$

$$\frac{200.2}{2530} + \frac{317.2}{611} = 0.079 + 0.519 = 0.598 < 1.0$$

This means that the member is stable over the length considered, i.e. no additional restraints to the compression flange are required between S_3 and A, as shown in Fig. 13.14. If the length of member being considered had proved to be inadequate, then additional restraint(s) would have been required or the effect of tension flange restraints (sheeting rails) could be taken into account (see clause G2(a)(1)).

Occasionally, it may not be possible to provide the necessary torsional restraints at S_4 and S_3 because diagonal braces intrude into the space required for some other function, such as door openings. Then the buckling resistance of the column member would need to be increased by other means, as a plastic hinge position must always be restrained. For example, select a more appropriate section with improved properties or encase the column member in concrete up to the underside of the haunch, i.e. the hinge position.

13.8.1.2 CHECK RAFTER BUCKLING IN EAVES REGION

First, check the stability of the haunched portion of the rafter (from the eaves connection to the haunch/rafter intersection) as this represents one of the most highly stressed lengths, and with its outstand flange (inner) in compression, this part of the rafter is the region most likely to fail due to instability. As it has already been decided to stay the inside corner of the column/haunch intersection (column hinge position), assume that the haunch/rafter intersection is also effectively torsionally restrained by diagonal braces, giving an effective length of 3.200 m, as indicated in Fig. 13.15. This diagram also gives the relevant moment distribution from the eaves to just beyond the point of contraflexure.

The depth of a haunch is usually made approximately twice the depth of the basic rafter section, as it is the normal practice to use a UB cutting of the same serial size as that of the rafter section for the haunch, which is welded to the underside of the basic rafter ($457 \times 191 \times 74$ UB). Therefore, assume the haunch has the same section as the rafter member and that the overall depth at the connection is 0.90 m (see Section 13.9).

Fig. 13.15 Member stability – haunched rafter region

It would appear that clause G.2(b)(2) is the most appropriate criterion to check the stability of the haunched portion, as there is at least one tension flange (intermediate) restraint between the effective end restraints, i.e.

$$L_t \leq L^* = \frac{L_k}{cn_t}$$

According to the code (clause G.3.5), the formula for L_k is appropriate[14] only if there is a plastic hinge associated with the length being considered, which is the case when the transient hinges develop.

$$L_k = \frac{[5.4 + 600(p_y/E)]xr_y}{\sqrt{[5.4(p_y/E)x^2 - 1]}}$$

$$= \frac{[5.4 + 600 \times 275/205\,000]33.9 \times 41.9 \times 10^{-3}}{\sqrt{[5.4(275/205\,000)33.9^2 - 1]}} = 3.256\,\text{m}$$

This value of L_k has to be modified by the shape factor (c), which accounts for the haunching of the restrained length (clause G.3.3) and n_t, which allows for the stress distribution across the length (clause G.3.6.1).

$$c = 1 + \frac{3}{(x-9)}(R-1)^{2/3}q^{1/2}$$

R = greater depth/lesser depth
q = haunched length/total length

Over the length being considered (3.200 m), then:

$R = 900/457 = 1.97$
$q = 3.200/3.200 = 1.00$

$c = 1 + 3(1.97 - 1.0)^{2/3}1.0^{1/2}/(33.9 - 9) = 1.118$

$$n_t = \sqrt{\frac{1}{12}\left[\frac{N_1}{M_1} + \frac{3N_2}{M_2} + \frac{4N_3}{M_3} + \frac{3N_4}{M_4} + \frac{N_5}{M_5} + 2\left(\frac{N_S}{M_S} - \frac{N_S}{M_E}\right)\right]}$$

N_i = applied factored moments at the ends, quarter points and mid-length of the length considered
M_i = plastic moment capacity at section i ($=p_y S_i$)
N_S/M_S = greatest of N_2/M_2, N_3/M_3, N_4/M_4
N_E/M_E = greater of N_1/M_1, N_5/M_5

The design strength is 275 N/mm² as the flange thickness is less than 16 mm. Though clause G.3.6.3 implies that the formula for n_t can be used for elastic conditions, M_i must always be the local plastic capacity. Any N_i which causes tension in the outstand flange is made zero. Also, the term $(N_S/M_S - N_E/M_E)$ is considered only if positive.

The plastic moduli are determined for the five cross-sections indicated on Fig. 13.16; the actual cross-sections considered are taken as being normal to the axis of the basic rafter (unhaunched) member. The plastic moduli together with other relevant information regarding the evaluation of the ratios N_i/M_i are given in the following table. The worse stress condition at the haunch/rafter intersection (location 5) is taken, i.e. that in the basic rafter immediately adjacent to the intersection.

Fig. 13.16 Member stability – haunch region

		1	2	3	4	5
Distance from apex	(m)	18.20	17.34	16.48	15.61	14.75
Factored moment	(kNm)	836.4	706.0	582.6	466.0	356.4
Plastic modulus	(cm³)	4082	3448	2909	2463	1660
Moment capacity	(kNm)	1123	948	800	67	456
Ratio	N/M	0.745	0.745	0.728	0.688	0.782

$$n_t = \sqrt{[0.745 + 3 \times 0.745 + 4 \times 0.728 + 3 \times 0.688 + 0.782 + 2\,(0.745 - 0.782)/12]}$$
$$= \sqrt{[0.728]} = 0.853$$
$$L^* = \frac{L_k}{cn_t} = 3.256/(1.118 \times 0.853) = 3.414\,\text{m} > 3.200\,\text{m}$$

This portion of the rafter is stable over the assumed restrained length of 3.200 m. Note: if a plastic hinge has developed at the haunch/rafter intersection, then a further design check is necessary, see Section 13.14.

An alternative method of assessing n_t

In Appendix B, a rapid method for giving an approximate assessment of n_t is outlined. The method enables the designer to establish quickly whether or not the more exact method just used in this example needs to be undertaken. If the approximate assessment indicates that the resulting permissible length is within **100 mm** of the restrained length being checked, then it is recommended that the exact check be made.

Taking the design example, calculate the approximate value of n_t (with reference to Appendix B):

$$n_t = \sqrt{\left\{\frac{1}{12M_e}\left[\frac{N_1}{R_1} + \frac{3N_2}{R_2} + \frac{4N_3}{R_3} + \frac{3N_4}{R_3} + \frac{3N_4}{R_4} + \frac{N_5}{R_5} + 2\left(\frac{N_S}{R_S} - \frac{N_E}{R_E}\right)\right]\right\}}$$

$M_e = 456.5 \text{ kNm}$

From Appendix B – plastic condition:

$R_1 = 2.555; R_2 = 2.155; R_3 = 1.755; R_4 = 1.495; R_5 = 1.00$

$$n_t = \sqrt{\left\{\frac{1}{12 \times 456.5}\left[\frac{836.4}{2.555} + \frac{3 \times 706.0}{2.155} + \frac{4 \times 582.6}{1.755} + \frac{3 \times 466.0}{1.495}\right.\right.}$$
$$\left.\left. + \frac{356.4}{1.00} + 2\left(\frac{706.0}{2.155} - \frac{356.4}{1.00}\right)\right]\right\}$$

$= 0.847$

$L^* = 3.200/(1.118 \times 0.847) = 3.379 \text{ m}$ (cf. 3.414 m by exact method)

and as $(3.379 - 0.100) = 3.279 \text{ m} > 3.200 \text{ m}$ there would be no need to check by exact method.

Note that the rapid method is applicable only for the typical British haunch (with middle flange) where the haunch depth is approximately twice that of the basic rafter section and the haunch cutting is from the same section size as the rafter member.

There are several ways in which the haunched region can be made stable:

- Redesign the member sizes in order to use a more torsionally stable rafter section.
- Increase thickness of compression flange – not only does it lower stresses but it also improves torsion stiffness of this flange; this can be achieved by using a different size for the haunch cutting, i.e. the same serial size but with increased weight, or alternatively using two plates welded together to form the haunch.
- Increase the haunch depth to lower stresses within haunch and reduce effect of the term cn_t.
- Add a lateral restraint within the haunched length, thereby reducing the restrained length to be checked.

The uniform rafter from the haunch/rafter intersection to the point of contraflexure (where outstand flange is just in compression) now needs to be checked for elastic stability by the method given in section 4, BS 5950. However, the point of contraflexure moves down the rafter as the moment distribution changes from the elastic to the plastic state. Therefore, it is advisable to consider that effective restraint occurs at distance beyond the point of contraflexure, equal to either the depth of the rafter or the distance to the next purlin up the slope, whichever is the lesser. The theoretical point of contraflexure, being the position of zero moment in the rafter, can rapidly be determined from:

$$0 = 5.4106x^2 - 29.642x - 416.33$$

Hence $x = 11.930$ m.

Therefore, with reference to Fig. 13.15, the notional length of this part of the rafter is $15.000 - 11.930 = 3.070$ m. Hence the effective length is $3.070 + 0.457 = 3.527$ m. The moment distribution is based on the notional length. Note that this portion of the rafter has two intermediate tension flange restraints (purlins) between the effective end restraints and therefore clause G.2(a)(1) can be used. However, in this case, the length between effective full restraints can be justified for checking the elastic stability of the uniform rafter, i.e. using the simpler procedure of clause 4.8.3.3.1 without the assistance of the intermediate partial restraints:

$$\frac{F_c}{P_c} + \frac{\overline{M}}{M_b} < 1$$

$F = 161.3$ kN (i.e. thrust in rafter $-R$)
$M_A^c = 356.4$ kNm (max. moment occurs at intersection)

As the 'point of contraflexure' represents the position of zero moment then

BS table 18 $\beta = 0$, hence $m = 0.57$ and

$\overline{M} = mM_A = 0.57 \times 356.4 = 203.1$ kNm
$\lambda = L/r_y = 3527/41.9 = 84$

BS table 27(b) $p_c = 173$ N/mm^2
$P_c = 173 \times 95.0/10 = 1644$ kN

BS table 13 $n = 1.0$
$u = 0.876$ (reference 6)
$x = 33.9$ (reference 6)
$\lambda/x = 84/33.9 = 2.48$

BS table 14 $v = 0.93$ (as universal section has equal flanges, then $N = 0.5$)
$\lambda_{LT} = 1.0 \times 0.876 \times 0.93 \times 84 = 68$

BS table 11 $p_b = 193$ N/mm^2
$M_b = 0.193 \times 1644 = 320$ kNm

$$\frac{161.3}{1644} + \frac{203.1}{320} = 0.098 + 0.635 = 0.733 < 1.0$$

The member is stable over the length P$_{11}$–PC and no further torsional restraints are necessary within this length of member.

13.8.1.3 CHECK RAFTER BUCKLING IN APEX REGION

Another highly stressed region is the length of rafter in which the 'apex' hinge occurs (see Fig. 13.17). Under dead + snow loading, the outstand flange is in tension, while the compression flange is restrained by the purlin/rafter connections.

Fig. 13.17 Member stability – apex region

Therefore, the buckling resistance of the rafter member between purlins in the apex region needs to be checked. Provided that the 'apex' hinge is the **last** hinge to form in order to produce a mechanism (which is true for low pitched portal frames under dead + snow loading), then adequate rotation capacity is not a design requirement, i.e. the hinge is required only to develop M_p, not to rotate – purely a strength condition.

This situation is not specifically covered by BS 5950 (1985), apart from a sentence in clause 5.5.3.1 which states that clause 5.3.5 (dealing with the spacing of restraints) need not be met. However, it is suggested, based on research evidence, that if the value of L_m (as defined by clause 5.3.5) is factored by 1.5 then this would represent a safe criterion for restraint spacing in the region of the last hinge position. That is, assuming that the purlins act as restraints because of their direct attachment to the compression flanges in the apex hinge region, then the purlin spacing should not exceed:

$$L_m = \frac{57 r_y}{\sqrt{\left[\frac{f_c}{130} + \left(\frac{p_y}{275}\right)^2 \left(\frac{x}{36}\right)^2\right]}}$$

$f_c = 161.3 \times 10/95.0 = 17.0 \, \text{N/mm}^2$
$x = 33.9$ reference 6

$$L_m = \frac{57 \times 41.9 \times 10^{-3}}{\sqrt{\left[\frac{17.0}{130} + \left(\frac{275}{275}\right)^2 \left(\frac{33.9}{36}\right)^2\right]}} = 2.368 \, \text{m}$$

As purlin spacing is 1.525 m (on slope), then no additional restraints are required. (Note that if the 'apex hinge' is not the last hinge to form, then L_m reverts to $2.368/1.5 = 1.578 \, \text{m} > 1.525 \, \text{m}$. If L_m had been less than 1.525 m then the purlin spacing would have to be reduced.

Now check the strength capacity of the rafter in the 'apex' region, using clause 4.8.3.2(b), i.e.

$$\left(\frac{M_x}{M_{rx}}\right)^2 + \frac{M_y}{M_{ry}} < 1$$

There is no minor axis moment, M_y, and M_{rx} is the reduced capacity of the rafter section about the major axis in the presence of axial load.
Remembering that the axial thrust of 161.3 kN includes the adequacy ratio, then:

$$M_{rx} = (1660 - 2480n^2)p_y \text{ (reference 6)}$$

$$f_c = 161.3 \times 10/95.0 = 17.0 \text{ N/mm}^2$$
$$p_y = 275 \text{ N/mm}^2$$
$$n = f_c/p_y = 17.0/275 = 0.06$$

$$M_{rx} = [1660 - 2480(0.06)^2]0.275 = 454.0 \text{ kNm}$$

This is slightly lower than the value of 456.5 kNm used in the analysis of the frame and to be strictly correct the analysis should be modified, such that $M_x = M_p = M_{rx} = 454.0$ kNm and hence the above strength capacity check would be just satisfied. However, as the difference in strength capacities is less than 1% and in view of the adequacy ratio being 14.1% above the design level, then it can be safely assumed in this case that the frame as designed is more than adequate, i.e. there is no need to carry out the minor adjustments to previous calculations and checks. It can be argued that the reduced plastic capacity for the rafter member should have been taken into account in the original equilibrium equations. The actual reduction in the adequacy ratio is dependent on the geometry and the relative strengths of the column and rafter sections, i.e. in this example the reduction is 0.2%.

This completes the member stability checks of the design frame for the dead + snow load condition, but this set of checks must be repeated for the dead + wind load case.

13.8.2 Stability checks for dead + wind loading condition

Figure 13.13(b) showed clearly that the dead + wind loading condition causes a reversal of moments in the frame compared with those obtained for the dead + imposed condition (Fig. 13.13a) and therefore the series of checks undertaken in Sections 13.8.1.1 to 13.8.1.3 needs to be repeated. Note that only one dead + wind condition is being considered in this analysis; nevertheless, when the wind direction is normal to the gables, a worse reversed moment condition might arise for the apex region.

13.8.2.1 CHECK COLUMN MEMBER BUCKLING

Under the dead + wind loading (Fig. 13.13b), the elastic bending moments in the right-hand column vary from 342 kNm at S_4 to 0 kNm at the base, causing the outer flange to go into compression. This moment of 342 kNm compared with the corresponding value of 789 kNm for the dead + imposed loading

(Fig. 13.13a), coupled with the fact that the outer flanges of the columns are restrained by the sheeting rails (see Fig. 13.14), indicates that this loading condition is not as severe as that investigated in Section 13.8.1.1. A check using section 4, BS 5950 (elastic condition) indicates the column member requires no further restraints.

13.8.2.2 CHECK RAFTER BUCKLING IN EAVES REGION

The wind loading condition causes the upper flanges in the eaves regions to sustain compression. These flanges are restrained by purlin cleats at 1.525 m intervals (see Fig. 13.15). As the magnitudes of the moments are significantly less than those for gravity loading, the application of section 4, BS 5950 would prove that the haunched rafters in these regions are more than adequate. No further restraints are necessary.

13.8.2.3 CHECK RAFTER BUCKLING IN APEX REGION

In the apex region, owing to the stress reversal conditions arising from the dead + wind case, the outstand (lower) flange is in compression and at present there are no restraints to that flange, though the 'tension' flange is restrained by the purlins. Therefore, this part of the rafter needs to be checked in detail as member buckling may prove to be more severe than that checked in Section 13.8.1.3, despite the moments being appreciably smaller. The unrestrained length between the two points of contraflexure is about $12.1 \times 1.525 = 18.5$ m (see Fig. 13.13b), hence:

$$\lambda = \frac{18500}{41.9} = 442$$

Clearly, this unrestrained length of 18.5 m is too slender, exceeding the limitation of 350 for slenderness for wind reversal (clause 4.7.3.2).
Therefore, restraints are required in the apex region. In considering the best location for the restraints the moment distribution for the dead + imposed load condition has a bearing on the decision. Also, it has to be remembered that the wind can blow in the opposite direction, i.e. right to left, therefore the restraints should be arranged symmetrically about the apex of the frame.

As the moment distribution for dead + imposed case in the apex region is fairly constant, there is a choice for the location of the restraints, i.e. at either first, second or third purlin position down for the apex purlin. If restraints are placed at the third purlin down for the apex purlin on either side of the apex, then the unrestrained lengths become 8.525 m, 9.150 m and 0.800 m from left to right in Fig. 13.18.

Consider the left-handed portion, 8.525 m long. Clause G.2.a(1), BS 5950 states that for checking the elastic stability of a uniform member which is restrained by intermediate restraints on the tension flange between effective torsional restraints:

$$\frac{F}{P_c} + \frac{\bar{M}}{M_b} < 1$$

Fig. 13.18 Member stability – apex region – reversed loading

However, the axial load (F) for the dead+wind condition is in tension and although axial tension should improve a member's buckling resistance, the code does not allow any benefit for this condition.

Conversely, clause 4.8.2 gives an erroneous impression that any axial tension would obviate a member buckling check, whereas this would depend on the relative magnitude of the bending moment and axial tension.

In the absence of clear guidance, a member buckling resistance check should be based on:

$$\frac{\overline{M}}{M_b} < 1$$

$$M_A = 142 \text{ kNm}$$

$$m_t = 1.0$$

clause G.3.4
$$\overline{M} = m_t M_A = 142 \text{ kNm}$$

The minor axis slenderness ratio for this particular stability check is defined by clause G.3.3, i.e.

$$\lambda_{TB} = n_t u v_t c \lambda$$
$$\lambda = \frac{8525}{41.9} = 203$$

The local moment capacity of a uniform member under elastic condition (load level = 1.141) is the yield moment ($p_y Z_x$), hence the expression for n_t (clause G.3.6) becomes:

$$n_t = \sqrt{\left[\frac{N_1 + 3N_2 + 4N_3 + 3N_4 + N_5 + 2(N_S - N_E)}{12 M_y}\right]}$$

The values of N_i can be evaluated from Fig. 13.18 and hence:

$$n_t = \sqrt{\left[\frac{0 + 3 \times 81 + 4 \times 134 + 3 \times 158 + 157 + 2(158 - 15)}{12 \times 0.275 \times 1460}\right]}$$

$$= 0.537$$

$u = 0.876$ (reference 6)
$x = 33.9$ (reference 6)
$\lambda/x = 204/33.9 = 6.02$

clause G.3.3

$$v_t = \sqrt{\left[\frac{\dfrac{4a}{h_s}}{1 + \left(\dfrac{2a}{h_s}\right)^2 + \dfrac{1}{20}\left(\dfrac{\lambda}{x}\right)^2}\right]}$$

a = half depth of purlin+half depth of member = 87+229 = 316 mm
h_s = distance between shear centres of flanges = 457.2 − 14.5 = 443 mm

$$\frac{a}{h_s} = \frac{316}{443} = 0.713$$

$$v_t = \sqrt{\left[\frac{4 \times 0.713}{1 + (2 \times 0.713)^2 + \dfrac{1}{20}(6.02)^2}\right]} = 0.767$$

$c = 1.0$ uniform member
$\lambda_{TB} = 0.537 \times 0.876 \times 0.767 \times 1.0 \times 203 = 73$
$p_b = 181$ N/mm^2
$M_b = 0.181 \times 1660 = 300$ kNm $< p_y Z_x$

$$\frac{\bar{M}}{M_b} = \frac{142}{300} = 0.473 < 1.0$$

There still remains the checking of the rafter between the new restraints 5 to 13 (see Fig. 13.18). A check would indicate that this part-member is stable for the dead + wind case.

13.8.3 Design of lateral restraints

During the various checks on member buckling undertaken in the previous sections, several positions along the frame have been assumed to be effectively restrained against both lateral and torsional displacements. Such restraints must be capable of carrying the lateral forces while being **sufficiently stiff** so that the member being braced is induced to buckle between braces.

Research evidence[15] has indicated that the magnitude of the restraining force in any **one** restraining element/brace before instability occurs is of the order of 2% of the squash load of the compression flange, i.e. $0.02 B T p_y$.

Though the restraining force is relatively small, it is **essential** that such a force (in the form of a brace) be supplied. Also, rafter braces may need to be designed for the additional force that results from their propping action to the purlins. Generally this force only becomes significant relative to the restraining force of 2% for small frames (< 25 m). Nevertheless, braces should be used in pairs, as illustrated in Fig. 13.19, as a single brace would cause the connected flange to move laterally due to the propping action in the brace. This action would increase the possibility of lateral buckling in the connected member.

Adequate stiffness is at least as important as the strength criterion of 2% squash load. In view of lack of sufficient experimental evidence, a limiting slenderness ratio of 100 is recommended for diagonal braces, illustrated in Fig. 13.19.

Fig. 13.19 Effective torsional restraints

Though individual design of the various column and rafter restraints would probably result in different sizes, it is more economic to design for the worst case and standardize on one size for all restraints. As the area of column flange is larger than that of the rafter, the design of the braces will be based on the condition appropriate to the restraint at the column/haunch intersection.

Therefore, assuming that the diagonal stays are approximately at 45° to the braced member:

$$\text{Length of brace} = \sqrt{2} \times (\text{depth of column section})$$
$$L = 1.4 \times 602 = 850 \text{ mm}$$

As an angle section is more effective as a strut (on a weight to weight basis) than a flat (thin rectangular section), an angle with a r_{vv} of at least $850/100 = 8.5$ mm will be selected. From the SCI guide[6] an appropriate angle, say a **45 × 45 × 4 Angle**, is chosen and is then checked against both strength and stiffness requirements.

The slenderness ratio of a discontinuous single angle strut with a single bolt at each end (clause 4.7.10.2(b)) is either

$$\lambda = 1.0L/r_{vv} \quad \text{or} \quad 0.7L/r_{xx} + 30$$
$$= 850/8.76 \quad \text{or} \quad 0.7 \times 850/13.6 + 30$$
$$= 97 \quad \text{or} \quad 74 < 100 \text{ Stiffness satisfied}$$

Lateral force $= 0.02BTp_y$

$$F = 0.02 \times 227.6 \times 14.8 \times 0.275 = 18.5 \text{ kN}$$

BS table 27(c)

$$p_c = 130 \text{ N/mm}^2$$

Clause 4.7.10.2(b) further states that for a single angle with a single fastener at each end, the compression resistance must not be greater than 80% of the compression resistance of the angle when treated as an axially loaded strut; hence for the lateral stays:

$$P_c = 0.80(0.130 \times 349) = 36.3 \text{ kN} > 18.5 \text{ kN Strength satisfied}$$

Use 45 × 45 × 4 Angle

There are alternative forms of restraint and the reader is directed to reference (16). Unlike this design example, difficulty may be experienced in giving lateral support exactly at a plastic hinge position. Should this situation arise then the hinge position may be regarded as being laterally restrained, provided the point of attachment of the brace to the **compression flange** is not more than $D/2$ from the assumed hinge position.

13.9 DESIGN OF CONNECTIONS

Apart from checking the adequacy of members against lateral torsional instability, the design of connections evokes much discussion with the result that there are several variations on how portal frame connections should be designed. Basically the connections have to perform as an elastic unit joining two main structural members together without loss of strength and undue distress such as gross deformations or plasticity. The basic method adopted for this design exercise[3] had been developed from theoretical considerations and experimental evidence. Connections based on this approach have been shown to perform satisfactorily. The method has since been modified to reflect the more recent research into the behaviour of end plate connections. In proportioning both the eaves and apex connections for the dead plus imposed case it should be remembered that the design is being undertaken at ultimate load level. Generally the eaves and apex joints are flush end plate connections, in which the bolt lever arms are increased by means of a haunch (see Figs. 13.20 and 13.21), thereby enhancing the moment capacity of the bolted connections. There are a number of different design criteria which need to be satisfied[3]. These will be explained during the process of designing the connections for the portal frame in this section. See also Section 13.14.

As mentioned, the design of portal frame connections is generally governed by the moments and forces resulting from the dead + imposed loading condition. However, if there is moment reversal due to another loading case (as in this example), then the connections have to be rechecked.

13.9.1 Design of eaves connection

From the portal frame analysis for the dead + imposed loading condition (Section 13.6), it can be seen that the factored moment and vertical shear, acting on the eaves connection, are 836.4 kNm (Fig. 13.13a) and 200.2 kNm (i.e. 1.141 × 351/2), respectively; also, there is an axial thrust of 195 kN from the rafter. Furthermore, the reverse moment condition needs to be checked; see Section 13.9.1.7. The initial design decisions to be made are the geometry of the end plate and the size of bolts to be used. From practical considerations (such as width of column and rafter flanges, discrete sizes of rolled plate sections), the end plate is made 220 mm. As the depth of the haunch is approximately twice that of the basic rafter member, the end plate is made 940 mm long, i.e. the end plate projects 20 mm beyond the flanges to allow for the top and bottom welds. The end plate thickness is determined from consideration of the flexural action imposed on the plate by the forces in the bolts. First, determine the weld sizes required for the end plate/rafter interface, before evaluating the bolt forces and end plate thickness.

13.9.1.1 WELD SIZES FOR END PLATE

The end plate is usually connected to the rafter section by means of fillet welds. A simple rule for proportioning the weld sizes at the ultimate load condition is to make the combined throat thicknesses of the welds equal to at least the thickness of the plate element being welded[3]. Therefore:

$$\text{Flange weld} = T_b/\sqrt{2}$$
$$= 14.5/1.41 \quad = 10.3\,\text{mm} \quad \textbf{Use 12mm FW}$$

$$\text{Web weld} = t_b/\sqrt{2}$$
$$= 9.1/1.41 \quad = 6.43\,\text{mm} \quad \textbf{Use 8mm FW}$$

The heavier flange weld must be continued down the web on the tension side of the connection for a minimum distance of 50 mm in order to avoid premature weld cracking in the vicinity of the root fillet of the rafter flange, owing to potential stress concentrations.

13.9.1.2 SIZE OF BOLTS

Today, high tensile bolts (grade 8.8) are used in moment connections. It should be noted that the use of the ultimate tensile capacity of these bolts is conditional on grade 10 nuts being used with the bolts[13] to prevent premature failure by thread stripping. Alternatively, clause 3.2.2 allows the use of non-preloaded HSFG bolts, which have deeper nuts and also a larger capacity than a grade 8.8 bolt of the same diameter. However, should the conditions be such that wind or crane vibrations might result in bolt loosening or fatigue, then it is advisable to use **preloaded** HSFG bolts, though other methods are available which prevent nut loosening.

Assume that the vertical pitch of the bolt rows is 90 mm, with the top row being positioned at 50 mm from top surface of the tension flange of the rafter (see Fig. 13.20). By minimizing the clearance between the top row of bolts next to the flange, the cross-bending in the end plate is reduced.

The appropriate load distribution in the bolt group to be used[3] is dependent on the ratio of the distance from the compression flange of the rafter to the penultimate row of the 'tension' bolts to that from the compression flange to the top row of bolts, i.e. $0.760/0.850 = 0.89$.

As the ratio is about 0.9 then the two top rows of tension bolts can be assumed to carry equal total load[3], with the remainder of the bolt loads varying linearly with their bolt distances (y_i) from the compression flange. Apart from the first row of bolts immediately inside the tension flange, the bolt loads are a combination of load due to the applied moment plus prying force. With this distribution, which assumes that the connection rotates about the compression flange, the **notional** bolt load (F_o) due to the applied moment flange is determined from:

$$F_o = M/[4.7d_e + 2(\Sigma y_i^2/d_e)]$$

Fig. 13.20 Details of eaves connection

The dimension d_e is the distance from the compression flange to a point midway between the two top rows of bolts, i.e.

$$d_e = (0.850 + 0.760)/2 = 0.805$$

Hence $F_o = 836.4/[4.7 \times 0.805 + 2(0.670^2 + 0.580^2)/0.805] = 145.9$ kN.

The maximum bolt force (F_b) occurs in the top two rows of bolts and is equivalent to $1.35 F_o$. The additional $0.35 F_o$ in the top row of bolts is due entirely to the applied moment, while in the second row of bolts it represents a realistic estimate of the prying force. If the notional bolt load F_o is limited to 0.6 of ultimate load capacity of the bolt P_{ult}, then the maximum bolt load F_b has a safety margin against sudden rupture of at least 1.2, i.e. $0.6 \times 1.35 \times 1.2 < 1.0$.

Assuming that no such conditions exist, then the design load capacity of a grade 8.8, 24 mm diameter bolt is

$$P_o = 0.6 \times 785 \times 353/10^3 = 166.3 \text{ kN} > 145.9 \text{ kN}$$

Checking the design capacity of a 22 mm diameter bolt would indicate that its capacity was insufficient, i.e. 142.7 kN < 145.9 kN.

Use eight 24 mm diameter bolts (grade 8.8) (tension region)

A commonly accepted assumption is that the vertical shear is taken by the bolts in the compression zone of a connection. The minimum number of bolts required to carry the vertical shear of 200.2 kN is:

$$200.2/(0.6 \times 166.3) = 2.01 \text{ (say 2 bolts)}$$

Use at least two 24 mm diameter bolts (grade 8.8) (compression region)

13.9.1.3 DETERMINATION OF END PLATE THICKNESS

The centres of the holes (A) can be fabricator dependent, but a good guide is to make the dimension equal to approximately 4.5–5 times the bolt diameter, say 110 mm. The end plate thickness (t_p) is calculated by assuming double curvature bending of the plate, from which the following expression has been derived:

$$t_p = \sqrt{\frac{4P_o m}{p_{yp} L_e}}$$

where m = effective span from centre of bolt to edge of weld
$= (A - t_b - 2 \times \text{weld size})/2$
$= (110 - 9.1 - 2 \times 8)/2 = 42.5$ mm
L_e = effective length, based on 30° dispersal
$=$ lesser of $(C + 3.5m)$ or $7.0m$
$= (90 + 3.5 \times 42.5) = 238.8$ mm or $7 \times 42.5 = 297.5$ mm
$t_p = \sqrt{[(4 \times 166.3 \times 42.5)/(0.265 \times 238.8)]} = 21.1$ mm

Note that the design strength of plate sections is 265 N/mm². With the modified method, a simple rule is to make the end plate thickness equal to 0.9 times the bolt diameter, i.e. 21.6 mm, which is supported in this instance by the preceding calculation. However, owing to the discrete sizes of plate rolled by the steel producers, the designer has to choose between 20 or 25 mm plate. In this case, the value of 21.1 mm is less than 24 mm, use a 20 mm thick end plate.

Use 220mm × 20mm plate × 940mm long

13.9.1.4 LOCAL PLASTICITY ADJACENT TO END PLATE

In advocating the design method outlined in this section it is accepted that as a connection approaches its ultimate capacity, local areas of plasticity would have developed adjacent to the end plate, both on the tension and compression sides. On the tension side, there is a diffusion of load from the tension flange into the web and across to the end plate. This diffusion, as well as residual stresses due to welding, are two of a number of factors which interact to produce large plastic strains in this region. It is certainly true that the load distribution in the bolts results from load emanating from the haunch web as well as from the tension flange via the end plate. Assuming that the bolt load distribution reflects approximately the actual distribution in the tension bolt group, then a conservative check can be undertaken for the rafter web in the tension zone:

$p_{yb} > 4F_o /(L_e t_b)$
$275 > 4 \times 145.9 \times 10^3/(238.8 \times 9.1) = 269$ N/mm²

This indicates that the tension zone in the rafter member is adequate, otherwise stiffening of the end plate would need to be considered. Similarly the haunch flange and web in the compression zone could also exhibit plasticity. It is suggested, based on available research evidence, that these relatively small areas of plasticity are acceptable, when compared with the larger yielded zones associated with the formation of plastic hinges, in the later stages of loading.

13.9.1.5 CHECK THE TENSION REGION OF THE COLUMN

The column flanges need to be checked for the effects of cross-bending due to the action of the bolts[3]. The design formulae given in reference (3) have been modified; that is, it is assumed that the effect of a hole has been compensated by the flexural action of the bolt which effectively replaces the missing plate material. These modified formulae are used in the following calculations. Stiffening is required if:

$$4.7 F_o > T_c^2 \left[\frac{C + w + w^*}{m} + \left(\frac{1}{w} + \frac{1}{w^*} \right)(m + n) \right] p_{yc}$$

$w = \sqrt{[m(m+n)]}$
$n = (B - A)/2 = (220 - 110)/2 = 55.0 \text{ mm}$
$w = \sqrt{[42.5(42.5 + 55.0)]} = 64.4 \text{ mm}$
$w^* = 70 < 2w$

$$4.7 \times 145.9 > 14.8^2 \left[\frac{90 + 64.4 + 70}{42.5} + \left(\frac{1}{64.4} + \frac{1}{70} \right)(42.5 + 55.0) \right] 0.275$$

686 > 493 kN Inadequate

Therefore stiffening of the flange is required in the tension region. Check that the stiffened flange is adequate[3], i.e.

$$4.7 F_o < T_c^2 \left[\frac{2v + w + w^*}{m} + \left(\frac{2}{v} + \frac{1}{w} + \frac{1}{2w^*} \right)(m + n) \right] p_{ye}$$

$$686 < 14.8^2 \left[\frac{68 + 64.4 + 70}{42.5} + \left(\frac{2}{34} + \frac{1}{64.4} + \frac{1}{140} \right)(42.5 + 45) \right] 0.275$$

686 < 716 kN   Adequate

If the stiffened flange had been inadequate, a backing plate could be used or the section size changed.

13.9.1.6 CHECK SHEAR IN THE COLUMN WEB PANEL

The factored moment acting on the connection produces a shearing action in the column web adjacent to the connection. Therefore, the shear capacity of the web (P_v) needs to be checked against the induced shear force (F_v) of:

$F_v = M/d_e = 836.4/0.805 = 1039 \text{ kN}$

The following design rule, which is slightly more correct than the guidance given in BS 5950, is based on research evidence[18]:

$P_v = 0.6 t_c (D - 2T_c) p_{yc}$
$ = 0.6 \times 10.6(602.2 - 2 \times 14.8)0.275 = 1001 \text{ kN}$
$F_v > P_v$

The column web has to be stiffened. As the inner column flange in the tension zone has also to be stiffened (Section 13.9.1.4), use the Morris stiffener, which combines both functions in one stiffening arrangement, (see Fig. 13.20). This form of shear stiffening has been shown to be both economic and structurally efficient[18]. Make the horizontal portion equal to 100 mm, thereby allowing easy bolt access for the erectors. Design the Morris stiffeners like diagonal stiffeners – they have to carry the excess shear force not taken by the column web, i.e. $1039 - 1001 = 38$ kN. The required area of stiffeners is obtained from:

$$A_s > (F_v - P_v)/(p_{ys}\cos\theta)$$
$$\tan\theta = d_e/(D_c - 2T_c - 100)$$
$$= 805/(602.2 - 2 \times 14.8 - 100) = 1.703$$
$$\cos\theta = 0.506$$
$$A_s > (1039 - 1001)/(0.265 \times 0.506) = 283 \text{ mm}^2$$

Use nominal sized stiffeners, say two 90 mm × 10 mm flats, which provide some 1800mm² of area.

Use two 90 mm × 10 mm flats, 6 mm FW

13.9.1.7 CHECK COMPRESSION ZONE

Web buckling A simple rule based on experimental evidence indicates that stiffening to the web is required if:

$$d/t > 52\varepsilon \text{ (cf. BS 5950)}$$
i.e $d/t = 602.2/10.6 = 56.8 > 52$

Therefore the web needs stiffening to prevent plate buckling. Generally, full depth web stiffeners are required in this position.

Web crushing The force being transmitted from the compression flange of the haunched rafter into the column web is:

$$F_c = M/d_e + F = 836.4/0.805 + 195 = 1234 \text{ kN}$$

Stiffeners required if:

$$F_c > P_c = [T_b + 5(T_c + \text{root fillet}) + 2t_p]\, t_c\, p_{yc}$$
$$= [14.5 + 5(14.8 + 12.7) + 2 \times 20]\, 10.6 \times 0.275 = 559.7 \text{ kN}$$
$$1234 > 559.7 \text{ Stiffener required}$$

Web stiffeners (placed on either side of the column web, opposite the rafter compression flange) are required to prevent both web buckling and crushing. The capacity of the stiffened column web in the compression zone is given by:

$$P_{vs} = A_s p_{ys} + 1.63 T_c (B_c\, t_c)^{1/2}\, p_{yc}$$

hence $A_s > [F_v - 1.63 T_c (B_c\, t_c)^{1/2}\, p_{yc}]/p_{ys}$
$$= [1234 - 1.63 \times 14.8(227.6 \times 10.6)^{1/2} 0.275]/0.265$$
$$= 3427 \text{ mm}^2$$

Use two 100 mm × 20 mm flats, 6 mm FW

Check the outstand edge of these compression stiffeners for buckling; stiffeners are adequate if:

$b/t < 7.5\varepsilon$
$b/t = 100/20 = 5.6 < 7.5\varepsilon$ Stiffeners adequate

13.9.1.8 CHECK FOR REVERSED MOMENT CONDITION

From Fig. 13.13(b), it can be seen that the eaves connection is also subjected to a reversed moment of 326 kN m, therefore the proposed connection details must be checked as to their suitability to sustain this moment. The following checks are based on the assumption that the size of the end plate and bolt diameter are not changed.

Moment capacity Assuming that the connection would rotate about the top flange, then the moment capacity of the connection as designed is:

$$M = 2 \times 166.3 \times 0.830 = 276 \; 326 \, \text{kNm}$$

i.e. moment capacity has to be increased. This is achieved by inserting two additional bolts in the bottom zone of the connection (see Fig. 13.20). Hence:

$$M = 2 \times 166.3 \, (0.830 + 0.740) = 522 > 326 \, \text{kNm}$$

Vertical shear The six bolts in the upper part of the connection are more than sufficient to cope with the vertical shear from the dead + wind case.

'Tension' region of column This particular region of the column is already reinforced by full depth web stiffness for a much larger force; it would be safe to assume that this region is adequate without a detailed check.

Shear in the column web panel As the reversed moment of 326 kNm is about 70% of the moment on which the original design was made, then the provisions for shear stiffening should prove adequate (see Fig. 13.20).

'Compression' zone of column The load to be transferred into the column web in the 'compression' zone (top of the column member) is:

$$\begin{aligned} F_c &= M/d_e + R \\ &= 326 \, / [(0.830 + 0.740)/2] - 70.7 = 345 \, \text{kN} \end{aligned}$$

Note that the axial load in the rafter for the dead + wind case is tension.
    It is known that the effectiveness of web stiffeners, when not placed in line with the application of the compression load, decreases rapidly as they are positioned further away from the load. In this case, it is decided to ignore any contribution that the horizontal part of the Morris stiffener may have on the bearing strength of the column web in the vicinity of the 'compression' zone.

With reference to Fig. 13.20, can be seen that the compression force from the haunched rafter flange is resisted only by about half the web length compared with that used in Section 13.9.1.6, i.e.

$$P_c = [T_b + 2.5(T_c + \text{root fillet}) + 2t_p]t_c\, p_{yc}$$
$$= [12.7 + 2.5\,(14.8 + 12.7) + 2 \times 20]\,10.6 \times 0.275 = 354\,\text{kN}$$

Nevertheless, the bearing capacity of the column web is adequate.
This completes the design of the eaves connection.

13.9.2 DESIGN OF APEX CONNECTION

The design of the apex connection is simpler than that of the eaves connection insofar as only the depth of haunch, geometry of end plate and bolt diameter need to be determined, i.e. no column is involved (see Fig. 13.21).

Fig. 13.21 Details of apex connection

The apex connection has to be designed for a moment of 416.3 kNm. (Note that if the apex moment has not been evaluated then a good estimate is to make it equal to the full plastic moment of the rafter section as the bending moment in this one is virtually constant, i.e. for this exercise it would be only 2% in error and would not affect the outcome.)

The vertical shear is theoretically zero (due to symmetrical loading and frame). Nevertheless, it is advisable to have a minimum of two bolts in the compression region. The connection details need to be checked if there is moment reversal due to other loading conditions, which in this example is 57 kNm.

13.9.2.1 PROPORTIONS OF APEX HAUNCH

The apex connection has to be haunched in order to increase the tension bolt group lever arm so that the connection has sufficient moment capacity. The actual depth of haunch is chosen to accommodate sufficient bolts within its depth to sustain the moment. This is a trial and error process. In this design case, it has been decided to make the overall depth of the connection 680 mm deep (see Fig. 13.21). To allow for the dispersion of the tension flange load to the bolts the length of haunch should be at least 1.5 times the depth of basic rafter section, i.e. $1.5 \times 457 = 685$ mm, or twice the depth of the haunch cutting, i.e. $2(680 - 457/\cos 10.41°) = 430$ mm. Therefore, make haunch length equal to 700 mm. Use the minimum weld size (6 mm FW) to connect the haunch cutting to the underside of the basic rafter member.

13.9.2.2 SIZING OF BOLTS

Having designed the eaves connection and established the bolt diameter to be 24 mm, it is economic to standardize on the size of bolts throughout the frame, i.e. check that 24 mm diameter bolts (grade 8.8) are suitable for the apex connection. Again, the ratio of the two largest lever arms of the tension bolts about the compression flange is evaluated, i.e.

$$0.530 / 0.620 = 0.85$$

Therefore, the bolt load distribution is assumed to be linear, i.e. it varies linearly with their distances from the compression flange, and the notional bolt load (F_o) can be determined, i.e.

$$\begin{aligned}F_o &= M/\,[2.7 y_{max} + 2\,(\Sigma y_i\,/y_{max})] \\ &= 416.3/[2.7 \times 0.620 + 2\,(0.530^2)/0.620] \\ &= 161.3 \text{ kN} < 166.3 \text{ kN}\end{aligned}$$

Though the four outermost tension bolts are adequate, it is advisable to position another row of bolts inside the depth of the basic rafter section, adjacent to its tension flange, to prevent local separation of end plates.

Use six 24 mm diameter bolts (grade 8.8) (tension region)
Use two 24 mm diameter bolts (grade 8.8) (compression region)

13.9.2.3 DESIGN OF END PLATE

As the rafter size is the same for both the eaves and apex connections, then make the width of the end plate the same as that for the eaves, i.e. >220 mm. Assume the same centres of holes (110 mm).

Use 220 mm × 20 mm plate × 720 mm long

As the basis of the calculations for the flange and web welds is identical to that determined for the eaves connection, then make:

Flange weld 12 mm FW
Web weld 8 mm FW

13.9.2.4 CHECK FOR REVERSED MOMENT CONDITION

Assuming that the size of the end plate and bolt diameter are unchanged, then the only check required is for the tension region. The moment capacity of the two bolts in this zone (Fig. 13.21) is:

$$M = 2 \times 166.3 \times 0.620 = 206 \, \text{kNm} > 57 \, \text{kNm}$$

Therefore the apex connection as detailed in Fig. 13.21 is satisfactory.

13.10 GABLE FRAMING

When it is specified that there are to be no extensions to a building in the future, and bearing in mind that the gable framing has to support only half the load carried by an intermediate main frame, then the gable arrangement shown in Fig. 13.22 is commonly used. Basically, the gable framing consists of inclined beam members (supporting the purlins and some gable sheeting), spanning between the vertical gable posts.

Fig. 13.22 Gable framing

In deciding the spacing of the gable posts, any dominant openings in the gables would have to be taken into account. However, in this example, it has been assumed there are no openings and it is proposed that the gable posts be positioned approximately 6 m apart, i.e. the span for which the sheeting rail was originally chosen. The gables are subject to a maximum wind pressure of $1.0 \times 0.59 \, \text{kN/m}^2$ (see Section 13.10.1.2). A quick check on the load capacity of the sheeting rails (Multibeam R145130) indicates that the proposed gable post spacing of 6.250 m for the four internal spans and $6.000 + 0.220$ (est.) = 6.220 m for the two outer spans is acceptable.

The in-plane stability for this relatively flexible gable framing is achieved by incorporating vertical gable bracing into the end bays of the gable (see Fig. 13.22a). The bracing members are designed as struts, resisting the side wind load acting on the corner gable posts. By triangulating the bracing as shown, additional wind load is induced into the edge members in the gable and bays.

13.10.1 Gable edge beams

For design purposes, consider the gable beams adjacent to the ridge of the building as being typical of the edge members. Such members are usually assumed to be simply supported, being designed to carry the gable cladding and wind loads, as well as the purlin loads (based on the revised spacing of 1.500 m on plan), back to the gable posts. The actual length of the member to be designed is 6.250/cos 10.41° = 6.355 m.

13.10.1.1 DESIGN LOADING

In most design situations the loads are usually relative simple to evaluate, but on occasions the time spent in specifying a loading regime precisely (and consequently the design forces) is not worth the disproportionate effort. In these circumstances, it is advantageous to make safe assumptions in order to effect a quick design solution. The assessment of the cladding weight acting on the edge beam and the wind load on that cladding represents one of those occasions (see Figs. 13.22 and 13.23a).

Fig. 13.23 Loading for gable edge beams

(a) Dimensions

(b) Dead + imposed loading (factored) (kN)

(c) Dead + wind loading (factored) (kN)

DESIGN OF SINGLE-STOREY BUILDING – PORTAL FRAME CONSTRUCTION

In order to simplify the calculations in this case, it is assumed that the edge member supports half the sheeting down to the sheeting rail 2, i.e. the effect of sheeting rail 1 is ignored. Also, it is assumed that the resulting distributed loads act uniformly along the member.

Thus, the various unfactored loads acting on the edge member are:

Dead load via each purlin $= (0.21 + 0.87)/2 = 0.55$ kN
Imposed load via each purlin $= (0.75 \times 1.500 \times 6.0)/2 = 3.38$ kN
(snow)
Wind load via each purlin $= -(1.4 \times 0.59 \times 1.500 \times 6.0)/2$
$= -3.72$ kN

Dead load (cladding +
self weight) $= 0.13(2.15 + 1.00) \times 6.250/4 + 2.3$ (est.)
$= 0.64 + 2.3 = 2.94$ kN

Wind load acting on gable sheeting The coexistent wind loading on the gable, associated with the uplift coefficient of 1.4 on the roof (for **wind on the side of the building**) is 0.8 suction (see Section 12.4.3 for explanation), hence the wind load on the relevant gable cladding is:

$$-0.8 \times 0.59 \, (2.15 + 1.00) \times 6.25/4 = -2.32 \text{ kN}$$

Axial load due to wind on gable Wind on the gable generates loads in the rafter bracing section (see Fig. 13.22b). As the edge members form part of this system, they have to carry axial load, the magnitude and nature of which depend on the particular wind condition occurring on the gable. For example, assuming that the wind load on the gable is $-0.8q$, then with reference to Fig. 13.26(a), the appropriate axial load in the edge member being considered is $-0.8 \times 65.29 = -52.7$ kN. Though there is frictional wind drag across the building, it is relatively small and is resisted by frame action and therefore ignored.

The other wind case which might prove more critical occurs when the wind blows on the gable end, producing $1.0q$ pressure on the sheeting, and a corresponding coefficient for the roof of $-1.0q$. The latter results in reduced uplift forces on the roof members, but the axial load becomes compression $(1.0 \times 65.29 \text{ kN})$. Also, the wind drag along the building is significant (see Fig. 13.26b), producing compression in the edge member $(1.0 \times 18.4 \text{ kN})$. Therefore, both wind cases have to be examined.

Dead + imposed load case (Fig. 13.23b) Factoring the loads by the appropriate partial load factors, the maximum moments acting on the member are calculated, i.e.

$$R_{LH} = (1.4 \times 0.55 + 1.6 \times 3.38)\frac{(0.250 + 1.725 + 3.250 + 4.750)}{6.25}$$
$$+ \frac{1.4 \times 3.02}{2}$$
$$= 6.18 \times 1.60 + 4.22/2 = 12.00 \text{ kN}$$

$$M_x = 12.00 \times 3.000 - 6.18 \times 1.500 - \frac{4.22 \times 3.000^2}{6.25 \times 2} = 23.7 \text{ kNm}$$

$$M_y = 0.0 \text{ kNm}$$

Dead + wind load case

(i) Wind suction on gable (−0.8q): again factoring the loads with reference to Fig. 13.23(c), the maximum design conditions associated with wind on the side of the building are calculated, i.e.

$$F_t = 1.4 \times 52.7 = 73.8 \text{ kN}$$

$$R_{LH} = (1.0 \times 0.55 - 1.4 \times 3.72)\frac{(0.250 + 1.750 + 3.250 + 4.750)}{6.25}$$
$$+ \frac{1.0 \times 2.94}{2}$$
$$= -4.66 \times 1.60 + 2.94/2 = -5.99 \text{ kN}$$

$$M_x = -5.99 \times 3.000 + 4.66 \times 1.500 - \frac{2.94 \times 3.000}{6.25 \times 2} = -11.7 \text{ kNm}$$

$$M_y = -3.25 \times 5.25/8 = -2.54 \text{ kNm}$$

(ii) Wind pressure on gable (1.0q): using the appropriate wind coefficients for this condition, then by similar calculations as in (i):

$$F_c = 117.2 \text{ kN}$$
$$M_x = -8.9 \text{ kNm}$$
$$M_y = +3.2 \text{ kNm}$$

13.10.1.2 MEMBER SIZE

As the member is loaded between the positional restraint provided by the simple connections, then $m = 1.0$ (BS table 13) and the member need only satisfy the following criterion for the two design cases:

$$\frac{M_x}{M_b} + \frac{M_y}{p_y Z_y} < 1.0$$

Unlike the purlin loading on intermediate portal frames where the load supported by the purlins is balanced about the vertical plane of the portal frame, the loading supported by the purlins attached to the edge member is not balanced. This represents a destabilizing condition and therefore $n = 1$ (BS table 13).

As the member has been assumed to be simply supported, then the effective length, $L_{Ex} = 1.0 \times 6.355 = 6.355$ m. For the dead + imposed load case, the top flange of the edge member is in compression, and is restrained at intervals by the purlins, therefore the effective length, $L_{Ey} = 1.0 \times 1.525 = 1.525$ m. However, the reverse is true for the dead + wind load case, i.e. the bottom flange, being in compression, is not restrained between the connections, therefore, $L_{Ey} = 1.0 \times 6.355 = 6.355$ m.

For lightly loaded gable edge beams, a channel section is commonly used, as it can be bolted directly (when suitably notched) onto the outside of the gable posts. This enables the cleats supporting the sheeting rails to be positioned in the same vertical plane without the use of special cleats. However, it has been decided to use a UC section which has better properties than a channel (weight for weight). The beams are to be positioned on the centre lines of the posts, which might result in special cleats for the sheeting rails. An alternative to using special cleats is to arrange the centres of the sheeting rails on the gables so that the rails are supported by the posts and not the edge members.

The design process can be shortened by making use of the tabulated values for the bending and compression resistances, given in the SCI guide[6], for the different hot-rolled sections for both grades 43 and 50 steels. Therefore, with reference to pp.32 and 176 of the guide[6], a **152 × 152 × 30 UC** is chosen and now has to be checked.

Dead + imposed load case

$$F = 0.0\,\text{kN}$$
$$M_x = 23.7\,\text{kNm}$$
$$M_y = 0.0\,\text{kNm}$$
$$L_{Ey} = 1.525\,\text{m}$$
$$\lambda = 40$$

From p.176, SCI guide, for the chosen section $M_b = 66.9\,\text{kN m}$, hence:

$$\frac{23.7}{66.9} + 0.0 = 0.354 < 1.0$$

Dead + wind load case (i)

$$F_t = 73.1\,\text{kN}$$
$$M_x = 11.7\,\text{kNm}$$
$$M_y = 2.54\,\text{kNm}$$
$$L_{Ey} = 6.355\,\text{m}$$
$$\lambda = 166$$

From p.176, SCI guide, $M_b = 35.7\,\text{kNm}$ and $p_y Z_y = 20.0\,\text{kNm}$, hence a safe estimate of the member buckling resistance is obtained by ignoring the tension:

$$\frac{11.7}{35.7} + \frac{2.54}{20.0} = 0.325 + 0.127 = 0.452 < 1.0$$

Dead + wind load case (ii)

$$F_c = 117.2\,\text{kN}$$
$$M_x = 8.9\,\text{kNm}$$
$$M_y = 3.2\,\text{kNm}$$
$$L_{Ey} = 6.355\,\text{m}$$
$$\lambda = 166$$

Using the compression resistance of $P_c = 220\,\text{kN}$ (p. 176, SCI guide) together with the values of M_b and $p_y Z_y$ already obtained, then

$$\frac{117.2}{220} + \frac{8.9}{35.7} + \frac{3.2}{20.0} = 0.533 + 0.249 + 0.160 = 0.942 < 1.0$$

Use 152 × 152 × 30 UC

Use the same section for all gable edge beams, and check that a more severe design condition does not exist. A typical beam–post intersection is detailed in Fig. 13.24).

13.10.2 Gable posts

The **central gable** post is to be designed as it has to sustain the worst design condition of all the posts. The posts are assumed to be simply supported between the base and the positional restraint provided by the rafter bracing (see Fig. 13.22).

For the wind suction condition (-0.8×0.59 kN/m^2), it would appear that the inner compression flange is unrestrained between the base and the positional restraint of the rafter bracing. The benefit of the sheeting rail restraint on the outer tension flange could be taken into account by using the stability clauses of appendix G, BS 5950, as was done in checking the wind condition for the main rafter, apex region (Section 13.8.2.3). In the latter case, the likelihood of the purlins being removed permanently is very remote. However, there is a greater possibility that the owner (who may be different from the original developer) may require other arrangements with respect to openings in the future.

Therefore, it is decided (for simplicity) to ignore this potential benefit from the rails. Nevertheless, it is felt that any openings would probably not extend above the eaves level, with the results that it is proposed to restrain laterally the inner flange of the five internal gable posts at eaves level, by bracing back to the sheeting rail, i.e. 5.10 m from the ground. Therefore, the design assumes that the gable posts are unrestrained up to the sheeting rail at 'eaves level', in which case the worse wind condition is the wind pressure of 1.0×0.59 kN/m^2.

13.10.2.1 DESIGN LOADING

The axial load includes the self weight of the rails and post, together with the calculated end reactions from the appropriate edge members, plus apex purlin loads and weight of the cladding and insulation (not included in end reactions of edge members). However, the axial load is usually relatively small, the main loading being the bending action induced into the post by the wind loading acting in the gable.

Dead + imposed load case

$$F_c = 1.4[(\text{rails+post}) + (\text{end reactions} + \text{apex purlins}) + (\text{cladding})]$$
$$= 1.4[0.6 \times 8.90 + (0.55 + 3.38 \times 1.6/1.4)(2 \times 1.64 + 1)$$
$$+ 0.13 \times 8.32 \text{ (av. ht.)} \times 6.25]$$
$$= 1.4(5.34 + 18.89 + 6.76) = 43.4 \text{ kN}$$
$$M_x = 0.0 \text{ kN m}$$
$$M_y = 0.0 \text{ kN m}$$

Dead + wind load case The wind loads used in these calculations are based on the wind pressure condition, i.e. 1.0×0.59 kN/mm^2. Note that in this combination of dead + wind, both the partial load factors are equal to 1.4, as uplift is not the condition being examined.

$$F = 1.4[0.6 \times 8.90 + (0.55 - 3.72)(2 \times 1.64 + 1) + 6.76]$$
$$= 1.4(5.34 - 13.57 + 6.76) = -2.1 \text{ kN (tension)}$$

Wind load on a typical sheeting rail is:

$$1.4\ (1.0 \times 0.59 \times 1.65 \times 6.25) \qquad = 8.52\ \text{kN}$$

Again, ignoring the small local effect from rail 1 (Fig. 13.23a), then by proportion the wind load on rail 2 is:

$$8.52[(2 \times 2.15 + 1.00\ )/6 + 1.65/2]/1.60 = 9.10\ \text{kN}$$

and similarly for the bottom rail, the wind load is:

$$8.52\ (1.65 + 0.15)/(2 \times 1.60) \qquad = 4.80\ \text{kN}$$

Therefore, the end reaction at the top of the post is:

$$R_{TP} = \frac{4.80 \times 0.15 + 8.52(1.80 + 3.45 + 5.10) + 9.10 \times 6.75}{8.90}$$

$$= 16.89\ \text{kN}$$

$$M_x = 16.89 \times 3.80 - 9.10 \times 3.30 - 8.52 \times 1.65 = 20.1\ \text{kNm}$$

$$M_y = 0.0\ \text{kNm}$$

13.10.2.2 MEMBER SIZE

By ignoring any restraint from the rails below eaves level, then:

$$L_{Ey} = 1.0 \times 5.10 = 5.10\ \text{m}: \qquad L_{Ex} = 1.0 \times 8.90 = 8.90\ \text{m}$$

With the experience gained in Section 13.10.1.2, it is advantageous initially to design the gable post for the dead + wind load cases, and then check its adequacy for the dead + imposed load case.

Referring to p. 134, SCI guide, it can be seen that a $254 \times 102 \times 28$ UB could be suitable. Therefore, check the adequacy of this section. The buckling resistance M_b of the universal section, for an effective length of 5.10 m, is 26.4 kN m, ignoring the effect of tension, hence checking the **dead + wind load case** gives:

$$\frac{20.1}{26.4} = 0.761 < 1.0 \qquad \text{Section is adequate.}$$

Use 254 × 102 × 28 UB

If the restraint afforded by the rails is taken into account, then the wind suction load case ($0.8 \times 0.59\ \text{kN/m}^2$) would be the worst design condition, i.e. causing the inner flange to go into compression, in which case a $254 \times 102 \times 25$ UB would probably prove satisfactory.

As there is only 43.4 kN load acting on the gable posts for the **dead + imposed load case**, then clearly the section is more than adequate, i.e. $43.4/914 = 0.047 < 1.0$. The compression resistance of 914 kN is obtained from p. 85, SCI guide[6]. Figure 13.24 shows a typical detail at the top of the gable post.

The same section size can be used for all other gable posts. However, a check should be made on the corner gable post, which, though supporting only half the load, is subject to wind loads acting simultaneously about the major and minor axes.

Fig. 13.24 Detail of gable beams and post

(a) Elevation
(b) Plan

13.10.3 Corner gable posts

It is common practice to arrange the outer flange of the corner post to be in the same vertical plane as the outer flange of the portal leg, i.e. the corner post is rotated through 90° compared with other gable posts. This means the worst design condition occurs when the wind direction is normal to the gable, resulting in a wind pressure of $1.0 \times 0.59 \, \text{kN/m}^2$ on the gable sheeting and a wind suction of $-0.2 \times 0.59 \, \text{kN/m}^2$ acting on the side walls.

The typical wind load on a gable sheeting rail acting on the corner gable post is:

$$1.4(1.0 \times 0.59 \times 3.1 \times 1.65) = 4.22 \, \text{kN}$$

The other rail loads are obtained by proportion, to give the loading pattern shown in Fig. 13.25. The end reaction at the top of the post about its minor axis for the loads acting normal to the web of the post is:

$$R_{TPy} = \frac{2.30 \times 0.15 + 4.22(1.80 + 3.45) + 3.50 \times 5.10}{5.95} = 6.78 \, \text{kN}$$

$$M_y = 6.78 \times 2.5 - 3.50 \times 1.65 = 11.2 \, \text{kN}$$

Also, the reaction about the post's major axis is:

$$R_{TPy} = \frac{0.46 \times 0.15 + 0.84 \times 1.80 + 0.80 \times 3.45 + 0.66 \times 4.90}{5.95}$$

$$= 1.22 \, \text{kN}$$

$$M_x = 1.22 \times 2.5 - 0.66 \times 1.45 = 2.09 \, \text{kN}$$

Fig. 13.25 Loading in corner gable post

As was the case with the internal posts, the effect of the axial load (which includes the load induced by bracing) is extremely small and can be ignored. Check that the section size used for the intermediate gable posts is satisfactory. Therefore, as the member is loaded between end restraints, $m = 1.0$ and the design criterion again becomes:

$$\frac{M_x}{M_b} + \frac{M_y}{p_y M_y} \leq 1.0$$

$$\frac{2.09}{27.3} + \frac{11.2}{0.275 \times 34.9} = 0.077 + 1.167 \not< 1.0 \qquad \text{Inadequate}$$

Try a 254 × 146 × 31 UB section. From pp. 29 and 134, SCI guide[6], $Z_y = 61.5 \text{ cm}^3$ and $M_b = 45.5 \text{ kNm}$, respectively, hence:

$$\frac{2.09}{45.5} + \frac{11.2}{0.275 \times 61.5} = 0.046 + 0.662 < 1.0$$

Clearly, there is more than enough reserve of strength to cater for the small axial load.

Use 254 × 146 × 31 UB

13.10.4 In-plane gable bracing

Referring to Figs. 13.22 and 13.25, it can be seen that the wind direction acting perpendicular to the side wall would produce the worse design condition for the diagonal bracing member, i.e.

Wind load acting at top of corner post $= 6.78 + 1.22$ $= 8.00 \text{ kN}$
Length of bracing member $= \sqrt{(6.20^2 + 5.95^2)} = 8.6 \text{ m}$
Wind load in bracing member $= 8.6 \times 8.00/6.2$ $= 11.1 \text{ kN}$

From p. 89 of the SCI guide, for an effective length of 8.6 m and a slenderness not exceeding 250, use a **114 × 3.6 CHS**.

13.11 OVERALL STABILITY OF BUILDING

The designer must always ensure the structural stability of the building. At this stage, both the portal frames and the gable framing have been designed for in-plane stability, particularly with respect to side wind loading. However, in order to provide stability to the building in its longitudinal direction, all frames need to be connected back to a braced bay. Generally, the **end bay(s)** of the building are braced, so that the wind loads acting on the gables can be transferred to the foundations as soon as possible, thereby not affecting the rest of the structure. Another function of a braced bay is that it ensures the squareness and verticality of the structural framework, both during and after erection.

The typical bracing system for a portal framed building usually takes the form of **rafter bracing** in the plane of the roof space (positioned as close to the top flange of the rafter without fouling the purlins), linked into a **vertical bracing** system (Fig. 13.22). These bracing systems are designed to cater for wind loading on the gable, plus the wind drag forces along the building. Figure 13.26(a) gives the forces in the rafter bracing due to a wind pressure of $1.0q$, assuming that half the load on the gable sheeting is taken by the bracing, while Fig. 13.26(b) gives the effect of drag on the rafter bracing. These forces are transferred via the rafter bracing to the vertical bracing system, which also transfers the wind drag forces from the side cladding. Figure 13.27(a) gives the unfactored loads in the vertical bracing transferred from the rafter bracing due to a unit wind pressure ($1.0q$) acting on the gable. Figure 13.27(b) indicates the unfactored loads in the vertical bracing due to the wind drag forces acting on the roof and sides. From these two sets of

Fig. 13.26 Unfactored loads on rafter bracing (kN)

(a) Wind pressure on gable

(b) Wind drag

Fig. 13.27 Unfactored loads in vertical bracing (kN)

(a) Wind pressure on gable

(b) Wind drag

loads, the factored loads for the different design cases can be deduced. The design calculations for both bracing systems for this example are not given, as they are similar to the appropriate detailed calculations given in Section 12.8. Also, to give longitudinal stability between the braced bay(s), **eaves ties** are required (see Section 12.8).

Those eaves ties give positional restraint to the top of the column members. In the normal design situation for portal frames (dead+imposed), unless there is some means of connecting the eaves ties to the inside corner of the haunch/column intersection, then this corner cannot be construed as being restrained by the eaves ties. Such restraint is provided by bracing back from the corner to a convenient sheeting rail. Sometimes, a single member is utilized to combine the two functions of eaves tie and gutter support.

13.12 DESIGN OF MAIN COLUMN BASE

The column base has to be designed for the vertical load, horizontal shear and zero moment ('pinned' base condition). The maximum factored horizontal shear which arises from the dead + imposed load case is 161.3 kN. The coexistent factored vertical load is 200.2 kN. However, the latter is basically the load from the roof, and any additional loading due to side cladding, insulation, liner, etc. has to be included, i.e.

- Weight of cladding
  (including insulation and liner) $0.13 \times 6.0 \times 5.95 = 4.6$
- Weight of side rails $\quad 5 \times 0.045 \times 6.0 = 1.4$
- Self weight of column $\quad 101 \times 5.95 \times 9.81/1000 = 5.9$ $\}$ 12.8 kN
- Weight of gutter[see reference 11, p. 224] $\quad 0.15 \times 6.0$
  $= 0.9$

Therefore, the total factored axial load is $200.2 + 1.4 \times 12.8 = 216$ kN.

13.12.1 Design of column base-plate

As the base carries no moment, then the common detail in these circumstances is either to place two holding down bolts along the neutral axis of the column section, at right angles to the column web, or to position four HD bolts just inside the section profile (see Fig. 12.34). The latter detail will be used as it affords a certain amount of moment resistance which could prove useful in case of a fire[19] and it helps erectors to position columns accurately. Thus, the base plate should be made wide enough for the plate to be welded to the column, i.e. 620 mm × 240 mm (see Fig. 13.28). Note that the grout hole in this relatively large base plate ensures that the grouting cement can flow easily under the entire base plate, thereby eliminating voids.

Fig. 13.28 Base details

Using a concrete mix for the foundation having a cube strength of $f_{cu} = 30\,\text{N/mm}^2$, the bearing pressure should not exceed $0.4 \times 30 = 12\,\text{N/mm}^2$ (clause 4.13.1):

$$\text{Bearing pressure} = 216 \times 10^3/(620 \times 240) = 1.45\,\text{N/mm}^2$$

As the projections of the base plate beyond the profile of the column section are minimal, then the formula given in clause 4.13.2.2 cannot be used as it would result in a small value for the plate thickness. Therefore it is recommended that the base plate thickness > flange thickness. Use a 20 mm thick base-plate (grade 43 steel); this is the minimum practical thickness used in the construction industry for this size of base.

The welds connecting the column member to the base plate need to transfer a horizontal shear of 161.3 kN. If a fillet weld were placed continuously around the profile of the column section then its length would be approximately 2 m, hence the required design strength of the weld is $161.3/2000 = 0.08\,\text{kN/mm}$, i.e. nominal size is required, use 6 mm FW.

Use 240 mm × 20 mm × 620 mm long
6 mm FW

13.12.2 Sizing of holding down bolts

Where the axial load is transmitted by the base plate (without moment) then nominal holding down bolts are required for location purposes; see Section 8.2 and reference (20). Assume **four** 24 mm diameter bolts, as smaller diameter bolts are more prone to damage. Nevertheless, these bolts may need to transfer the horizontal shear of 161.3 kN into the concrete foundation block if the bond between steel base and grout fails:

Shear/bolt $\qquad\qquad\quad = 161.3/4 \qquad = 40.3\,\text{kN}$
Shear capacity of bolt $= 0.160 \times 353 = 56.5\,\text{kN}$

Although the portal frame bases have been assumed to be pinned, the base detail does generate a certain amount of fixing moment at the base plate/concrete block interface. It is recommended that the HD bolts be designed to resist 10% of M_{ps}, i.e. 78.9 kNm. It is not recommended to use this partial base fixity in the design of the portal frame as there is no guarantee that the concrete block/soil interface would sustain the moment, i.e. rotation of the concrete block would negate any partial fixity. There are exceptions, e.g. piled foundations, when any built-in base fixity would be maintained.

Assuming the lever arm between the column flange and the outer bolts is 550 mm, the tension that would be induced in each bolt is:

Tension/bolt $\qquad\qquad\quad = 78.9/(0.550 \times 2) = 71.2\,\text{kN}$
Tension capacity of bolt $= 0.195 \times 353 \qquad = 68.8\,\text{kN}$

The tension capacity of the 24 mm diameter bolts (grade 4.6) is inadequate, therefore use 27 mm diameter bolts (grade 4.6):

$$\frac{F_s}{P_s} + \frac{F_t}{P_t} \leq 1.4$$

$$\frac{40.3}{73.4} + \frac{71.2}{89.5} = 0.594 + 0.796 = 1.35 \leq 1.4$$

Use four 27 mm diameter bolts (grade 4.6 steel)

It is advisable to use double nuts to prevent loosening of the nuts and the possibility of thread stripping.

13.13 DESIGN OF FOUNDATION BLOCK

The design of the foundations for any structure is very dependent on the ground conditions that exist on site. It is important that the engineer has this data available or some reasonable basis for formulating the foundation design. In this case, a site investigation has indicated that the soil conditions are such that they can support a bearing pressure of 150 kN/m². This pressure is a **permissible** value and is applicable to serviceability conditions, i.e. working load level. Therefore, the following design cases need to be checked:

(A) $1.0w_d + 1.0w_i$
(B) $1.0w_d + 1.4w_w$
(C) $1.0w_d + 1.0w_i + 1.0w_w$

First, determine the serviceability loads acting on the foundation block for the design case (A):

$$\begin{aligned}
\text{Load ex. roof} &= \alpha[1.0(w_d + w_i)L]/(2\cos 10.41°) \\
&= 1.141[1.0(1.60 + 4.44)37.0]/(2 \times 0.984) \\
&= 1.141(30.1 + 83.5) = 134.5 \text{ kN} \\
\text{Column loading} &= 12.8 \text{ kN} \\
F_c &= 134.5 + 12.8 = 147.3 \text{ kN}
\end{aligned}$$

This vertical load should be used in conjunction with the appropriate horizontal shear, which is evaluated by multiplying the factored horizontal load by the ratio of the unfactored vertical load to the factored vertical load, i.e.

$$F_h = 161.3 \times 134.5/(351/2) = 123.6 \text{ kN}$$

To be strictly correct, the horizontal shear obtained from an elastic analysis using unfactored loads should have been used, i.e. 121.4 kN. However, for portal frames with shallow pitched roofs, the two values are almost identical[1], e.g. 123.6 kN compared with 121.4 kN. Therefore, the adjusted value from the plastic analysis is acceptable. The design case (A) loading is shown in Fig. 13.29.

Fig. 13.29 Resultant forces on base

As shown in Section 12.11, design case (B) tends to govern the design of the foundation block. In order to maximize the wind uplift condition (dead + wind), α is taken as unity, hence:

$$F_v = 30.1 - 1.4[0.59 \times 6.0 \times 18.5(1.4 \times 0.75 + 0.6 \times 0.25)] + 12.8$$
$$= +42.9 - 110.0 = -67.1 \text{ kN}$$

Again, using the horizontal shear from the plastic analysis, adjusted for serviceability conditions, i.e.

$$F_h = 157.2(30.1 - 110.0)/[2.654(1.4 \times 30.1 - 110.0)] = -69.7 \text{ kN}$$

Figure 13.29(b) shows the loading for design case (B).

In the design of the foundation block for the column members in the previous chapter (Section 12.11) the block was made square in shape. However, with the column member size being larger in this example, it might prove more economic to use a rectangular shaped base. Again, use mass concrete of sufficient depth to spread the vertical load at 45° through the concrete block to the substrata. Therefore, try initially a foundation block of 2.2 m × 1.8 m × 1.0 m proportions, which weighs 2.2 × 1.8 × 1.0 × 23.7 = 78.2 kN (see Fig. 13.28b). Hence, the bearing pressure at the concrete/soil interface for the maximum vertical load condition is $(147.3 + 78.2) / (2.2 \times 1.8) = 57 \text{ kN/m}^2$ which is satisfactory.

The combination of vertical and horizontal loads on the base should be considered in design. The proposed foundation block now has to be checked to show the resultant soil bearing force lies within the middle third of the base length, i.e. not more than $L/6$ (=0.367 m) from the base centre-line. Both load combinations (A) and (B) are now examined. Anticipating the magnitude of the overturning moments, the column is positioned at 0.7 m from the centreline of the proposed block (see Fig. 13.29a). This causes the resultant forces to act to the concrete/soil interface for the two cases:

Case (A) $\dfrac{123.6 \times 1.0 - 147.3 \times 0.7}{78.2 + 147.3} = 0.075 \text{ m}$ (satisfactory)

Case (B) $\dfrac{69.7 \times 1.0 - 67.1 \times 0.7}{78.2 - 67.1} = 2.05 \text{ m}$

Case (B) is not satisfactory: had the margin been small, then one could either increase the size of the block or allow some tension at the concrete/soil interface, i.e. cause the resultant to act outside the $L/6$ dimension. However, in this particular case it would be economic to eliminate the moment due to the horizontal shear by tying the foundation block to the floor slab (assuming the slab is at the same level). That is, the floor acts as a tie and absorbs the horizontal shear. Reinforcement would need to be incorporated at floor slab level in both the slab and the block. Detailed design can be found in good concrete design textbooks.

A special block will be required for the penultimate frame to counterbalance the additional wind uplift from the vertical bracing system (see Fig. 13.27 and Section 12.11).

13.14 OTHER CONSIDERATIONS

Although this example deals only with the design of a single-storey, pitched portal frame, the designer should be aware of other considerations which might affect the final design, see Section 12.12. Furthermore, the portal frame can take many forms, see reference (3), be subjected to non-uniform loading and have several spans. Each frame needs to be designed carefully taking full cognizance of the available published information, the main criteria being strength, stiffness and economy.

Though the strength of a portal frame can be readily checked by a plastic or elastic analysis, it is implicit that the stresses in the haunched region are not greater than the appropriate design strength. The in-plane stiffness of the frame is more complex, and apart from the guidance given in BS 5950 the reader is directed to references (11) and (12) particularly for multi-span frames. Member stability and design of connections continue to be researched and again the reader should keep up to date with current design knowledge with regard to these subject areas for example, reference 21.

In broad terms, economy is achieved by minimizing the material and fabrication costs. The former might be compromised because of member stability considerations. The latter is dependent on the resources a particular fabricator has available, e.g. there might be a limit on the size of plate that can be punched and therefore it becomes economical for that fabricator to use stiffened thin end-plates, instead of unstiffened thicker end-plates.

STUDY REFERENCES

Topic	References
1. Comparative costs	**Horridge J.F. & Morris L.J.** (1986) Comparative costs of single-storey steel framed structures, *Structural Engineer*, vol. 64A (no. 7), pp. 177–81
2. Plastic design	**Baker J.F., Horne M.R. & Heyman J.** (1965) *The Steel Skeleton* vol. 2, *Plastic Behaviour and Design*. Cambridge University Press
3. Plastic design	**Morris, L.J. & Randall, A.L.** (1979) *Plastic Design*. Steel Construction Institute
4. Plastic design	**Horne M.R. & Morris L.J.** (1981) *Plastic Design of Low-Rise Frames*. Collins

5.	Plastic design	**Horne M.R. & Chin M.W.** (1966) *Plastic Design of Portal Frames in Steel to BS 968*. BCSA Ltd
6.	Section properties	(1987) *Steelwork Design* vol.1, Section properties, member properties. Steel Construction Institute
7.	Snow drifting	BS 6399 *Loading for Buildings* Part 3: *Imposed roof loads* (1988)
8.	Wind loading	British Standards Institute CP3 Chapter V Part 2
9.	Stiffness method	**Coates R.C., Coutie M.G. & Kong K.C.** (1988) *Structural Analysis*. Van Nostrand Reinhold
10.	Frame stability	**Horne M.R.** (1977) Safeguard against frame instability in the plastic design of single storey pitched roof frames. In Proc. Conf. *The behaviour of slender frames*. The City University, London
11.	Frame stability	**Davies J.M.** (1990) Inplane stability in portal frames. *Structural Engineer*, vol. 69 (no. 8), pp. 141–7
12.	Frame stability	**Davies J.M.** (1991) The stability of multibay portal frames, *Structural Engineer*, vol. 69 (no. 12), pp. 223–9
13.	Column stability	**Horne M.R.** (1979) *The Plastic Design of Columns*, issued as supplement to reference (3). Steel Construction Institute
14.	Rafter stability	**Horne M.R., Shakir Khalil H. & Akhtar S.** (1979) *The Stability of Tapered and Haunched Beams*, Proc. ICE, vol. 67 (no. 9), pp. 677–94
15.	Lateral restraint	**Morris L.J. & Nakane K.** (1983) Experimental behaviour of haunched members. In Morris L.J. (ed.) *Instability and Plastic Collapse of Steel Structures*, Proc. of Int. Conf., pp. 547—59. Granada Publishing
16.	Lateral restraint	**Morris L.J.** (1981) A commentary on portal frame design, *Structural Engineer,* vol. 59A (no. 12), pp. 394–404
17.	Bolt strength	**Godley M.H.R. & Needham F.H.** (1982) Comparative tests on 8.8 and HSFG bolts in tension and shear, *Structural Engineer*, vol. 60A (no. 3), pp. 94–9
18.	Column web panel	**Morris L.J. & Newsome C.P.** (1981) Bolted corner connections subjected to an out-of-balance moment – the behaviour of the column web panel. In Howlett, Jenkins and Stainsby (eds.) *Joints in Stuctural Steelwork,* Proc. Int. Conf. Pentech Press.
19.	Fire boundary	(1980) *The Behaviour of Steel Portal Frames in Boudary Conditions*. Steel Construction Institute
20.	Holding down bolts	(1980) *Holding Down Systems for Steel Stanchions,* BCSA publication 8/80
21.	Moment connections	(1966) *Joints in Steel Construction*. Steel Construction Institute/BCSA Ltd

14
DESIGN OF AN OFFICE BLOCK – COMPOSITE CONSTRUCTION

In earlier chapters, the design of individual elements such as beams, columns and composite floors has been described. Complete multi-storey structures consist of a number of these elements fitted together to form a framework. In addition to the design of individual elements, the engineer must ensure that the complete structure is stable under all loading conditions. For example, the structure must be capable of withstanding some horizontal loading either actual, e.g. wind (see Section 2.3), or notional (see Section 10.2). As emphasized in Section 1.5, when bringing together structural elements into a framework the designer must ensure proper load paths, i.e. reactions from one element form loads on the supporting elements, and so on until the loads are transferred to the foundations.

14.1 LAYOUT AND BASIC CHOICES

An eight-storey block, for general office occupancy, is to be designed in structural steelwork for a site on the outskirts of Newcastle upon Tyne. The principal dimensions are shown in Fig. 14.1. The arrangement of each floor is similar, allowing the steelwork layout to be the same on each floor, and on the roof as well (with minor modifications).

Fig. 14.1 Office block floor plan

The designer has to make basic choices with regard to:

- floor construction
- frame construction
- stair construction
- resistance to wind loading
- architectural details
- integration of structure with building services

These choices will be made taking into account:

- the economy of construction, which may require specialist advice, e.g. from quantity surveyors;
- the speed of construction, which may require liaising with contractors;
- details of possible finishes, which will generally be decided in conjunction with an architect.

All these factors affect the final cost and quality of the building, and the design team must produce a combination of these which is satisfactory to the client. A review of the factors affecting multi-storey steel frame construction is given by Mathys[1].

14.1.1 Floor construction

The floor construction could be *in situ* reinforced concrete, precast concrete, or composite construction. For speed of construction a composite flooring using a profiled steel formwork is chosen. This form of construction has been discussed in Section 9.6, and type CF60 by PMF[2] has been chosen for the present design, as shown in Table 14.1.

Table 14.1. CF60 – 1.2 mm with LWAC (Courtesy of PMF Ltd.)
Maximum spans in metres

Concrete thickness in mm	Imposed loading in kN/m²							
	3.0	4.0	5.0	6.0	7.0	8.0	9.0	10.0
120	3.80	3.80	3.80	3.60	3.35	3.10	2.90	2.70
140	3.60	3.60	3.60	3.60	3.55	3.40	3.20	3.05
160	3.45	3.45	3.45	3.45	3.45	3.45	3.40	3.35
180	3.30	3.30	3.30	3.30	3.30	3.30	3.30	3.30
200	3.20	3.20	3.20	3.20	3.20	3.20	3.20	3.20
220	3.00	3.00	3.00	3.00	3.00	3.00	3.00	3.00

This floor is capable of spanning up to 3.3 m with a lightweight aggregate concrete, and design and construction details are given by Lawson[3]. For a fire rating of 1 hour, mesh reinforcement type A193 is recommended[4], giving a cross-section for the floor construction as shown in Fig. 14.2.

Fig. 14.2 Floor construction

14.1.2 Frame construction

The design of a multi-storey steel frame may use the method known as rigid design (clause 2.1.2.3) or simple design (clause 2.1.2.2). In rigid design, the connections are assumed capable of developing the required strength and stiffness for full continuity. In simple design the connections are assumed not to develop significant moments, i.e. beams are designed assuming they are simply supported. The choice between the two forms of construction is generally economic, and is outside the scope of this chapter. The present design assumes simple construction.

As discussed in Section 9.1, it is advantageous to make the slab and beams act compositely, and such an arrangement is possible with profiled steel sheeting. Fire protection is required for the steel beams and a lightweight system such as Pyrotherm is chosen (see Section 15.4).

While it is possible to design the columns to act compositely with a concrete casing, it may be preferred not to involve the process of shuttering and *in situ* casing. In the present design, lightweight casing for fire protection is used, of the same type as for the beams.

14.1.3 Stair construction

A number of methods of stair construction are possible, some of which influence the speed of construction in general, and the access of operatives during construction.

Generally, concrete construction is chosen rather than an all steel arrangement, owing to the complexity of the steelwork fabrication. The concrete may be *in situ* or precast, or a combination of both. The choice of method may affect the supporting steelwork arrangement, and possible alternatives are shown in Fig. 14.3. For the present design, flights and half landings are supported separately.

14.1.4 Resistance to wind loading

The horizontal loading due to wind may be resisted either by frame action, in which all the beams and columns act together, or by designing specific parts of the structure to resist these forces. In rigid frame design, wind loading would be included as one of the load systems, and the frame analysed accordingly. This is discussed further in Section 14.8.

The alternative to frame action is to transfer the wind forces to wind towers, shear walls or bracing located at specific points in the structure. These

Fig. 14.3 Stair construction

(Left diagram: Flights and landings span together)
(Right diagram: Flights and landings span separately)

wind resisting parts of the structure may be constructed in either steel (as a framework described in Section 10.3) or in concrete (as shear walls or shafts). In the present design, a wind bracing framework has been chosen (Section 14.7), with a brief comparison as to the effects of frame action (Section 14.8). The final choice is based on both economic and architectural considerations, as above six to eight storeys the use of a wind frame becomes cost-effective, but its presence in the structure may affect both the façade and the building layout.

14.1.5 Architectural details

All details of the steelwork frame affect the appearance and layout of the building and the design team must be aware of the results of each other's actions. Some further choices relate to external façade construction, internal partitions, floor and ceiling finishes.

In the present design, a precast wall unit (below sill level) with glazing above is chosen, giving loadings as in Section 14.2. The same unit is used at roof level as a parapet. Internal partitions are not defined in position, and an allowance for movable lightweight partitions is made in the imposed floor loading. A screened floor finish is allowed for, together with a lightweight suspended ceiling.

Imposed loads[5] and wind loads[6] are obtained from the appropriate British Standard.

14.2 LOADING

14.2.1 Roof and floor loading

Roof loading: CF60 slab   3.0 kN/m^2
Roof finishes   1.8 kN/m^2
Total dead load = 3.0 + 1.8   = 4.8 kN/m^2

Imposed load   1.5 kN/m^2

Floor loading:
	CF60 slab	3.0 kN/m²
	Floor finishes	1.2 kN/m²
	Total dead load = 3.0 + 1.2	= 4.2 kN/m²
	Imposed load	5.0 kN/m²
	Partitions	1.0 kN/m²
	Total imposed load = 5.0 + 1.0 =	6.0 kN/m²

14.2.2 Stair loading

Flights:
Precast concrete	5.5 kN/m²
Finishes	0.8 kN/m²
Total dead load = 5.5 + 0.8	= 6.3 kN/m²
Imposed load	4.0 kN/m²

Landings:
Precast concrete	3.5 kN/m²
Finishes	1.2 kN/m²
Total dead load = 3.5 + 1.2	= 4.7 kN/m²
Imposed load	4.0 kN/m²

14.2.3 Wall unit and glazing

Roof parapet:	Precast unit	2.0 kN/m
Floor wall unit:	Precast unit	2.0 kN/m
	Glazing	0.3 kN/m
	Total dead load = 2.0 + 0.3	= 2.3 kN/m

14.2.4 Wind loading

The following notation and method may be found in reference (6).

Basic wind speed V (Newcastle upon Tyne)	46 m/s
Topography factor S_1	1.0
Ground roughness (outskirts of city)	Type (3)
Building size (max. dimension 40 m)	Class B
Factor S_2 (increases with height)	
Statistical factor S_3 (50-year exposure)	1.0
Design wind speed $V_s = S_1 S_2 S_3 V$ m/s	
Dynamic pressure $q = 0.613 V_s^2$ N/m²	
Force coefficient $C_f = 1.3$ (for $l/w = 3.6$, $h/b = 0.8$)	

The wind speed and pressure vary with height, and the appropriate values are shown in Fig. 14.4. The wind pressures may be resolved in forces at each floor level, which are also shown in Fig. 14.4, giving values for one bay width of 5 m only.

Fig. 14.4 Wind loading

Fig. 14.5 Steelwork arrangement

14.3 ROOF BEAM DESIGN

A suitable arrangement of beams for all floors is shown in Fig. 14.5. Roof beams are denoted R1, R2, etc., and typical floor beams (floors 1 to 7) are denoted T1, T2, etc. Using the composite slab (type CF60) spanning 3.3 m maximum, secondary beams type 1 to 4 must be provided to support the slab, at a spacing not greater than 3.3 m. These beams are supported on main beams 5 and 6, which are in turn supported by the columns. In the region of the stair wells, special beams may be required such as 9 and 10, and in the vicinity of lift shafts there will be additional requirements and loadings affecting beams 11 to 14.

(a) Roof beam R1 – 152 × 89 × 16 UB

The design generally follows the recommendations of BS 5950: Part 3.1[7] and clause numbers are given for guidance where appropriate. With reference to Figs. 14.6 and 14.7:

Fig. 14.6

Effective breadth B_e (= span/4) = 1250 mm
Concrete cube strength = 30 N/mm^2
Steel design strength = 275 N/mm^2
Beam spacing = 3.25 m

Fig. 14.7

Fig. 14.8

Loading: dead, $4.8 \times 5.0 \times 3.25 = 78$ kN
own weight, $0.16 \times 5.0 = 1$ kN
fire casing, $0.2 \times 5.0 = \underline{1\ \text{kN}}$
80 kN

imposed $1.5 \times 5.0 \times 3.25 = 24$ kN

Design loading (u.d.l.) $1.4 \times 80 + 1.6 \times 24 = 150$ kN

Shear force F_v $150/2 = 75$ kN
Moment M_x $150 \times 5.0/8 = 94$ kNm

Reactions (unfactored):
$R_d = 40$ kN
$R_i = 12$ kN

(See Fig 14.8.)

clause B.2.1

Force in concrete $R_c = 0.45 f_{cu} B_e (D - D_p)$
$= 0.45 \times 30 \times 1250 \times (120 - 60)\ 10^{-3} = 1010$ kN

Force in steel $R_s = p_y A = 275 \times 20.5 \times 10^{-1} = 564$ kN

Neutral axis is in the slab, as $R_c > R_s$ and therefore

$x_p = 564/(0.45 \times 30 \times 1250) = 33$ mm

Moment of resistance $= R_s (D/2 + D_s - x_p/2)$
$M_c = 564 \times 10^3/(152.4 + 120 - 33/2)\ 10^{-3} = 101$ kNm
$M_x/M_c = \mathbf{0.93}$

clause 4.1

Modular ratio[6] appropriate for deflections $= 10$
(one-third is long term)

$r = A/(D_s - D_p) B_e$
$= 20.5 \times 10^2 / (120 - 60)1250 = 0.0273$

Equivalent second moment of area

$I_g = A(D + D_s + D_p)^2 / 4(1 + \alpha r) + B_e(D_s - D_p)^3 / 12\alpha + I_x$
$= 20.5 \times 10^2 (152.4/2 + 120 + 60)^2/[4(1 + 10 \times 0.0273)]$
$\quad + 1250(120 - 60)^3/(12 \times 10) + 838 \times 10^4$
$= 3700 \times 10^4\ \text{mm}^4 = 3700\ \text{cm}^4$

clause 6.1.3.5

Deflection (based on unfactored imposed load, 24 kN)

$5 \times 24 \times 5^3/(384 \times 205 \times 3700 \times 10^{-5}) = \mathbf{5.1}$ mm

Limit of deflection $= 5000/360 = 13.9$ mm

clause 5.4.6

Use 19 mm diameter studs 100 mm high as shear connectors with design strength (Q_k) of 90 kN.

clause 5.4.7.2

Reduction due to profile introduces a factor,

$= 0.85/\sqrt{N} \times b_r/D_p \times (h - D_p)/D_p$
$= 0.85/1 \times 113/60 \times (100 - 60)/60 = 1.07$

which should not exceed 1.

clause 5.4.3

$N_p = F_p/(0.8 Q_k)$
$= 564/(0.8 \times 90) = 8$

Note that the lower value of R_c or R_s is taken and constitutes the actual force in the steel and concrete.

Use a total of **16 studs** along the beam.

clause 5.6.4   Longitudinal shear transfer (through the concrete) is not a likely problem where the sheeting is attached by shear connectors to the beam.

Vertical shear capacity P_v
$$= 0.6 \times 275 \times 4.6 \times 152.4 \times 10^{-3} = 116 \text{ kN}$$
$F_v/P_v = \mathbf{0.65}$

(b)   Roof beam R2 – 152 × 89 × 16 UB

The design is as beam R1.
Beam spacing = 2.75 m

Loading: dead	$4.8 \times 5.0 \times 2.75 =$	66 kN
own weight		1 kN
fire casing		1 kN
		68 kN

imposed $1.5 \times 5.0 \times 2.75 =$ 21 kN

Design loading $= 1.4 \times 68 + 1.6 \times 21 = 129$ kN
Moment $M_x = 129 \times 5.0/8$ $= 81$ kNm
Reactions (unfactored):
$R_d = 34$ kN
$R_I = 11$ kN

(c)   Roof beam R3 – 152 × 89 × 16 UB

The design is as beam R1.
Beam spacing = 2.25 m

Loading: dead	$4.8 \times 5.0 \times 2.25 = 54$ kN
own	weight + casing = 2 kN
	56 kN

imposed $1.5 \times 5.0 \times 2.25 = 17$ kN

Reactions (unfactored):
$R_d = 28$ kN
$R_i = 8$ kN

(d)   Roof beam R4 – 152 × 89 × 16 UB

clause 4.6   For an edge beam, the effective breadth, $B_e = 0.8 \times 5000/8 = 500$ mm

Loading: dead	$4.8 \times 5.0 \times 3.25/2 = 39$ kN
own weight + casing	= 2 kN
parapet 2.0×5.0	= 10 kN
	51 kN

imposed $1.5 \times 5.0 \times 3.25/2 = 12$ kN

DESIGN OF AN OFFICE BLOCK – COMPOSITE CONSTRUCTION

$$\text{Design loading} = 1.4 \times 51 + 1.6 \times 12 = 91 \text{ kN}$$
$$\text{Moment} \quad M_x = 91 \times 5.0/8 \quad\quad = 57 \text{ kNm}$$
Reactions (unfactored):
$$R_d = 25 \text{ kN}$$
$$R_i = 6 \text{ kN}$$

clause B.2
$$R_c = 0.45 \times 30 \times 500 \times (120-60) \times 10^{-3} = 405 \text{ kN}$$
$$R_s = 275 \times 20.5 \times 10^{-1} \quad\quad = 564 \text{ kN}$$

Neutral axis is in the steel as $R_s > R_c$

clause B.2.2
$$M_c = [405(120+60)/2 + 564 \times 152.4/2]10^{-3} = 79 \text{ kNm}$$
$$M_x/M_c = \mathbf{0.76}$$
$$N_p = 405/(0.8 \times 90) = 6$$

Use a total of **12 studs** along beam.

(e) Roof beam R5 – 254 × 146 × 37 UB

Roof beam carries reactions from two type R1 beams as well as some distributed load (see Fig. 14.9).

Point load:	dead	2×40	$= 80$ kN
	imposed	2×12	$= 24$ kN
Distributed load:	own weight	0.37×6.5	$= 2$ kN
	casing	0.3×6.5	$= 2$ kN

$$\text{Design loading (point)} = 1.4 \times 80 + 1.6 \times 24/8 = 150 \text{ kN}$$
$$\text{Design loading (distributed)} = 1.4 \times 4 \quad\quad = 6 \text{ kN}$$
$$\text{Moment} \quad M_x = 150 \times 6.5/4 + 6 \times 6.5/8 = 248 \text{ kNm}$$
$$\text{Shear force} \quad F_v = (150+6)/2 \quad\quad = 78 \text{ kN}$$

Reactions (unfactored):
$$R_d = 42 \text{ kN}$$
$$R_i = 12 \text{ kN}$$

(See Fig. 14.10.)

Fig. 14.9

Fig. 14.10

clause 4.6   Effective breadth, $B_e = 6500/4$ $= 1625$ mm

clause B.2.1
$$R_c = 0.45 \times 30 \times 1625 \,(120-60) = 1300 \text{ kN}$$
$$R_s = 275 \times 47.5 \times 10^{-1} = 1310 \text{ kN}$$
$$M_c = [1300(120+60)/2 + 1310 \times 256/2]10^{-3} = 251 \text{ kNm}$$
$$M_x/M_c = \mathbf{0.87}$$
$$r = 47.5 \times 10^2/[1625\,(120-60)] = 0.0487$$

Following the calculation for beam R1:

$$I_g = 17500 \text{ cm}^4$$
$$\text{Deflection} = 24 \times 6.5^3/48 \times 205 \times 17500 \times 10^{-5} = 3.8 \text{ mm}$$

clause 5.4.3
$$N_p = 1300/(0.8 \times 90) = 18$$

Use a total of **36 studs** along beam.

$$P_v = 0.6 \times 275 \times 6.4 \times 256 = 270 \text{ kN}$$
$$F_v/P_v = \mathbf{0.29}$$

(f) Roof beam R6 – 254 × 146 × 37 UB

This beam carries reactions from two type R3 beams as well as some distributed load.

Point load: dead $\quad 2 \times 28 = 56$ kN
$\qquad\qquad$ imposed $2 \times 8 = 16$ kN
Distributed load $\qquad\qquad = 4$ kN
Reactions (unfactored):
$\qquad R_d = 30$ kN
$\qquad R_i = 8$ kN

Design is the same as beam R5.

(g) Roof beam R7 – 254 × 146 × 37 UB

This beam carries reaction from one type R1 beam as well as distributed load.

Point load: $\qquad$ dead $\qquad\qquad\qquad = 40$ kN
$\qquad\qquad\qquad$ imposed $\qquad\qquad = 12$ kN
Distributed load: own weight + casing $\;= 4$ kN
$\qquad\qquad\qquad$ parapet $2.0 \times 6.5 \quad = 13$ kN
Design loading (point): $1.4 \times 40 + 1.6 \times 12 = 75$ kN
Design loading (distributed): $1.4 \times 17 \quad = 24$ kN
Moment $M_x = 75 \times 6.5/4 + 24 \times 6.5/8 \quad = 141$ kNm

Reactions (unfactored):
$\qquad R_d = 28$ kN
$\qquad R_i = 6$ kN

clause 4.6 $\qquad$ Effective breadth, $B_e = 0.8 \times 6500/8 \qquad = 650$ mm

clause B.2.1
$$R_c = 0.45 \times 30 \times 650 \,(120-60) \qquad = 526 \text{ kN}$$
$$R_s = 275 \times 47.5 \times 10^{-1} \qquad\qquad = 1306 \text{ kN}$$
$$M_c = 526\,(120+60)/2 + 1306 \times 256/2 = 214 \text{ kNm}$$
$$M_x/M_c = \mathbf{0.66}$$
$$N_p = 526/(0.8 \times 90) = 8$$

Use a total of **16 studs** along beam.

(h) Roof beam R8 – 254 × 146 × 37 UB

This beam carries one reaction from type R3 beam as well as distributed load.

Point load: dead   = 28 kN
               imposed   = 8 kN
Distributed load: own weight + casing = 4 kN
                      parapet   = 13 kN
Reactions (unfactored):
$R_d = 22$ kN
$R_i = 4$ kN

Design is the same as beam R7.

Roof beams over the stair will support additional loads where stairs exit on to the roof. The typical stair beams are included as beams T9 and T10 in Section 14.4. Beams in the lift shaft area will be designed to suit detailed layouts and loadings for the lift motor room, etc., and are not included in this example.

14.4 TYPICAL FLOOR BEAM DESIGN

The same arrangement is used for the steelwork on the typical floor (floors 1 to 7 inclusive) as that used for the roof. Some variation may be needed in the vicinity of the stair wells and the lift shaft.

In general, the number of beam sizes used is kept to a minimum to ease ordering and fabrication. Two sizes only were used for the roof steelwork. Four sizes will be used for the typical floor, which is, of course, repeated seven times.

The design calculations follow the layout in Section 14.3.

(a) Typical floor beam T1 – 254 × 102 × 25 UB

Beam spacing 3.25 m

Loading: dead $4.2 \times 5.0 \times 3.25 = 68$ kN
           own weight $0.25 \times 5.0 = 1$ kN
           casing $0.2 \times 5.0 = \underline{1 \text{ kN}}$
                                        70 kN
           imposed $6.0 \times 5.0 \times 3.25 = 98$ kN

Fig. 14.11 (254 kN UDL over 5.0 m span)

With reference to Fig. 14.11:

Design loading: $1.4 \times 70 + 1.6 \times 98 = 254$ kN
Shear force $F_v = 254/2$ $= 127$ kN
Moment $M_x = 254 \times 5.0$ $= 159$ kNm
Reactions (unfactored):
$R_d = 35$ kN
$R_i = 49$ kN

Effective breadth, $B_e = 1250$ mm (see Fig. 14.12)

$R_c = 0.45 \times 30 \times 1250(120-60)10^{-3} = 1010$ kN
$R_s = 275 \times 32.2 \times 10^{-1}$ $\qquad\quad = 886$ kN
$M_c = 1010(120+60)/2 + 886 \times 257/2 = 205$ kNm
$M_x/M_c = \mathbf{0.78}$
$r = 0.0429$
$I_g = 12\,040$ cm^4
Deflection $= 5 \times 98 \times 5.0^3/(384 \times 205 \times 12\,040 \times 10^{-5}) = 6.5$ mm
$N_p = 1010/(0.8 \times 90) = 14$

Fig. 14.12 (1250; 254 × 102 × 25 UB)

Use a total of **28 studs** along beam.

$P_v = 0.6 \times 275 \times 6.1 \times 257 = 259$ kN
$F_v/P_v = \mathbf{0.49}$

(b) Typical floor beam T2 – 254 × 102 × 25 UB

Beam spacing 2.75 m

Loading: dead $\quad 4.2 \times 50 \times 2.75 = 58$ kN
own weight + casing $\quad = \underline{\;2\text{ kN}}$
$\qquad\qquad\qquad\qquad\qquad\qquad\quad 60$ kN
imposed $6.0 \times 5.0 \times 2.75 = 83$ kN

Reactions (unfactored):
$R_d = 30$ kN
$R_i = 41$ kN

Design is the same as beam T1.

(c) Typical floor beam T3 – 254 × 102 × 25 UB

Beam spacing 2.25 m
Loading: dead $\quad 4.2 \times 5.0 \times 2.25 = 47$ kN
own weight + casing $\quad = \underline{\;2\text{ kN}}$
$\qquad\qquad\qquad\qquad\qquad\qquad\quad 49$ kN
imposed $6.0 \times 5.0 \times 2.25 = 68$ kN

Reactions (unfactored):
$R_d = 25$ kN
$R_i = 34$ kN

Design is the same as beam T1

(d) Typical floor beam T4 – 254 × 102 × 25 UB

Loading: dead $4.2 \times 5.0 \times 3.25/2 \quad = 34$ kN
own weight + casing $\quad = \;\;2$ kN
wall + glazing $\quad 2.3 \times 5.0 = \underline{12\text{ kN}}$
$\qquad\qquad\qquad\qquad\qquad\qquad\quad 48$ kN
imposed $6.0 \times 5.0 \times 3.25/2 = 49$ kN
Design loading: $1.4 \times 48 + 1.6 \times 49 = 146$ kN

Moment $M_x = 146 \times 5.0/8 = 91$ kNm
Reactions (unfactored):
$$R_d = 24 \text{ kN}$$
$$R_i = 25 \text{ kN}$$

$$B_e = 500 \text{ mm}$$
$$R_c = [0.45 \times 30 \times 500(120-60)]\,10^{-3} = 405 \text{ kN}$$
$$R_s = 275 \times 32.2 \times 10^{-1} = 886 \text{ kN}$$
$$M_c = [405\,(120+60)/2 + 886 \times 257/2]10^{-3} = 150 \text{ kNm}$$
$$M_x/M_c = \mathbf{0.61}$$
$$N_p = 405/(0.8 \times 90) = 6$$

Use a total of **12 studs** along beam.

(e) Typical floor beam T5 – 356 × 171 × 57 UB

This beam carries two reactions from two type T1 beams as well as some distributed load.

Point load: dead 2×35 = 70 kN
              imposed 2×49 = 98 kN
Distributed load: own weight 0.57×6.5 = 4 kN
              casing 0.4×6.5 = 3 kN

Fig. 14.13 (See Fig. 14.13.)

Design loading (point): $1.4 \times 70 + 1.6 \times 98 = 255$ kN
Design loading (distributed): 1.4×7 = 10 kN
Shear force $F_v = (255 + 10)/2$ = 133 kN
Moment $M_x = 255 \times 6.5/4 + 10 \times 6.5/8 = 422$ kNm
Reactions (unfactored):
$$R_d = 38 \text{ kN}$$
$$R_i = 49 \text{ kN}$$

$$B_e = 1625 \text{ mm}$$
$$R_c = 1300 \text{ kN}$$
$$R_s = 275 \times 72.2 \times 10^{-1} = 1986 \text{ kN}$$
$$M_c = [1300\,(120+60)/2 + 1986 \times 358.6/2]10^{-3} = 473 \text{ kNm}$$
$$M_x/M_c = \mathbf{0.89}$$
$$r = 0.0741$$
$$I_g = 38\,200 \text{ cm}^4$$
Deflection $= 98 \times 6.5^3/(48 \times 205 \times 38\,200 \times 10^{-5}) = 7.2$ mm
$$N_p = 1300/(0.8 \times 90) = 18$$

Use a total of **36 studs** along beam.

$$P_v = 0.6 \times 275 \times 8.0 \times 358.6 = 473 \text{ kN}$$
$$F_v/P_v = \mathbf{0.28}$$

(f) Typical floor beam T6 – 356 × 127 × 33 UB

This beam carries two reactions from two type T3 beams as well as distributed load.

$$
\begin{aligned}
\text{Point load: dead } 2 \times 25 &= 50 \text{ kN} \\
\text{imposed } 2 \times 34 &= 68 \text{ kN} \\
\text{own weight} + \text{casing} &= 4 \text{ kN} \\
\text{Design loading (point): } 1.4 \times 50 + 1.6 \times 68 &= 179 \text{ kN} \\
\text{Design loading (distributed): } 1.4 \times 4 &= 6 \text{ kN} \\
\text{Moment } M_x = 179 \times 4.5/4 + 6 \times 4.5/8 &= 205 \text{ kNm}
\end{aligned}
$$

Reactions (unfactored):
$R_d = 27$ kN
$R_i = 34$ kN

$$
\begin{aligned}
B_e &= 4500/4 = 1125 \text{ mm} \\
R_c &= 0.45 \times 30 \times 1125 \, (120-60) = 911 \text{ kN} \\
R_s &= 275 \times 41.8 \times 10^{-1} = 1150 \text{ kN} \\
M_c &= [911(120+60)/2 + 1150 \times 348.5/2]10^{-3} = 282 \text{ kNm}
\end{aligned}
$$
$M_x/M_c = \mathbf{0.73}$
$N_p = 911/(0.8 \times 90) = 13$

Use a total of **26 studs** along beam.

(g) Typical floor beam T7 – 356 × 127 × 33 UB

Span 6.5 m

$$
\begin{aligned}
\text{Point load:} \quad \text{dead} &= 35 \text{ kN} \\
\text{imposed} &= 49 \text{ kN} \\
\text{Distributed load: own weight} + \text{casing} &= 4 \text{ kN} \\
\text{wall} + \text{glazing, } 2.3 \times 6.5 &= 15 \text{ kN} \\
\text{Design loading (point): } 1.4 \times 35 + 1.6 \times 49 &= 127 \text{ kN} \\
\text{Design loading (distributed): } 1.4 \times 19 &= 26 \text{ kN} \\
\text{Moment } M_x = 127 \times 6.5/4 + 26 \times 6.5/8 &= 228 \text{ kNm}
\end{aligned}
$$

Reactions (unfactored):
$R_d = 27$ kN
$R_i = 25$ kN

$$
\begin{aligned}
B_e &= 650 \text{ mm} \\
R_c &= 526 \text{ kN} \\
R_s &= 275 \times 41.8 \times 10^{-1} = 1150 \text{ kN} \\
M_c &= [526 \,(120+60)/2 + 1150 \times 348.5/2]10^{-3} = 248 \text{ kNm}
\end{aligned}
$$
$M_x/M_c = \mathbf{0.92}$
$N_p = 526/(0.8 \times 90) = 8$

Use a total of **16 studs** along beam.

DESIGN OF AN OFFICE BLOCK – COMPOSITE CONSTRUCTION

(h) Typical floor beam T8 – 356 × 127 × 33 UB

Span 4.5 m

Point load (T3): dead = 25 kN
imposed = 34 kN
Distributed load: own weight + casing = 4 kN
wall + glazing 2.3 × 4.5 = 10 kN
Design loading (point): $1.4 \times 25 + 1.6 \times 34$ = 89 kN
Design loading (distributed): 1.4×14 = 20 kN
Moment $M_x = 89 \times 4.5/4 + 20 \times 4.5/8$ = 111 kNm
Reactions (unfactored):
$R_d = 19$ kN
$R_i = 17$ kN

$B_e = 450$ mm
$R_c = 364$ kN
$R_s = 1150$ kN
$M_c = 233$ kN m
$M_x/M_c = \mathbf{0.48}$
$N_p = 364/(0.8 \times 90) = 6$

Use a total of **12 studs** along beam.

(i) Typical floor beam T9 – 254 × 146 × 37 UB

This is a non-composite beam in the stair well supporting precast flight and landing units.

Span 5.0 m

Loading: dead $6.3 \times 5.0 \times 2.5/2$ = 39 kN
$4.7 \times 5.0 \times 2.0/2$ = 24 kN
own weight + casing = 4 kN
———
67 kN
imposed $4.0 \times 5.0 \times 4.5/2$ = 45 kN
Design loading: $1.4 \times 67 + 1.6 \times 45 = 166$ kN
Shear $F_v = 166/2 = 83$ kN
Moment $M_x = 166 \times 5.0/8 = 104$ kNm
Reactions (unfactored):
$R_d = 34$ kN
$R_i = 23$ kN

clause 4.2.3   Shear resistance $P_v = 0.6 \times 275 \times 6.4 \times 256 = 270$ kN

$F_v/P_v = \mathbf{0.31}$

clause 4.2.5   Moment capacity $M_c = 275 \times 485 \times 10^{-3} = 133$ kNm

$M_x/M_c = \mathbf{0.78}$

Deflection $= 5 \times 45 \times 50^3/(384 \times 205 \times 5560 \times 10^{-5}) = \mathbf{6.4\ mm}$
Limit of deflection $= 5000/360 = 13.9$ mm

(j) Typical floor beam T10 – 356 × 171 × 57 UB

This is a non-composite beam in the stair well supporting a reaction from beam T9 as well as some distributed load. Lateral restraint along the beam is provided only by beam T9. The design of such unrestrained beams is discussed in greater detail in Section 3.2.

Span 6.5 m

Point load (T9):	dead	= 34 kN
	imposed	= 23 kN
Distributed load:	own weight + cladding	= 4 kN
	wall + glazing, 2.3 × 6.5	= 15 kN

(See Fig. 14.14.)

Design loading (point): $1.4 \times 34 + 1.6 \times 23$ = 84 kN
Design loading (distributed): 1.4×19 = 27 kN
Maximum shear $F_v = 84 \times 4.5/6.5 + 27/2$ = 72 kN
Moment $M_x = 84 \times 2.0 \times 4.5/6.5 + 27 \times 6.5/8 = 138$ kNm
Reactions (unfactored):
At A: $R_d = 20$ kN
$R_i = 7$ kN
At B: $R_d = 28$ kN
$R_i = 16$ kN

Fig. 14.14

BS table 9
BS table 13
BS table 16

Effective length $L_E = 4.5$ m
$m = 1.0$
$n = 0.94$
$\lambda = 4500/39.2 = 115$
$\lambda/x = 115/28.9 = 4.0$

BS table 14

$v = 0.86$
$\lambda_{LT} = 0.94 \times 0.88 \times 0.86 \times 115 = 82$

BS table 11

$p_b = 161$ N/mm^2
$M_b = 161 \times 1010 \times 10^{-3} = 163$ kNm
$M_x/M_b = \mathbf{0.85}$
$P_v = 0.6 \times 275 \times 8.0 \times 358.6 = 473$ kN
$F_v/P_v = \mathbf{0.15}$

14.5 COLUMN DESIGN

Loads for each column must be calculated, and in the present design three columns are selected as typical: an external (side) column C1; a corner column C2; and an internal column C6. Further columns could be designed individually if desired.

Column loads are best assembled from the unfactored beam reactions, with dead and imposed loads totalled separately. The **imposed** loads may be reduced where a column supports more than one floor. The reduction is 10% per floor until a maximum of 40% is reached, and 40% thereafter. This applies to buildings up to 10 storeys, and is detailed in BS 6399[8]. Column loads must include an allowance for self weight and fire casing.

The design load condition is $1.4W_d + 1.6W_i$.

(a) Column loads

COLUMN C1

Reactions R4, R5, T4, T5 are taken from Section 14.3(d), etc., and are tabulated overleaf. With reference to Fig. 14.15, typical calculations for column loading are given:

Column length: 7th floor – roof:

Total $\quad W_d = 25 + 25 + 42 + 4 \quad = 96\,\text{kN}$
Total $\quad W_i = 6 + 6 + 12 \quad\quad\quad\;\; = 24\,\text{kN}$
Design load $\;F = 1.4 \times 96 + 1.6 \times 24 = 173\,\text{kN}$

Column length: 6th–7th floors:

Total $\quad W_d = 96 + 24 + 24 + 38 + 4 = 186\,\text{kN}$
Total $\quad W_i = 24 + 25 + 25 + 49 \quad = 123\,\text{kN}$
Reduced $\; W_i = 0.9 \times 123 \quad\quad\quad\quad\; = 11\,\text{kN}$
Design load $\;= 1.4 \times 186 + 1.6 \times 111 = 438\,\text{kN}$

In the same way, each table may be completed, see below.

Fig. 14.15

Column length	Beams	Reactions R_d (kN)	R_i (kN)	Own weight (kN)	Totals W_d (kN)	W_i (kN)	Reduction (%)	Reduced design W_i (kN)	load F (kN)
COLUMN C1									
7–R	R4	25	6	4					
	R4	25	6						
	R5	42	12		96	24	0	24	173
6–7	T4	24	25	4					
	T4	24	25						
	T5	38	49		186	123	10	111	438
5–6	ditto	86	99	4	276	222	20	178	671
4–5	ditto	86	99	4	366	321	30	225	872
3–4	ditto	86	99	4	456	420	40	252	1040
2–3	ditto	86	99	4	546	519	40	311	1260
1–2	ditto	86	99	4	636	618	40	371	1480
G–1	ditto	86	99	6	728	717	40	430	1710
COLUMN C2									
7–R	R4	25	6	4					
	R7	28	6		57	12	0	12	99
6–7	T4	24	25	4					
	T7	27	25		112	62	10	56	246
5–6	ditto	51	50	4	167	112	20	90	378
4–5	ditto	51	50	4	222	162	30	113	492
3–4	ditto	51	50	4	277	212	40	127	591
2–3	ditto	51	50	4	332	262	40	157	716
1–2	ditto	51	50	4	387	312	40	187	841
G–1	ditto	51	50	6	444	362	40	217	969

Column length	Beams	Reactions R_d (kN)	R_i (kN)	Own weight (kN)	Totals W_d (kN)	W_i (kN)	Reduction (%)	W_i (kN)	Reduced design load F (kN)
COLUMN C6									
7–R	R2	34	10	4					
	R2	34	10						
	R5	42	12						
	R6	30	8		144	40	0	40	265
6–7	T2	30	41	4					
	T2	30	41						
	T5	38	49						
	T6	27	34		273	205	10	185	678
5–6	ditto	125	165	4	402	370	20	296	1040
4–5	ditto	125	165	4	531	535	30	375	1340
3–4	ditto	125	165	4	660	700	40	420	1600
2–3	ditto	125	165	4	789	865	40	519	1930
1–2	ditto	125	165	4	918	1030	40	618	2270
G–1	ditto	125	165	6	1049	1195	40	717	2610

Fig. 14.16

(b) Column C1 design: Ground – 1st floor – 203 × 203 × 86 UC

Over this length, the column carries an axial load of 170 kN. The reaction from beam T5 at first floor level is eccentric to the column. With reference to Fig. 14.16:

clause 4.7.6
$$e = 100 + 222.3/2 = 211 \text{ mm}$$
$$R_{T5} = 133 \text{ kN}$$

clause 4.7.7 The nominal moment is divided between column lengths above and below the first floor equally, assuming approximately equal column stiffnesses:

$$M_x = 133 \times 0.211/2 = 14 \text{ kNm}$$

BS table 6 $\quad p_y = 265 \text{ N/mm}^2$
BS table 24 $\quad L_E = 0.85 \times 4.7 = 4.00 \text{ m}$
$\lambda = L_E/r_y = 4000/53.2 = 75$
BS tables 25, 27c $\quad p_c = 167 \text{ N/mm}^2$
$P_c = p_c A_g = 167 \times 110 \times 10^{-1} = 1840 \text{ kN}$

clause 4.7.7 $\quad m = 1.0$
clause 4.7.7 $\quad \lambda_{LT} = 0.5L/r_y = 0.5 \times 4700/53.2 = 44$
BS table 11 $\quad p_b = 244 \text{ N/mm}^2$
$M_b = p_b S_x = 244 \times 979 \times 10^{-3} = 239 \text{ kNm}$

clause 4.8.3.3 Overall buckling check (simplified):

$$F/P_c + mM_x/M_b \leq 1$$

$$1710/1840 + 1.0 \times 14/239 = \mathbf{0.98}$$

Using the same method, other lengths of the column may be designed and the results tabulated. Where floor beams providing directional restraint are substantial, and are not required to carry more than 90% of their moment

Column length	Size	F (kN)	M_x (kNm)	λ	p_c (N/mm²)	P_c (kN)	λ_{LT}	p_b (N/mm²)	M_b (kNm)	Check
G–1	203 × 203 × 86 UC	1710	14	75	167	1840	44	244	239	0.98
1–2	203 × 203 × 71 UC	1480	14	60	201	1830	35	265	213	0.87
2–3	203 × 203 × 60 UC	1260	14	61	199	1510	36	271	177	0.91
3–4	203 × 203 × 46 UC	1040	13	62	197	1160	36	271	135	0.99
4–5	ditto	872	13	satisfactory						
5–6	ditto	671	13	satisfactory						
6–7	152 × 152 × 30 UC	438	12	82	157	600	48	243	60	0.93
7–R	ditto	173	14	satisfactory						

capacity, the effective length may be taken as 0.7 of actual length for column lengths above the first floor. The design strength, p_y, is 265 N/mm² for column sections greater than 16 mm thick and 275 N/mm² for sections less than 16 mm thick.

It is the normal practice, in the interests of economy, for columns to be fabricated in two-storey lengths, and assembled on site using splices. It is not therefore good practice to change column size for every storey, and in this case the same size (203 × 203 × 86 UC) would probably be used between ground and second floors, and one size (203 × 203 × 60 UC) between second and fourth floors.

(c) Column C2 design: Ground – 1st floor – 203 × 203 × 60 UC

Over this length the column carries an axial load of 968 kN. The reactions from beams T4 and T7 at the first floor are both eccentric to the column. With reference to Fig. 14.17:

For T7: $e = 100 + 209.6/2 = 205$ mm
For T4: $e = 100 + 9.3/2 = 105$ mm

$R_{T7} = 1.4 \times 27 + 1.6 \times 25 = 78$ kN
$R_{T4} = 1.4 \times 24 + 1.6 \times 25 = 74$ kN
$M_x = 78 \times 0.205/2 = 8.0$ kNm
$M_y = 74 \times 0.105/2 = 3.9$ kNm

BS table 6 $\quad p_y = 275$ N/mm²
BS table 24 $\quad L_E = 0.85 \times 4.7 = 4.00$ m
$\lambda = 4000/51.9 = 77$
BS tables 25, 27c $\quad p_c = 167$ N/mm²
$P_c = 167 \times 75.8 \times 10^{-1} = 1260$ kN

clause 4.7.7 $\quad m = 1.0$
$\lambda_{LT} = 0.5 \times 4700/51.9 = 46$
BS table 11 $\quad p_b = 248$ N/mm²
$M_b = 248 \times 652 \times 10^{-3} = 161$ kNm

clause 4.8.3.3 Overall buckling check (simplified):

$$F/P_c + mM_x/M_b + mM_y/p_yZ_y \leq 1$$

$968/1260 + 1.0 \times 8.0/161 + 1.0 \times 3.9/(275 \times 199 \times 10^{-3}) = \mathbf{0.89}$

Fig. 14.17

Using the same method other lengths of the column may be designed and the results are tabulated below.

Column length	Size	F (kN)	M_x (kNm)	M_y (kNm)	λ	p_c (N/mm^2)	P_c (kN)	λ_{LT}	p_b (N/mm^2)	M_b (kNm)	Check
G–1	203 × 203 × 60 UC	968	8.0	3.9	77	167	1260	46	248	161	0.89
1–2	203 × 203 × 46 UC	841	7.9	3.9	62	197	1160	36	271	135	0.86
2–3	ditto	716	7.9	3.8	satisfactory						
3–4	ditto	591	7.9	3.8	satisfactory						
4–5	152 × 152 × 37 UC	491	7.1	3.8	81	159	754	48	243	75	0.90
5–6	152 × 152 × 30 UC	378	7.0	3.8	82	157	600	48	243	60	0.94
6–7	ditto	246	6.2	3.4	satisfactory						
7–R	ditto	99	8.7	4.6	satisfactory						

As for column C1, two-storey lengths at the same size are preferred.

Fig. 14.18

(d) Column C6 design: Ground – 1st floor – 254 × 254 × 132 UC

Over this length the column carries an axial load of 2610 kN. The reactions from the beams at first floor level are eccentric, but will tend to balance each other, see Fig. 14.18. The difference between the reactions from T5 to T6 will, however, give a net moment about the major axis. Note that the effect of the absence of imposed load on any beam (pattern loading) is not taken into account, and all beams are considered fully loaded (clause 4.7.7).

Column length	Size	F (kN)	M_x (kNm)	λ	p_c (N/mm^2)	P_c (kN)	λ_{LT}	p_b (N/mm^2)	M_b (kNm)	Check
G–1	254 × 254 × 132 UC	2610	3.7	60	195	3300	35	265	496	0.80
1–2	254 × 254 × 89 UC	2270	3.6	48	217	2470	28	265	326	0.93
2–3	254 × 254 × 73 UC	1930	3.5	49	222	2060	29	275	272	0.95
3–4	ditto	1600	3.5	satisfactory						
4–5	203 × 203 × 60 UC	1340	3.5	61	199	1510	36	271	177	0.91
5–6	203 × 203 × 46 UC	1040	3.6	62	197	1160	36	271	135	0.92
6–7	152 × 152 × 37 UC	678	3.5	81	159	754	48	243	75	0.95
7–R	152 × 152 × 30 UC	265	4.1	82	157	600	48	243	60	0.51

As previously, two-storey lengths at the same size are preferred.

14.6 CONNECTIONS

The design of typical beam to column connections is given in Section 3.7(g). A typical connection for the present design is detailed for beam T5 to column C6. With reference to Fig. 14.19:

Beam reaction (Section 14.4e) = 133 kN
Moment = 133 × 0.05   = 6.7 kNm

Use **9 no. 20 mm bolts grade 4.6**
Use **2 no. 90 × 90 × 10 angle cleats**

Fig. 14.19

(a) Column bolts

	Shear/bolt	$133/6 =$ **22.2 kN**
clause 6.3.2	Shear capacity	$P_s = 160 \times 245 \times 10^{-3}$ = **39.2 kN**
clause 6.3.3	Bearing capacity of bolts $P_{bb} = 20 \times 9.4 \times 435 \times 10^{-3}$ = **81.8 kN**	

where 9.4 is the column flange thickness.

(b) Beam bolts

Vertical shear/bolt $= 133/3 = 44.3$ kN
Horizontal shear due to eccentric bending moment,
$$Md_{max}/\Sigma d^2 = 6.7 \times 0.10/(2 \times 0.10^2) = 33.5 \text{ kN}$$
Resultant shear/bolt $= (44.3^2 + 33.5^2)$ $= 55.5$ kN
Shear capacity (double shear) $P_s = 160 \times 2 \times 245 \times 10^{-3}$ $= 78.4$ kN
Bearing capacity of bolt $P_{bb} = 20 \times 8.0 \times 435$ $= 73.1$ kN

(c) Angle cleat

Shear area of cleats $= 2 \times 0.9 (300 \times 10 - 3 \times 22 \times 10) = 4210 \text{ mm}^2$
Shear capacity $P_v = 0.6 \times 275 \times 4210 \times 10^{-3}$ = **695 kN**
Shear force $F_v = 133$ kN

Fig. 14.20

Similar connections may be designed for all other beams. Where a beam is designed for composite action, such as T1, T2, T3 and T4, no load is considered to be transferred to the column by the slab, and the cleat and bolts should carry all the beam reaction.

Splices connect the ends of each section of column together so that loads are transmitted between them satisfactorily. Such connections are proportioned in accordance with empirical rules as shown in *Steel Designers' Manual*[9]. Typical splice details are given by Needham[10] and SCI[11].

14.7 WIND BRACING

As discussed in Section 14.1.4, and previously in Section 10.3, the wind loading may be designed to be carried by a wind bracing. It is commonly convenient to locate the wind bracing at stair/lift wells where the diagonal members may be hidden by brickwork. In some situations, such as industrial frameworks, it may be satisfactory to leave the wind bracing exposed.

An arrangement for the bracing is shown in Fig. 14.20. The stair wells provide four frames in the lateral direction (two frames in the longitudinal direction) as shown.

(a) Loading and forces

Force above ground floor level is the sum of the forces shown in Fig. 14.4 = 187.3 kN (for one 5 m bay).

Force on each lateral wind bracing,
$W_w = 187.3 \times 8.4 = 375$ kN

Fig. 14.21

Moment of wind forces about ground level is the sum of the forces × heights shown in Fig. 14.4 = 3550 kNm
Moment on each internal wind bracing
$$M_w = 3550 \times 8/4 = 7100 \text{ kNm}$$

Considering the part of the frame between ground and first floors and analysing as a pin-jointed frame (see Fig. 14.22):

$$R_v = 7100/4.5 = 1580 \text{ kN}$$
$$R_h = 375 \text{ kN}$$
$$F_{C13} = 1580 \text{ kN compression}$$
$$F_{diagonal} = 375/\cos 46° = 540 \text{ kN tension}$$
$$F_{C7} = 1580 - 540 \sin 46° = 1190 \text{ kN tension}$$

As discussed in Section 10.2, cross-bracing allows a tension only design for the diagonals. For this arrangement wind from either direction produces tension in the appropriate diagonal, but tension or compression in the columns.

Fig. 14.22

(b) Column C7 (G–1) – 254 × 254 × 167 UC

The forces in column C7 will include dead and imposed loads similar to C6 (Section 14.5):

$$W_d = 1049 \text{ kN}$$
$$W_i = 717 \text{ kN}$$
$$W_w = 1580 \text{ kN compression or } 1190 \text{ kN tension}$$

Load combinations for maximum compression:

BS table 2

either $1.4 W_d + 1.4 W_w = 1.4 \times 1049 + 1.4 \times 1580 = 3680 \text{ kN}$
or $1.2 [W_d + W_i + W_w]$ = 1.2
$(1049 + 717 + 1580) = 4020 \text{ kN}$

Load combination for maximum tension:

$$1.0 W_d + 1.4 W_w = 1.0 \times 1049 - 1.4 \times 1190 = -617 \text{ kN}$$

$$F_c = 4020 \text{ kN}$$
$$M_x = 3.7 \text{ kN m (see note below)}$$
$$\lambda = 4000 / 67.9 = 59$$

BS table 27c
$$p_c = 197 \text{ N/mm}^2$$
$$P_c = 197 \times 212 \times 10^{-1} = 4180 \text{ kN}$$

$$\lambda_{LT} = 0.5 \times 4700/67.9 = 35$$
BS table 11
$$p_b = 265 \text{ N/mm}^2$$
$$M_b = 265 \times 2420 \times 10^{-3} = 641 \text{ kNm}$$

Overall buckling check:

clause 4.8.3.3 $4020/4180 + 3.7/641 = \mathbf{0.97}$

Note that $M_x = 3.7$ kN m is used as in Section 14.5. This value could be reduced to take account of the lower values of γ_f used here (1.2 W_d + 1.2 W_i in place of 1.4 W_d + 1.6 W_i in Section 14.5), but this would have little effect.

(c) Column C13 (G–1) – 254 × 254 × 167 UC

Taking the dead and imposed loads as similar to those in column C1:

$W_d = 728$ kN
$W_i = 430$ kN
$W_w = 1580$ kN compression **or** 1190 kN tension

Maximum compression:

$1.4\,[W_d + W_w] = 1.4\,(728 + 1580) = 3230$ kN
$1.2\,[W_d + W_i + W_w] = 1.2\,(728 + 430 + 1580) = 3290$ kN

Maximum tension:

$1.0\,W_d + 1.4\,W_w = 1.0 \times 728 - 1.4 \times 1190 = -938$ kN

$F_t = 938$ kN
$F_c = 3290$ kN
$M_x = 10.5$ kNm (see note after Section 14.7(b))
$P_c = 4180$ kN
$M_b = 641$ kNm

Overall buckling check:

$$3290/4180 + 10.5/641 = \mathbf{0.80}$$

(d) Diagonal (G–1) – 203 × 89 channel

The force due to wind only $W_w = 540$ kN tension
$1.4\,W_w = 756$ kN tension

Net area of web (allowing two no. 24 diameter holes across section)

clause 4.6.3
　　　　　　　　　　　　　$= 203.2 \times 8.1 - 2 \times 24 \times 8.1 = 1260$ mm^2
Area of flanges　　　$= 3790 - 1260 = 2530$ mm^2
Multiplier　　　　　　$= 3 \times 1260/(3 \times 1260 + 2530) = 0.60$
Effective area　　　$A_e = 1260 + 2 \times 2530 \times 0.60 = 2780$ mm^2

clause 4.6.1　　Tension capacity　$P_t = A_e\,p_y = 2780 \times 275 \times 10^{-3} = 765$ kN

$$F/P_t = 756/765 = \mathbf{0.99}$$

The design of all members in the bracing system follows the method outlined. The bracing system in the direction at right angles is designed in a similar manner.

14.8 WIND RESISTANCE BY FRAME ACTION

Previous design codes (e.g. BS 449) permitted a simplified frame action for wind resistance, and design methods for this appear in, e.g. *Steel Designers' Manual*[12]. The method makes a number of assumptions regarding shear distribution and points of contraflexure. Although these methods once enjoyed wide application, they are no longer sanctioned under BS 5950: Part 1.

Frame action to resist wind loading requires the frame elements to be connected by rigid joints, and the design is thus controlled by section 5 of BS 5950: Part 1. Plastic or elastic design is permitted, but the horizontal loads must be applied to the whole frame and forces analysed accordingly (clause 5.4.2).

An alternative design method using simple connections for vertical loading, but recognizing their stiffness in the design for wind loading, has been examined by Nethercot[13]. This is shown to give some advantages, but may lead to some overstressing.

Frame action for wind resistance has the disadvantage economically of more complex connections, as well as increased member sizes generally. These costs must be offset against the saving of the wind bracing in any economic comparison.

STUDY REFERENCES

Topic	References
1. Steel buildings	**Mathys J.H.** (1987) Multistorey steel buildings – a new generation, *Structural Engineer*, vol. 65A, no. 2, pp. 47–51
2. Profiled sheeting	(1985) Profiles for composite flooring, *Profiles for Concrete*. Precision Metal Forming Ltd
3. Composite slabs	**Lawson R.M.** (1989) Composite slabs, *Design of Composite Slabs and Beams with Steel Decking*, pp. 5–9. Steel Construction Institute
4. Fire resistance	(1986) Fire resistance of composite slabs with steel decking. CIRIA Special publication 42
5. Loading	BS 6399 *Loading for Buildings* Part 1: *Dead and imposed loads* (1984) Part 2: *Wind loads*
6. Wind loading	British Standards Institute CP3 ChapterV Part 2.
7. Composite beams	BS 5950 *The Structural Use of Steelwork in Buildings* Part 3.1: *Design of composite beams (1990)*
8. Imposed load reduction	Reduction in total imposed floor loads. BS 6399 *Loading for Buildings* Part 1: *Dead and imposed loads* (1984), clause 5
9. Column splices	(1972) Design of connections, *Steel Designers' Manual*, 4th edn. pp.707–78. Blackwell
10. Column splices	**Needham F.H.** (1980) Connections in structural steelwork for buildings, *Structural Engineer*, vol. 58A, no. 9, pp. 267–77
11. Column splices	(1992) *Joints in Simple Constuction* vol. 2. Steel Construction Institute
12. Frame action	(1972) Wind on multistorey buildings, *Steel Designers' Manual*, 4th edn, pp. 847–67. Crosby Lockwood Staples
13. Frame action	**Nethercot D.A.** (1985) Joint action and the design of steel frames, *Structural Engineer*, vol. 63A, no. 12, pp. 371–9

15

DETAILING PRACTICE AND OTHER REQUIREMENTS

15.1 FABRICATION PROCESSES

The designer needs to have an understanding of the processes involved in the fabrication and erection of structural steelwork. This understanding is necessary to ensure that:

(a) all the details shown by the designer are capable of fabrication;
(b) the effects of the fabrication processes on the design are allowed for, e.g. corrosion traps, plate distortion in cropping and bending;
(c) the details shown do not involve unnecessarily complex, time-consuming and hence costly processes;
(d) the responsibilities of the fabricator are clear, e.g. what assembly of cleats is required prior to delivery to site;
(e) the details chosen should allow a safe means of erection.

The processes involved in structural steelwork fabrication, and the requirements of good design, are described by Taggart[1]. Other publications are available giving fuller descriptions of steelwork fabrication[2].

The processes may be summarized as follows.

15.1.1 Surface preparation and priming

Surface preparation is usually carried out either by blast cleaning or by use of mechanical tools. In blast cleaning, an abrasive material is projected at high speed at the surface to be cleaned. The abrasive material can be metallic ('shot' blasting) or non-metallic such as slag or other minerals ('sand' blasting).

Alternatively, preparation may be carried out by a variety of mechanical tools, such as wire brushes and sanders, or by mechanical chisels and needle guns. These are usually less effective than blast cleaning but may be used in smaller fabrication works and on site prior to the final painting.

Priming of the steel surface is carried out immediately after cleaning with the surface clean and free from moisture. A number of different primers are available, and their use should take account of the processes which are to follow. In particular, some primers may give rise to hazardous fumes during subsequent cutting and welding. In addition, some primers may interfere with the welding processes which are to be used.

15.1.2 Cutting and drilling

The steel sections or plates are cut to length and size by guillotining, sawing or flame cutting. Guillotining is a process of shearing steel plates to the required length and width, and cropping is a similar process, but which may be applied to steel sections. The method may be limited in its use for a particular fabrication by minor distortion and burring which require subsequent correction.

Sawing may be carried out by circular saws, hacksaws or bandsaws. Clean, accurate, straight cutting may be achieved.

The thermal cutting process ('flame' cutting) involves a number of different systems which may be process controlled, and used to produce steel plates cut to a predetermined profile.

Drilling of the required holes in steelwork may be carried out using single and multi-spindle machines which may be set to produce a pattern of holes determined by a template. Punching is also used for making holes, but this has a limited use owing to embrittlement of the edge of the hole and possible edge cracking. Punched holes should not, for example, be permitted in connections which are designed to develop yield lines.

15.1.3 Bending and forming

In more complex steelwork assembly, bending of sections and plates to specified shapes may be needed. Sections can be bent to circular and other profiles as required, but with the local radius of curvature limited by the proportions of the sections. Presses are used in bending plates to form sections of specified shape. Examples are shown in Fig. 15.1.

Fig. 15.1 Bending and forming

15.1.4 Welding

Many welding processes are available, but metal arc welding is the one which is normally permitted for steelwork fabrication. Manual metal arc welding is used for attaching end plates, cleats, etc., to steel members, while automatic gas shielded processes are used for the fabrication of plate girders. Different types of weld (Fig. 15.2) are used in different situations and further details may be found in BS 5315 and specialist literature[2,4].

Fillet welds Butt weld (single V) Butt weld (double bevel)

Fig. 15.2 Types of weld

15.1.5 Inspection and protection

Inspection of the steelwork is carried out at several points in the fabrication processes. The most important of these are at entry and after welding. Before commencement of fabrication, the steel sections or plates are checked for straightness, and where steel is subject to stresses applied perpendicular to the direction of rolling, the material should be examined by ultrasonic techniques to detect hidden defects. Defects discovered at a later stage can prove costly to rectify, and may involve rejection of a finished item. Welding is tested on either a sample basis (say 5%) or fully. The methods used may be ultrasonic or radiographic or may involve the use of dyes or magnetic particles. The choice of method will depend on the quality control required, accessibility, and the relative importance of the weld to the overall structure. Tension welds are normally tested to a greater frequency, and sometimes all welds in tension are required to be tested.

Final surface protection of the steelwork is carried out both in the fabrication works and on site. It will involve the retreatment of damaged primer and the application of a variety of finishing paints from oil and resin based paints to polyurethanes and chlorinated rubbers. In addition, special finishes using metallic coatings are available where additional protection is advisable. For some special types of structure galvanizing is selected as the means of preventing corrosion (see also Section 15.5).

15.2 STEELWORK DRAWINGS

In practice, the production of steelwork drawings is to ensure that the original concepts for the structure shown in the calculations and sketches are translated into complete instructions for fabricating and erecting the steel framework. Individual firms maintain varying practices for detailing structural steelwork, but will include some or all of the items described in this section. Some guidance regarding detailing practice is given in a BCSA publication[3].

For the student the preparation of scale drawings will assist in:

(a) visualizing the structure being designed;
(b) bringing a recognition of relative size, e.g. the slenderness of a particular member, or the relationship of a bolt hole to the member size and edge distance;
(c) adopting a discipline in providing complete information both in drawing and in calculation.

15.2.1 Construction drawings

The first drawing produced by the designer is one showing the overall relationship of the steel framework to the other building components. This general arrangement drawing will have plans, elevations and sections showing clearly the relative positions of floor slabs, cladding, walls, windows, foundations, etc., to the steel framework. Multi-level plans as shown in Fig. 15.3 are useful in condensing much information into one plan view. For the designer, the general arrangement is essential to ensure that all the client's requirements have been included, that the steel framework is fully compatible with the other building components, and that the structure is shown to be inherently stable with discernable load paths to the foundations.

15.2.2 Steelwork general arrangement

The layout of all the steelwork members and their relationship to each other must be shown on one drawing. This will be used by the steelwork erectors to assemble the framework with its connections on site. For the simplest of structures, it may be possible to show this information on the construction drawings, but more usually a special drawing of the steelwork layout is necessary.

This layout (Fig. 15.4) will show only the steelwork with principal dimensions and grid lines. It will incorporate a numbering system for each steelwork member and may give member sizes and other details. This latter information is, however, usually placed in a steelwork schedule on the layout drawing, or on a separate sheet (Fig. 15.5). In addition, this general arrangement drawing will need to show the information required to enable the connections to be designed and detailed. This takes the form of end reactions (and moments where appropriate), beam levels and eccentricities. Forces and moments must be clearly indicated as factored or unfactored.

15.2.3 Fabrication drawings

The fabrication process requires drawings of the steelwork members in full detail showing precise sizes, lengths, positions of holes, etc. While in practice these drawings are often produced by the fabricators themselves, it is useful for the student to draw fabrication details.

Fig. 15.3 Construction drawing

(a) Plan
(b) Side view
(c) End view

Fig. 15.4 Steelwork arrangement

310 STRUCTURAL STEELWORK DESIGN TO BS 5950

Drawing	Ref.	Section	Size	No. off	Length (mm)	Remarks
471/2	A1	UB	533×210×92	4	6750	
"	A2	UB	ditto	4	6750	Handed A1
"	A3	UB	305×127×37	6	3520	
"	B1	UC	203×203×60	6	3520	
"	C1	UC	203×203×86	2	3710	
"	C2	RSC	152×76	18	2890	
471/3	E1	UB	533×210×92	2	5210	
"	E2	UB	ditto	2	5210	Handed E1
,	E3	UB	305×127×37	2	4790	
,	E4	UC	203×203×60	4	3520	Docket attached

Fig. 15.5 Steelwork schedule

This ensures that the design concepts are practicable and develops the discipline of conveying these concepts with precision.

An example of a beam fabrication drawing is shown in Fig. 15.6. The information required will include:

- size of member and steel grade;
- precise length, allowing for clearance at each end;
- size of notches and any other special shaping;
- size and position of bolt holes, but not bolt sizes;
- welds types, size and length where appropriate;
- parts (such as cleats) to be connected during fabrication, and in this case only the bolt sizes are appropriate;
- notes giving number of members required, any handed (mirror image) members required, reference to painting preparation and specification.

Fig. 15.6 Beam fabrication

15.2.4 Connection details

Reference has been made to connection details previously (Sections 3.6, 6.6 and 8.3) and the overall need for neat and balanced solutions to design problems has already been emphasized (Section 1.1). Guidance regarding the arrangement of connections and their design is given by Pask[5] and Needham[6] for beam and column construction with I- and H-sections. The British Steel CIDET publication[7] gives guidance for the use of hollow sections, including connection arrangements. Morris[8] deals specifically with connections for single-storey structures.

Drawings of connections may originate with the designer, particularly when a special arrangement has been assumed in the calculations, and a sketch detail has been given. It is important that sketches of such details are conveyed to the fabricators when they are responsible for connection design. Typical connection details showing bolt sizes, packs and clearances similar to those referenced above[5,6] may be needed where site erection is not the responsibility of the fabricators.

It is useful for the student to detail some of the assembled connections required by the design, in order to become aware of the difficulties a particular arrangement may cause, and its effect on design capacity. Where the fabricator is required to design the connections, details of forces and moments (factored or unfactored) must be provided (see Section 15.2.2).

The precise behaviour of particular connections is complex and the subject of ongoing research. The performance of connections is generally defined by the three permitted design methods. These design methods are defined as **simple** design, **semi-rigid** design and **fully rigid** design. The 'simple' method is based on the assumption that beams are simply supported and therefore implies that beam-to-column connections must be sufficiently flexible so as to restrict the development of end fixity. Any horizontal forces have to be resisted by bracing or other means.

The design code permits the use of both elastic design and plastic design within the context of fully rigid design. This method, based on full continuity at the connections, gives the greatest rigidity and economy (in terms of weight of steel) for a given framework. Whether or not a fully rigid design produces more economic structures in terms of cost is continually being debated. Any horizontal loads are resisted by rigid frame action. 'Moment' or 'rigid' connections required by this design method must be capable of carrying the design bending moment, shear force and axial load, while maintaining more or less the angle between connected members, i.e. the required connection should behave 'rigidly'.

In previous chapters, simple design has been assumed in most cases, and simple connections have been designed. In Chapter 13, however, fully rigid design has been assumed and moment connections have been designed accordingly.

15.2.5 Movement joints

Movements occur in buildings owing to changes of temperature, moisture content, foundation arrangements, etc., and in buildings of unusual shape or size it may be necessary to accommodate these movements by provision of a joint. For single-storey construction, the provision of an expansion joint should be considered when the length of the structure exceeds about 150 m. For multi-storey construction, expansion joints may be required at lesser lengths (say 70 m), but in addition settlement joints may be necessary at major height changes, and for unusual plan shapes.

The simplest joint is provided by dividing the structure at the joint and placing columns on either side of the joint. Joint details involving members sliding on a bearing, or the use of slotted holes, are less desirable due to their greater complexity and uncertain lifespan.

15.3 COST CONSIDERATIONS

Previous discussion (Section 1.1) showed briefly that a minimum weight of steelwork would not necessarily produce the minimum cost structure. Clearly, simplicity in fabrication and repetition of member types affects total cost significantly.

In multi-storey construction, comparative costs will be influenced by the choice of floor system, especially the use of composite steel deck floors (Chapter 9). In addition, simplicity of layout and connections has its effect on the speed of construction, and hence on the considerable cost of servicing the capital involved in such a building project. The arrangements for wind bracing and staircases also have cost effects, and these and other factors are discussed in reference (9).

Comparative costs of single-storey construction will depend on both the spacing and span of frames as well as the form of structure chosen. The most common forms of structure are included in a cost comparison by Horridge and Morris[10]. The charts produced may be used for settling the basic layout and choosing the best spacing and span for a single-storey structure. They also provide guidance on the cost effects of the choice of a particular form for the roof structure.

15.4 FIRE PROTECTION

The protection of a building structure from the effects of fire is required by regulations to provide adequate time prior to collapse in order to:

- allow for any occupants to leave
- allow for fire fighting personnel to enter if necessary
- delay the spread of fire to adjoining property

To achieve these requirements it is often necessary to cover the bare steelwork with a protective coating. This may be simply concrete cast around the steel with a light steel mesh to prevent spalling, or any one of a number of proprietary systems. These may be sprayed on to the steel surface, or may take the form of prefabricated casings clipped round the steel section. Examples of each of these are shown in Fig. 15.7. Details of a wide range of these systems are set out by Elliott[11,12].

Systems of fire protection are designed and tested by their manufacturers to achieve the fire resistance periods specified in the Building Regulations. These periods are somewhat inflexible and a more fundamental design approach is possible using structural fire engineering[12,13]. In this design

Fig. 15.7 Fire protection — Concrete cased — Sprayed — Preformed casing approach, an assessment of the maximum atmosphere temperature is made from the fire load, ventilation and other conditions. The heating curve of the steel member is then estimated, allowing for the location of the steel and its protection. Finally, the effects of temperature on the structural capacity of the steelwork are determined.

Special requirements apply to steel portal frames where they form part of the fire barrier to adjacent buildings. A method of designing portal frames to ensure the integrity of the boundary wall in severe fires is given in reference (14).

15.5 CORROSION PROTECTION

The detailing of steelwork can affect the manner and speed of corrosion. Care should therefore be taken to minimize the exposed surface, and to avoid ledges and crevices between abutting plates or sections which may retain moisture. Protective coatings are dependent for their effectiveness on their type, quality and thickness, but most of all on the degree of care taken in the preparation of the steel and in the application of the coating.

The mechanisms of corrosion form a special study area, and this is basic to the proper protection of steelwork. This area, together with information on coatings, surface preparation, inspection and maintenance, is discussed in a CIRIA report[15]. The choice of a protective system usually involves consultation with experts in this area. The cost of protection varies with the importance of the structure, its accessibility for maintenance, and the frequency at which this can be permitted without inconvenience to the users.

STUDY REFERENCES

Topic	References
1. Steelwork fabrication	**Taggart R**. (1986) Structural steelwork fabrication, *Structural Engineer,* vol. 64A, no. 8, pp. 207–11
2. Steelwork fabrication	**Davies B.J. & Crawley E.J.** (1980) *Structural Steelwork Fabrication* vol. 1. BCSA Ltd
3. Detailing practice	(1979) *Metric Practice for Structural Steelwork.* BCSA Ltd
4. Welding	**Pratt J.L.** (1979) *Introduction to the Welding of Structural Steelwork.* Steel Construction Institute
5. Connections	**Pask J.W.** (1982) *Manual on Connections.* BCSA Ltd
6. Connections	**Needham F.H.** (1980) Connections in structural steelwork for buildings, *Structural Engineer,* vol. 58A, no. 9, pp. 267–77
7. Connections	Design of joints under static loading, *Construction with Hollow Steel Sections,* pp. 129–52. British Steel Corporation (Tubes Division)

8.	Connections	**Morris L.J.** (1980) A commentary on portal frame design, *Structural Engineer*, vol. 59A, no. 12, pp. 384–404
9.	Steelwork costs	**Gray B.A. & Walker H.B**. (1985) *Steel Framed Multi-storey Buildings. The economics of construction in the UK*. Steel Construction Institute
10.	Steelwork costs	**Horridge J.F. & Morris L.J**. (1986) Comparative costs of single-storey steel framed structures, *Structural Engineer*, vol. 64A, no. 7, pp. 177–81
11.	Fire protection	**Elliott D.A.** (1974) *Fire and Steel Construction*. Steel Construction Institute
12.	Fire protection	**Elliott D.A.** (1983) *Introduction to the Fire Protection of Steelwork*. Steel Construction Institute
13.	Fire engineering	**Kirby B.R.** (1985) *Fire Resistance of Steel Structures*. British Steel Corporation
14.	Fire engineering	(1980) *The Behaviour of Steel Portal Frames in Boundary Conditions*. Steel Construction Institute
15.	Corrosion protection	**Haigh I.P.** (1982) *Painting Steelwork*. CIRIA report no. 93

APPENDIX A

PLASTIC SECTION PROPERTIES

With reference to Fig. A1.

Area $A = A_s + A_p$

Neutral axis at position of equal area above and below it, hence:

$$A_s/2 + d_p t = A_s/2 - d_p t + A_p$$
$$d_p = A_p/2t$$
$$S_s = 2A_f d_f + td^2/4$$
$$\begin{aligned}S_x &= A_f(d_f - d_p) + A_f(d_f + d_p) + t(d/2 - d_p)^2/2 \\ &\quad + t(d/2 + d_p)^2/2 + A_p(D/2 + T_p/2 - d_p) \\ &= 2A_f d_f + td^2/4 + td_p^2 + A_p(D/2 + T_p/2 - d_p) \\ &= S_s + td_p^2 + A_p(D/2 + T_p/2 - d_p)\end{aligned}$$

ELASTIC SECTION PROPERTIES

With reference to Fig. A2.

Area $A = A_s + A_p$

Neutral axis at centroid, hence:

$$Ad_e = A_p(D/2 + T_p/2)$$
$$d_e = A_p(D + T_p)/2A$$
$$I_x = I_s + A_s d_e^2 + A_p(D/2 + T_p/2 - d_e)^2$$
$$Z_x = I_x/(D/2 + d_e)$$

Note: the formula for S_s ignores the effect of the root fillets associated with rolled I-sections; exact values can be obtained from the SCI publication *Steelwork Design*, vol.1, *Section properties, member capacities*. These formulae apply only when the neutral axis of the combined section lies in the web depth. If the neutral axis lies within the flange of the I-section, the various section properties would need to be determined from first principles.

Fig. A1

Fig. A2

APPENDIX B

A RAPID METHOD FOR ASSESSMENT OF n_t FACTOR

The following method gives an assessment of n_t (the stress distribution factor) for a typical British haunch detail where a UB cutting of the same section as the rafter is used as the haunch, producing a haunch depth of approximately twice that of the basic rafter section. The resulting value of n_t can be used in the various formulae dealing with member buckling, BS 5950.

$$n_t = \sqrt{\left\{\frac{1}{12M_o}\left[\frac{N_1}{R_1} + \frac{3N_2}{R_2} + \frac{4N_3}{R_3} + \frac{3N_4}{R_4} + \frac{N_5}{R_5} + 2\left(\frac{N_S}{R_S} - \frac{N_E}{R_E}\right)\right]\right\}}$$

where

M_o = plastic/elastic moment capacity of basic section ($p_y S_x$ or $p_y Z_x$)
N_i = applied factored moments at quarter points (kN m)
R_i = coefficient applied to uniform rafter moment capacity to produce an estimate of moment capacity at point i

$\quad = [1.45 + 0.90(p^* - 0.20) + 0.70(p^* - 0.45)]$ **plastic**
$\quad = [1.60 + 0.65(p^* - 0.20) + 0.50(p^* - 0.45)]$ **elastic**

$$p^* = \frac{\text{length from intersection to point } i \text{ within haunch}}{\text{total length of haunch}}$$

NOTES

(a) All compound terms have to be positive, otherwise make zero.
(b) $R_i = 1.0$ when N_i is located in the uniform part of the rafter.
(c) If quarter point coincides with intersection, $R_i = R_{inter} = 1.0$.
(d) If intersection lies between two quarter points then N_i nearest to intersection in the uniform section becomes N_{inter} with $R_{inter} = 1.0$.
(e) Particular values of (N_i/R_i) *cannot be used* in conjunction with the appropriate modulus to evaluate individual stresses.
(f) If $L^* (=L_k/cn_t)$ is less than actual length plus 100 mm, then an exact check of n_t must be made; see Section 13.8.1.2.

Examples

These examples are made with reference to plastic conditions; the corresponding R_i values for elastic conditions are noted in parentheses.

1. Special case, when restrained length equals haunch length, with reference to Fig. B1:

   $R_1 = [1.45 + 0.90(1.00 - 0.20) + 0.70(1.00 - 0.45) = \mathbf{2.555}$ (2.395)
   $R_2 = [1.45 + 0.90(0.75 - 0.20) + 0.70(0.75 - 0.45) = \mathbf{2.155}$ (2.107)
   $R_3 = [1.45 + 0.90(0.50 - 0.20) + 0.70(0.50 - 0.45) = \mathbf{1.755}$ (1.820)
   $R_4 = [1.45 + 0.90(0.25 - 0.20) + 0.70(0.25 - 0.45) = \mathbf{1.495}$ (1.632)
   $R_5 = R_{inter} = \mathbf{1.00}$ (1.00)

2. With reference to Fig. B2:

   $R_1 = [1.45 + 0.90(1.00 - 0.20) + 0.70(1.00 - 0.45) = \mathbf{2.555}$ (2.395)
   $R_2 = [1.45 + 0.90(0.67 - 0.20) + 0.70(0.67 - 0.45) = \mathbf{2.027}$ (2.015)
   $R_3 = [1.45 + 0.90(0.33 - 0.20) + 0.70(0.33 - 0.45) = \mathbf{1.567}$ (1.684)
   $R_4 = \mathbf{1.00}$ (1.00)
   $R_5 = \mathbf{1.00}$ (1.00)

3. With reference to Fig. B3:

   $R_1 = [1.45 + 0.90(1.00 - 0.20) + 0.70(1.00 - 0.45) = \mathbf{2.555}$ (2.395)
   $R_2 = [1.45 + 0.90(0.63 - 0.20) + 0.70(0.63 - 0.45) = \mathbf{1.963}$ (1.970)
   $R_3 = [1.45 + 0.90(0.25 - 0.20) + 0.70(0.25 - 0.45) = \mathbf{1.495}$ (1.632)
   $R_4 = R_{inter} = \mathbf{1.00}$ (1.00)
   $R_5 = \mathbf{1.00}$ (1.00)

Fig. B1

Fig. B2

Fig. B3

INDEX

Adequacy factor, 232, 236
Angle section
  design, 188, 255
  cleats, 35, 40
  compound, 216
  deflection, 49
  moment capacity, 49
  purlins, 42
  shear capacity, 48
  side rails, 42, 48
Apex connection
  see portal frame
Axial compression, 81, 84, 195

Base plate
  see column bases
Beam
  laterally restrained, 29, 32
  local moment capacity, 30, 84
  member buckling, 31
  secondary, 36
  unrestrained, 29, 36, 55
Bearing pressure, 95, 99, 219, 276
Bending strength, 31
Bolt
  sizes, 32
  capacities, 257
  grade 8.8, 257
  holding down, 95, 100, 219, 276
  HSFG, 69, 257
  loads, 35, 98, 102, 103
Braced bays
  load, 170, 196
Bracing, 65, 113–116, 119
  connections, 214
  forces, 113, 190
  gable, 115, 205, 265
  in-plane, 273
  multi-storey, 114, 119
  rafter, 196–202, 274
  single-storey, 115
  systems, 64, 114, 115, 265
  vertical/side, 115, 116, 119, 196–204
British Standard
  BS 2573, 52
  BS 4360, 8, 10
  BS 5400, 6, 123, 159
  BS 5950, 5, 6, 13, 123
  BS 6399, 7, 16, 53
Brittle fracture, 6, 100, 159

Buckling
  modes of failure, 29, 82–83
  parameter, 29
  resistance, 29, 31, 38, 45, 51, 58
  simple design, 30
  see also member buckling

Capacity, 9
Channel section
  buckling resistance, 45
  deflection, 47, 49
  moment capacity, 45, 49
  shear capacity, 44, 48
Chord
  see lattice girder
  see truss
Cladding, 42, 165, 166, 171
Classification of sections, 10, 31, 181
  compact, 31, 186
  plastic, 30, 181
  semi-compact, 31, 186
  slender, 31, 181
Coexistent shear, 30
Cold-formed sections, 42, 171, 228
Column member, 80, 191, 217, 243, 296
  axial compression, 81
  bases, 94–96, 99, 218, 275
  battened, 81
  brackets, 96–98, 101–103
  design, 85
  laced, 81, 91
  local capacity, 84, 194
  member buckling, 85, 194
  slenderness, 82
Compact section
  see classification
Composite
  beams, 106, 110, 286, 291
  construction, 105, 281
  deflection, 109, 112
  floor/roof slab, 105, 110
  local shear, 108
  moment capacity, 106, 111
  shear capacity, 106, 111
  shear connectors, 107, 111
  slab, 110
  transformed section, 109, 287
Compression resistance, 69, 183, 186, 188
Concrete slab, 18, 32

Connections
  apex, *see* portal frame
  design of bolts, 32, 35, 39, 101–103
  eaves, *see* portal frame
  eccentric, 70, 83
  lattice girder, 205–212
  site splices, 212
  truss, 70, 71, 75
  web cleats, 34, 39, 300
Continuous spans, 49
  loading, 18
Corrosion protection, 224, 313
Cost consideration, 164, 312
Crane girder, 52
  bracing, 116
  bracket, 102, 103
  braking load, 53, 55
  crab; crabbing, 53
  diaphragm, 58
  dynamic effects, 17, 53
  loading on, 17, 54
  overhead, 53
  surge, 52, 55
  surge girder, 61–63
  web bearing, 60, 62
  web buckling, 59, 62
  wheel loading, 52, 55

Dead load
  see load
Deflection, 10, 34, 39, 47, 49, 51, 60, 62, 76, 124, 159, 223
  limits, 11, 31
Design examples
  bracing, 116, 119
  column – industrial building, 85
  crane girder bracket, 101
  composite beam, 110
  continuous spans, 18
  gable wind girder, 116
  gantry girder, 17, 55, 61
  laced column, 91
  lattice girder, 76
  multi-span purlins, 49
  multi-storey wind bracing, 119
  plate girder, 125, 129, 135, 145
  portal frame loading, 21
  purlins, 43
  restrained beam, 32
  side rails, 47

slab base, 99
truss with sloping rafter, 71
unstrained beam, 36
Design of complete buildings
single-storey building,
see lattice girder
see portal frame
multi-storey building,
see composite construction
Design requirements, 5
Design strength, 10, 30
Detailing practice, 305
Diagonal braces, 255
Drawings, 12, 307
connection details, 310
construction, 308
fabrication, 308
general arrangement, 308
Durability, 6

Eaves connection, 256
Eaves tie, 205, 274
Economy, 5, 164
Effective area, 69, 187
Effective length, 29, 244
Elastic modulus, 30
calculation, 315
End plate, 256
hole positions, 256, 264
thickness, 258, 264
welds, 257
Equivalent moment factor, 30, 31, 38, 184, 244, 249, 253
Equivalent slenderness, 30
Erection, 4, 50, 312

Fabrication processes
bending and forming, 306
cutting and drilling, 4, 306
painting, 305, 307
surface preparation, 305, 313
welding, 307
Fatigue, 6
Fire; fire protection, 167, 223, 312
Force
coefficients, 21, 24, 169
components, 44
Foundation block, 219, 277
Frame stability, 242
Frictional drag forces, 170

Gable
edge beams, 266
framing, 205, 265
posts, 66, 115, 197, 205, 270
Gantry girder
loading, 17
Gusset plate, 70

Gutter, 167, 191, 275

Haunched members
see portal frame
Holding down bolts
see bolts

Imposed load
see load
Influence lines
moment, 19, 54
shear, 20

Lacings, 93
Lateral restraint, 29, 38, 45, 49, 243, 245, 252, 255
Lateral torsional instability, 29, 31, 46, 243
Lattice girder, 61, 66, 76, 163, 174
analysis, 66, 176–180
bottom chord, 185
bracing, 196
forces, 199
rafter, 199–201
vertical, 200–204
wind, 199–202
column bases, 218
connections, 207–214
diagonal members, 188
eaves tie, 196, 205
foundation block
load cases, 219
wind uplift, 219–221
gable posts, 196, 205
general details, 164, 205
load cases, 176, 192, 219
longitudinal tie, 186, 199, 204, 214
main column, 191
member forces, 176–180
overall stability, 196
preliminary decisions, 165
purlins and sheeting rails, 170
self weight assessment, 167, 190
site splices, 180, 211
top chord, 181
wind suction, 185
Layout of calculations, 11
drawings, 12
references, 12
results, 12
sketches, 11
Load, 32, 36, 66, 222
coefficients, 24
combinations, 7, 16, 24, 25, 86, 176, 192, 219, 235
crane wheel, 54, 55
dead, 7, 15, 167, 229

distribution, 7
imposed, 16
pattern, 19, 20
snow, 7, 16, 167
transfer, 7
wind, 16, 167, 235, 283
Local buckling, 31, 181, 231
Local capacity, 30, 84, 89, 91, 182, 194
Longitudinal tie, 183, 203, 212

Member buckling, 85, 89, 92, 182, 194, 243, 245, 250, 252
apex region, 252–254
haunched rafter, 245
Modulus
elastic, 30, 315
plastic, 30, 315
Moment
capacity, 30, 33, 37, 45, 49, 50, 57, 62, 89, 127, 129
coefficients, 21, 24
Morris stiffener
see shear stiffening
Movement joints, 311
Multi-storey office block
see composite construction

Notional load, 113, 257, 281

Other considerations, 155, 222, 279
brittle fracture, 159
corrosion, 224
deflections, 223
erection, 159
fatigue, 159
fire, 223
general, 224
loading, 222
temperature, 159
transportation, 159
Overall buckling
see member buckling
Overall stability of buildings, 196, 273–279

Partial safety factors, 7
Plastic analysis, 26, 229–242
Plastic modulus, 232
calculation, 315
Plastic moment capacity, 30, 229, 231
axial load effect, 232, 244, 253
Plate girder, 121
anchor force, 144
aspect ratio, 123, 135–138
design, 124
end post, 145, 151–155

moment capacity, 127, 129
other considerations, 155
shear resistance, 126, 130
stiffened, 135, 145
stiffeners,
    end bearing, 123, 133, 142, 151
    intermediate, 123, 135, 139, 148
    load carrying, 123, 131, 140, 149
tension field action, 143–145
unstiffened, 124, 129
web panel, 123
welds, 127, 132, 134, 140, 142, 143, 154
Plate sizes
  plates, 122
  wide flats, 122
Portal frame, 226
  adequacy factor, 232, 236
  apex connection, 263, 264
  bracing systems, 265, 273
  collapse mechanism, 230, 233–241
  column base, 275
  column design, 243
  computer analysis, 237
  design of frame, 229, 240
  eaves connection, 256–263
  end plate design, 256–259
  foundation block, 277
    load cases, 277
  gable connections, 272
  gable framing, 266–273
  gable posts, 270–273
  haunch length, 234, 241
  lateral restraints, 255
  loading, 21, 229
  member buckling, 243–254, 272
  member sizes, 232, 240
  overall stability, 273–279
  pinned bases, 229, 275
  plastic moment distribution, 234, 237, 238, 241
  preliminary member size, 230
  purlin and side rails, 228
  rafter design, 231, 240
  reversed wind condition, 235
  shear stiffening, 260

Pressure coefficients, 22, 168–170
Profiled sheeting, 105, 283
Purlin, 42, 43, 66
  design, 43, 170–173, 228
  effective lengths, 41
  multi-span, 49
  table, 172

Rafter bracing, 201, 265
Reduced design strength, 182, 215
Resistance, 9
Restraints, 38, 61, 82, 116, 182, 191, 205, 243–245, 250, 254
Rotation capacity, 250

Sag rods, 43, 173
Semi-compact section, 31, 186
Serviceability limit state, 6
Shape factor, 237
Shear
  capacity, 30, 33, 37, 44, 48, 57, 61, 106, 126, 261
  connectors, 105, 107, 108
  resistance, 130, 136–138, 146–148
  stiffening, 261
Sheeting/side rails, 42
  spacing, 171, 228, 240
  table, 172
Single-storey structure
  see lattice girder
  see portal frame
Slender section, 31, 181
Slenderness, 82, 181–184
  correction factor, 29
  equivalent factor, 30
  minor axis, 29
  truss members, 67–69
Snap-through instability, 242
Stairs, 284, 285
Steel grades, 10
Steel section properties, 4
Strength, 9
Strength reduction factor, 182, 215
Stress, 9
Stress distribution factor, 247
  rapid assessment of, 248, 317
Structural integrity, 6
Sway stability, 113, 242

Tee-section design, 73, 181–185
Tension capacity, 69, 187, 189
Tension flange restraint, 246
Torsional index, 29
Transportation, 180, 224
Truss, 65, 66, 71, 114
  analysis, 66, 72, 77, 117
  chord, 77, 118
  compression resistance, 69, 74
  connections, 70, 75, 78
  continuous strut, 68
  deflection, 76
  diagonal members, 78, 118
  discontinuous strut, 68
  hipped, 65
  lattice girder, 65, 76
  mansard, 66
  slenderness of member, 67
  tension capacity, 69
  vierendeel, 66
Tubular members, 76, 166, 201–205

Vertical bracing, 114, 198, 203, 273
Vibration, 6

Web
  bearing, 31, 60, 62, 128
  buckling, 31, 59, 62, 128, 131, 140, 149
Weld
  capacity, 32
  group design, 207–210
  sizes, 32
Wind
  bracing, 113–115, 196–204, 265, 274, 301
  design case, 22, 167, 192–194, 198–200, 235
  drag force, 170, 200, 273
  force coefficients, 169
  girder, 115, 199, 205
  pressure coefficients, 16, 22, 168, 169
  speed, 21, 168
  suction, 43, 46, 185
  frame action, 226, 283, 303

Yield moment, 234